Practicing Utopia

Practicing Utopia

An Intellectual History of the New Town Movement

ROSEMARY WAKEMAN

The University of Chicago Press
Chicago and London

Rosemary Wakeman is professor of history and director of the Urban Studies Program at Fordham University.

The University of Chicago Press, Chicago 60637
The University of Chicago Press, Ltd., London
© 2016 by The University of Chicago
All rights reserved. Published 2016.
Printed in the United States of America

25 24 23 22 21 20 19 18 17 16   1 2 3 4 5

ISBN-13: 978-0-226-34603-8 (cloth)
ISBN-13: 978-0-226-34617-5 (e-book)
DOI: 10.7208/chicago/9780226346175.001.0001

Library of Congress Cataloging-in-Publication Data

Wakeman, Rosemary, author.
  Practicing utopia : an intellectual history of the new town movement / Rosemary Wakeman.
    pages cm
  Includes bibliographical references and index.
  ISBN 978-0-226-34603-8 (cloth : alk. paper) — ISBN 978-0-226-34617-5 (e-book)
   1. New towns—History—20th century.   2. New towns—Philosophy.
3. City planning—History—20th century.   4. Regional planning—History—20th century.   I. Title.
  HT169.55.W35 2016
  307.76′80904—dc23

                                                                2015029536

CONTENTS

ILLUSTRATIONS

ACKNOWLEDGMENTS

The history of the new town movement has taken me to an abundance of places, both real and virtual. The result is that I have many people to thank for their guidance and help, ideas and resources. I am greatly indebted to the archivists and librarians at the various institutions where I gathered materials. The staff at the Centre de documentation de l'urbanisme in La Défense outside Paris patiently fulfilled my unending requests and dug out documents during my summer research trips. The new towns collection at the CDU is outstanding, as are the materials assembled by the Programme interministériel d'histoire et d'évaluation des villes nouvelles françaises. Thank you to the staffs at the Bibliothèque nationale, Paris; the Bundesarchiv, Berlin; and the New York Public Library for their invaluable help in locating sources, and especially to Jay Barksdale of the New York Public Library for the privilege of the Allen Room. At the International New Towns Institute in Almere, the Netherlands, Jean-Paul Baeten allowed me to shuffle through boxes of materials he was only just then archiving. I am immensely grateful to the library staffs and webmasters at a host of new towns, from Cergy-Pontoise to Eisenhüttenstadt and Akademgorodok; they made gathering historical resources a pleasure.

Although any errors that remain in the text are entirely mine, this book has been a collective endeavor. Friends and colleagues at the Institut d'urbanisme de Paris, especially Thierry Paquot and Laurent Coudroy de Lille, have sustained this project from the beginning. Clément Orillard and Loïc Vadelorge contributed their expert insights on the new town phenomenon. Thank you to Christoph Bernhardt at the Leibniz Institute for Regional Development and Structural Planning, Erkner, Germany, for his perceptive comments. The best illustrations of East German new town plans are from the extraordinary IRS archive. Colleagues and doctoral fellows in the Center for Metropolitan Studies at the Technical University in Berlin contributed ideas at various stages of research, and my thanks go especially to Dorothee Brantz and Alexander Nützenadel. Colleagues at the Central European University in Budapest critiqued an early presentation on East European new towns. Thank you to Gabor Gyani for his suggestions, and to Livia Szelpal for her invaluable research help with Hungarian new towns.

Florian Urban read early chapter drafts, and Janet Ward and Jeff Diefendorf added their comments on East German new towns for my essay that appeared in their collected work *Transnationalism and the German City* (New York: Palgrave Macmillan, 2014). Presentations on various chapters at the University of Toronto; the University of Utrecht, Netherlands; the University of Maastricht, Netherlands; the Center for Metropolitan Studies, Berlin; the University of Paris at Marne-la-Vallée; and the University of Toulouse, France, were welcome opportunities for discussing new towns, and I profited immensely from the feedback and insights gleaned on these occasions. Likewise, presentations at the New York Public Library and Columbia University's Seminar on the City were crucial to understanding the new town movement, as were discussions with Bob Beauregard, Owen Gutfreund, and Guy Ortolano. Carola Hein and Sandor Horvath, Michelle Provoost and Pierre Weidknnet, and Rhodri Windsor-Liscombe generously contributed suggestions and resources.

A Senior Fellowship from the European Institutes for Advanced Study in 2012–13 enabled me to write a complete draft of the book. During that sabbatical year, colleagues at the Netherlands Institute for Advanced Study in the Humanities and Social Sciences, Wassenaar, provided continuous support and ideas—especially Gabor Demszky and Arie Molendijk, who read and provided helpful comments on the manuscript. A Fordham Faculty Fellowship in 2012–13 also supported this research. I extend my thanks as always to colleagues and students at Fordham University for their interest and suggestions—especially Asif Siddiqi for his help with Soviet ZATOs, and the small band of students attending the seminar on utopia who helped flesh out ideas. Thank you as well to Charlotte Labbe and the staff at Fordham Library's Interlibrary Loan service for their ability to find any document, anytime, anywhere, and to Erwin Nolet and Dindy van Maanen at the library of the Netherlands Institute for Advanced Study in Wassenaar who performed the same expert wizardry.

In acknowledging the sources for images of new towns printed in this book, I am brought to a community of fellow travelers for whom I have the utmost appreciation. The joy of researching contemporary history is that with luck, you can actually communicate with the subjects of your inquiry. I give sincere thanks to the late Charles Correa, and to Dennis Crompton, Jerzy Kozłowski, Randhir Sahni, and Derek Walker for their help in securing images of their work. Thank you as well to Louise Avery at the Kitimat Museum and Archives, Kitimat, British Columbia; Neil Baxter at the Royal Incorporation of Architects in Scotland; Ross Brown at the University of Strathclyde, Glasgow; Lucas Buresch at the Rockefeller Archive Center, Sleepy Hollow, New York; Joanne Burman of the British Pertroleum Archive at the University of Warwick, Coventry, England; and Cynthia Davidson at Eisenman Architects, New York City. Claire Galopin at the Institut d'aménagement et d'urbanisme d'Île-de-France, Paris, helped locate photographs, as did Aryn Glazier of the Dolph Briscoe Center for American History at the University of Texas, Austin; Ulrike Heinisch at the Urban History Institute in Frankfurt am Main; Paweł Jagło at the Nowa Huta branch of the Historical Museum of Crakow; and Chandra Kumar at the World Bank Group Archives in Washington, DC. Images were contributed by Maureen Irish at the North Ayrshire Council, Irvine, Scotland; Hanne Sue Kirsch at the Soleri Archives, Mayer, Arizona; Dmitriy Kobzov at Bratsk State University, Irkutsk Oblast, Russia; Josabeth Leidi at the Stockholm City Archive; Steven Leclair at the National Research Council of Canada, Ottawa, Ontario; and Klas Lundkvist at the Stockholm City Museum. My thanks to Dwight Magee at Rio Tinto Alcan, Cleveland; Jonathan Makepeace at the Royal Institute of British Architects, London; Wiebke McGhee at the North Lanarkshire Archives, Motherwell, Scotland; Dr. Lynne McGowan at the University of Liverpool, England; and Catherine McIntyre at the Milton Keynes City Discovery Centre, Milton Keynes, England. Mark Monforte at the Irvine Company, Irvine, California, generously contributed images, as did Matthew Parrent at Gruen Associates, Los Angeles; Giota Pavlidou at the Constantinos A. Doxiadis Archives, Athens, Greece; Shelley Power, agent for the Archigram Archive; Sari Piirainen at the Espoo City Museum, Esbo, Finland; and Chise Takahashi at Tange Associates, Tokyo. Thank you to Robert Tennenbaum for his photographs of Columbia, Maryland, as well as to the Columbia Association; Irina Tikhonova at the V. S. Sobolev Institute of Geology and Mineralogy, Siberian Branch of the Russian Academy of Sciences, Novosibirsk, Russia; Łukasz Trzcinski at Fundacja Imago Mundi, Cracow, Poland; Bernard Venis for his personal photos; Gabriele Urban at the Eisenhüttenstadt City Archive, Germany; Adrienne Voller at the Kwinana Public Library, Kwinana, Australia; and Morgan Yates at the Automobile Club of Southern California in Los Angeles.

Tim Mennel and Nora Devlin at the University of Chicago Press provided invaluable support for this project. I thank them for making this book a success. Discussions with friends Michèle Collin and Thierry Baudouin, Francis Nordemann and Emeline Bailly kept me on track. My husband, Tom Wakeman, found himself going to new town places he never knew existed, and did so with aplomb. This book is a result of his abiding patience and support, which always inspire me.

Introduction

There is nothing new about new towns. They have been constructed since the beginning of recorded history. The ancient Phoenicians named their colonies Qart Hadasht, or New City, applying the term to their bases in what later became Carthage in North Africa and Cartagena in Spain. The ancient Romans were of course master town builders as they marched their way across a vast empire, and we have been living in their new towns ever since. During the Middle Ages, dozens of fortified new towns were constructed along territorial frontiers. And in his proposal to break up congested sixteenth-century Milan, Leonardo da Vinci suggested that ten new towns be built on the city's outskirts.

We could easily add a long list of town-building endeavors to these examples, since theoretically speaking, every town and city was at some point "new." The adjective is so vague that it can seem meaningless as a category of understanding.[1] What we can say is that the label *new town* has been applied at specific moments in history to denote deliberate and highly symbolic acts of territorial control and settlement. A new town was not unlike a flag planted in the soil. Commanding the design and construction of towns across a territory was an unmistakable sign of hegemony. There was nothing spontaneous or unrehearsed about these places. New towns were calculated acts. They required a near-magical template that could perform a switch in power and authority. Each town acted as a brand, an image of the future, a step into a new era. How a town was designed and laid out, how it was built, how society would function inside it, could be deciphered by anyone living within or passing through its precincts.

It is from these perspectives that this book investigates the new town movement in the mid- to late twentieth century. The period from 1945 to 1975 was a golden age of new towns. Throughout Europe and North

America, and beyond in the Middle East, Australia, Africa, and Asia, new towns popped up in extraordinary numbers. It is nearly impossible to know exactly how many new towns were built during this period, because the definition of a "new town" was not exact, nor was it the same everywhere. In fact, investigating the complex meaning of *new town* is part of my purpose in the pages that follow: it found expression as *satellite towns, new communities, new cities, worker cities, cities of science, garden cities.* This last—Ebenezer Howard's formulation—has received intense treatment in planning histories, and indeed virtually all the late twentieth-century new towns claimed the garden city as their birthright. But little has been written about the magnitude and character of the subsequent new town crusade. By the second half of the twentieth century, it dominated thinking about urban utopia and succeeded in building hundreds, if not thousands, of settlements. The magnitude of construction, its extraordinary geographic extent, the immensity of the discourse and rhetoric about new towns, are what made it into an unmistakable *movement.*

New towns of this era were a tool for reconstruction and resource extraction, for population resettlement and territorial dominion. They were a solution to the afflictions of the big city, a testing ground for regional planning and transportation systems, for living in nature and living in outer space. There was no end to such blueprints for the future. They were designed and created by governments in different national contexts, by private corporations, by different societies with different political agendas. Despite the vague terminology and the variety of experiences, the unremitting effort to build new towns had a common aim: each project was a campaign to construct, literally, an entirely new world.[2] The examples that immediately come to mind are the sweeping architectural statements of Le Corbusier's Chandigarh in India and Oscar Niemeyer and Lúcio Costa's socialist utopia of Brasilia in Brazil. But beyond these two celebrity capital cities lay new town projects that are arguably less well known and yet defined the aspirations of the late twentieth century: Tapiola in Finland, Milton Keynes in England, Cergy-Pontoise in France, Navi Mumbai in India, Nowa Huta in Poland, Islamabad in Pakistan, Irvine in the United States. The movement defied the East-West dichotomy of the Cold War and the polarization of north and south between rich and poor nations. Wherever they were located, whatever their size, whether famous or forgotten, all these new town projects shared a utopian rhetoric and conception. In a phrase, the new town was a marvelous glimpse at tomorrow.

The adjective *new* implied a model or prototype of the future. Architects and planners poured their talents into figuring out this ideal. The energy that went into designing and constructing new towns is the best indication of their symbolic power. Decisions about them were made at the highest political levels. The deluge of publications about the new town phenomenon from the Second World War through the 1970s is also proof of this passionate laboring over the invention of schema, the design and planning of cities from scratch. Their definition, nomenclature, and classification obsessed urban practitioners. All this meant grappling not only with the meaning of *new* but also with the meaning of *town*. The latter word in particular distinguished these projects from large-scale residential subdivisions such as the Levittowns built near New York City and Philadelphia or the infamous *grands ensembles* surrounding Paris or the *Neubaugebiet* (development area) districts of Germany. New towns were qualitatively different. They were attempts to prefabricate, literally and figuratively, complete urban totalities: they were holistic, whole-cloth, complete places. It was this grand intention that made delineating their definition and features so fundamental, and what made them utopian in form.

They offered a range of housing types, but they also applied cutting-edge urban design, paradigmatic concepts of centrality, community, and neighborhood, and new metropolitan regional perspectives. They put forth ingenious flows of people, transportation, water, and waste. They offered jobs, schools, and services. They were used as test sites for comprehensive master planning, for consumerism and the shopping mall, for recreation and leisure ideals, and for what constituted the civic domain. They were laboratories for the application of systems analysis and computer forecasting, and for the dreamscapes of the architectural avant-garde. These were huge megaprojects imagined as a perfected social order—a flawless paradise in which families would find every modern convenience and enjoy happiness and harmony. Some measure of autonomy and self-sufficiency was integral to their identity. They were a complete urban configuration as a course of action for the future. Some new towns were successful. Others degenerated into dystopian statements of modern brutalism. But for their time, they all set groundbreaking standards for the quality of urban life.

The Hermeneutics of Utopia

Utopia is always made from the grist of history. The wildest dreams about what is yet to come are derived from the here and now, the raw materials of the period in which fantasies are construed.[3] Utopian reveries illuminate the historical conditions of their possibility. They ask questions. For David Harvey, "The question of what kind of city we want cannot be divorced from that of what kind of social ties, relationship to nature, lifestyles, technologies and aesthetic values we desire." We seek to change ourselves by changing the city.[4] The new town was this fantastic transformational object, different from what had gone before. It was the miraculous frontier between two worlds, the present and the future. The result was a time warp. The new town sped up progress into an imminent age. It was a tremendous leap into a bright, new, limitless world that was happy and good. These places were not just abracadabra, nor did architects and planners simply dream up arbitrary worlds. Rather, the new towns that captured the late twentieth-century imagination provided the coordinates of desire, although even the category of imagination has a plurality of meanings. They were fantasy acts that taught what constituted the good life, how people should live, and what cities should be like. New towns were the material structure for knowing what to want in the second half of the twentieth century.

Especially during the reconstruction years after the Second World War and then the heyday of developmental modernism in the 1960s and early 1970s, these fantasies seemed attainable. It was a period of intense transformation. Indeed, utopian urban projects tend to appear in times of dramatic historical upheaval, and postwar society clearly qualified. When things fall apart, utopian energies are released, so there was a tenor of plenitude, optimism, and hopefulness during these years. The sense of immediacy was palpable. The point was to fulfill the hopes for the future, to accelerate progress and development. Although it may appear to us as no more than a whimsical chase, in the late twentieth century utopianism had genuine, sincere meaning. Reformers did not imagine it as impractical or impossible—it was a way of thinking, a mentality, a philosophical attitude. They believed that a radical transformation of the urban condition could be achieved. For instance, the plan for the new town of Nordweststadt outside Frankfurt begins with a quote published in the *Frankfurter Allgemeine Zeitung*: "We have the intellectual and material means to order our environment. With so many technical skills and so much scientific knowledge we need no longer put up with contaminated water, polluted air, disruptive noise, a mutilated landscape and a chaotic society."[5] There was a profusion of utopian projects and manifestos devoted to imagining cities differently.[6] The solution to urban problems lay in starting anew. New towns were the anodyne for all urban ills—a swift single shot at solving everything wrong about the past. They were a clarion call against the injustices of urban life, against the congestion and rudities that for reformers made big cities such a misery. At a 1972 conference on new towns at UCLA, anthropologist Margaret Mead claimed that they were an experiment in "life style":

If new towns would give us an opportunity to develop social forms and architectural styles that would make life more human again, then they could offer a great contribution. . . . The time seems to be ripe for experiment; those who could participate are clamoring for new life styles and recognizing that, without a well-designed structural basis in town planning, regional planning, and new architectural forms, the kind of life style they would like to see will be impossible.[7]

These were Promethean objectives. The transition to a better urban society required not just policy and planning but visionary thinking. Writing in 1960, Karl Mannheim described utopia as a "state of mind" that tends "to shatter either partially or wholly, the order of things prevailing at the time."[8] Even so, the operation of the utopian impulse cut both ways: it could also reinforce the established power structure. Fredric Jameson makes the point that utopia has two lines of descent: the one aims at founding a new society, the other is a host of "tempting swindles of the here and now," from liberal reform to commercial pipe dreams.[9] New towns displayed this sort of utopian hermeneutics. They were both a reflection on and a critique of mid- to late twentieth-century society. A steadfast belief in physical determinism was shared across the architectural and planning professions. An ideal social atmosphere could be achieved by carefully planning all the physical elements of the city. Designing the physical fabric would change individual behavior, social relations, civic life, and community. The assumption was that the ideal city could be mass-produced for a mass cultural age. Life would be balanced and harmonious.

The precise meaning of terms such as *balance, harmony,* and *quality of life*—mantras evoked repeatedly in explaining the benefits of new towns—remained ambiguous. Certainly, a decent home to raise a family, a living wage at a local job, clean air and water, and access to education and health benefits were uppermost in everyone's minds, especially in the immediate postwar years. New towns were deeply embedded in social welfare reform, even into the 1960s, when such programs as the Johnson administration's Great Society agenda were launched in the United States. The idea of the new town also awakened self-identity and self-improvement, a sense of belonging in neighborhood and community, the enjoyment of a balanced blend of work and leisure. By the 1960s, it additionally meant freedom, cruising down the highway in flashy cars, imbibing the new consumerism and the showy world of pop culture. A melting pot of architectural avant-gardes flowered in the freewheeling atmosphere of that decade.[10] These visionaries concocted new cities for the Space Age, played around with the stuff of lifestyle and leisure, and tinkered with technological gadgetry in conceiving a whole new way to live. They designed flying cars and sky-borne trains, heroic urban structures that pulled away from the earth's surface and expressed the vibrancy and optimism of the future. These were all associated with material progress, modernization, and the out-of-this-world possibilities for transforming urban life.

The hermeneutics of utopia provides a powerful way of reading the written and visual artifacts of the new town movement. A phenomenal quantity of publications, plans, statistical surveys, and case study investigations was written about new towns. Debates, conferences, and colloquia took place. Extensive architectural and design drawings, cybernetic logic diagrams, and a stunning photographic and cinematic record of new towns were produced. Together they reveal the complicated fascination with a romanticized urban ideal. Conjuring up a whole new world was an enchanted feat. The promotional machinery accompanying the new town movement was replete with slogans, imagery, and spellbinding drama about what the future could hold. Steeped in rational planning techniques and systems analysis, new towns were also mass-marketed products of the media and publicized as utopian theater. They entered into popular imagery as "happy cities" of harmony and bliss. The materiality of utopia can be found in the thousands of planning and policy documents produced in these years, and also in the ephemera of advertising and promotional films, dramatic articles in the popular press, and photographs of daring New People living in the urban future. It is a wide and open field of evidence.

My argument throughout this book is that the new town movement of the mid- to late twentieth century represented a rich corpus of ideas and influences that carried forward the inheritance of urban utopianism. This output about future cities was instrumental in both framing the social imaginary about how to live and shifting popular attitudes toward modernization and ultimately toward technocratic planning. It was an erudition shared widely within international planning culture—even more, an attitude or deportment about progress and the future. This book attempts to recover this discourse and practice of utopian aspiration.

Modernization and Official Sanction

To a real degree, the terminology around the new town in the second half of the twentieth century was about official sanction. A particular site was designated a new town as a formal endorsement of development by the state, by private corporations, by international development agencies, or most likely some combination of the three. New towns were capital- and infrastructure-intensive megaprojects. They included mass housing, mass transportation, highways and roads, water and sanitation, electricity and communications, the gigantic infrastructure of natural resource extraction. Yet the chronicle of a new town can at times be elusive. Sometimes it is not even easy to find one in a local landscape. The new towns' obscurity is made worse because they emerged in a variety of countries for different purposes. Moreover, the rhetoric about them can sound monotonous and overblown, and not worthy of much attention. Nonetheless, the new town *movement* is best seen as a significant intellectual undertaking and a meaningful artifact of the modernization era.

Temporality under modernization was accelerated. Maintaining this momentum required a total mobilization of resources. The label of *new town* signified massive flows of public and private money into infrastructure, and into urban and regional planning. New town projects were an economic stimulus package, and a way of mopping up surplus capital and labor after the crisis of depression and war. To borrow a phrase from James Scott, these were "state spaces."[11]

The state and its development-minded allies took responsibility for organizing the welfare and betterment of society, and assumed that society was a complex machine that only technocratic experts could operate. They turned urban society into an object of cognitive apparatus and rationalizing interventions. Newness was a defining feature of technocratic power and mastery. New towns were used to resettle refugees, to militarize and populate frontier territory, to tap unexploited natural resources and develop national infrastructure, to redistribute population, to tackle housing shortages and provide employment opportunities, to offer a better life to the citizenry.[12] In other words, new towns were not merely stage sets. They produced and reproduced state power, nation building, and modernization. The new town was an attempt to transform living conditions and old ways of life—and people—into something new. It was an occasion to display and legitimize, to territorialize the authority and beneficence of the state in a complete package. The residents of new towns were a spunky lot, no matter which new town they happened to live in. There was a sense of verve about them. They were young; many were highly skilled. They were pioneers, willing to shed the past and surge forward as citizens of the future. They were the New Men and Women of the modern age.

The power and agency of this modernization regime are revealed in the careful planning, the stereotypic schemes and their reiteration across space. The literal meaning of utopia is "no place," and new towns certainly suffer a reputation as identical nonentities out there somewhere in the mind-numbing land of suburbia. This is the predominant postmodern viewpoint, one that soundly rejects utopian thinking. Something isn't just utopian anymore: it is suspiciously utopian. This book steers away from this perspective by attempting to explore the currents of utopian thought and how they worked for specific purposes. The modernization regime believed fervently in the role of utopian illumination as vital to new possibilities. It had a positive role to play.

Utopianism also invoked the radical political energies, the technologism and futuristic expectations, of the 1960s. New towns were transparent spaces free from the past and the plagues of old cities, free from delusion and from political conflict. They promised happiness, self-determination, and a full civic life. Progress was the rational unfolding of these universal ideals. The constitutive elements of the city as well as the people who lived there were understood as abstract transposable types. Urban residents were categorized by race, class, age, income level. Urban infrastructure and services were fitted neatly into classifications and typologies. Liberated from the past and from political turmoil, the "population" of new towns could occupy itself with consumer and leisure activities. According to Michel Foucault, the surveying and redistribution of populations, the regulatory technologies of government and its developmental arms, the creating of typologies and hierarchies, were all the disciplinary apparatus of Western modernity. It was a model ready for international diffusion by experts. New town planners and architects swam in this broad current of technocratic know-how—developmental modernism and the institutional nexus that carried it out.

Yet modernism and modernization were never truly international. New towns spread across national borders with profound modifications. It is the irony of the modernization regime that it was implemented within local contexts. New towns were a standardized prescription: they all shared recognizable spatial forms. However, each national new town program was proclaimed to be different from those elsewhere in the world. Capitalist new towns were declared to be the opposite of communist new towns (and vice versa): new towns in what were deemed "underdeveloped" countries, in modernist parlance, were said to be unlike those in developed countries. Each state stood by its own vision of balanced and equitable regional development. New town utopias as a strategy of modernization were also deeply associated with the rhetoric and posturing of the Cold War. These political dimensions go almost entirely unmentioned in conventional histories of postwar urban planning. Nonetheless, utopias are always materialized through political and economic practices. The imaginary of urban utopia exuded the moral ordering and militarism of the Cold War years.

In both the Eastern and the Western blocs of the Cold War, new towns were designed for "containment," and for reining in political unrest and counteracting communist (or capitalist) influence. They were a tactic for militarizing, modernizing, and controlling territory on the Cold War frontier. Foucault pointed to the interplay of macrocosm and microcosm in the relationship between town, sovereignty, and territory. The drive for control over a territorial domain found its complement in colossal infrastructure and resource extraction projects, and in planned settlements. These structures were a communication of political power[13] on both sides of the Iron Curtain. As a strategy for political stabilization, new towns were then exported to countries within and beyond the US and Soviet spheres of influence. They functioned in the international arena as neocolonial agents abetting the transfer of Western ideals of progress, which were then incorporated directly into the social sciences and urban planning. New town projects were imbued with promises of freedom, democracy, and prosperity and laden with Western notions about family and community. They were sweeping statements of Western modernization.

However, recent scholarship has upended this story and the notion that modernism and modernization established an institutional order and a framework of power regardless of place. There exist instead a variety of modernisms that respond to local urban circumstance and are shaped by the exigencies of urban history and culture. Creative adaptations of the modernist paradigm are highlighted in this perspective. In the words of Dilip Parameshwar Gaonkar, "A people 'make' themselves modern, as opposed to being 'made' modern by alien and impersonal forces." They rise to meet modernism, negotiate it, and appropriate it in their own fashion,[14] asserting their individuality against the great generality of the modern age.

Modernization, then, produced difference as much as it produced uniformity. It brought about diversity and complexity in a myriad of urban contexts. The selection of new towns in this book, the crisscrossing of geography and planning cultures, is meant to showcase this richness. While new town architects and planners circulated what they deemed an awe-inspiring discourse on the urban future, it was habitually altered and appropriated by local practitioners. Scholarship about the forces of modernization has focused on the apolitical subjectivity of the citizen-resident and the active denial of any possibility of agency or political action. And indeed, new towns were usually controlled by "development authorities" that oversaw every aspect of their construction and progress. Yet new town schemes as externalities applied everywhere homogeneously can tell us only so much unless we take into acount the local culture as well as the sociocultural tensions implicit to any city or urban region. In the case of the French new town program, which is considered to be among the most blatant cases of state authoritarianism, new town pioneers reveled in the sense of community and local participatory democracy. "We were all involved in local politics . . . there was a spirit of community that even guided the EPA [development authority] in their participatory management style: the neighborhoods worked like a cooperative," remembered a resident of the new town of Evry.[15] It was a brave new world.

Dispersal and Suburbia

Late twentieth-century new towns have largely escaped not only study but even acknowledgment as places. One obvious reason for this oversight is their peripheral location: they are out in the sticks. Another is that they generally have reputations as drab and boring. For most people, if you've seen one new town, you've seen them all. This makes them easy to ignore, especially since urban history has traditionally focused on the great Western capital cities, and even then on the experience of modernity in their most spectacular centralized spaces. But by the mid-twentieth century, it was exactly the central city that came under withering condemnation for its dreariness and squalor. The overcrowding, the wretched housing and sordid neighborhoods, had turned progressive reformers into militant agitators against "old cities." Their fiery rebukes bordered on the fanatical and made nineteenth-century urban reformers seem meek by comparison. The vocabulary of "urban nightmare," "crisis," and "tragedy" introduced seemingly every twentieth-century document about planning. In the aftermath of the Second World War, the "housing crisis" was unrelenting. The "population boom" threatened apocalypse. Environmental disaster would soon follow. The solution to all this impending doom was sweeping slum clearance projects in the inner cities and the dispersal of the population. For the most part, scholars have interpreted the outcome of this policy as uncontrolled suburban sprawl. The new town movement suggests the need to overturn this conviction. Brand-new towns were exciting because they offered people, especially young families, a modern future. Planned development in a regional configuration was highly articulated, politically produced, and implemented on a far wider scale than historians acknowledge. Retrieving this history is one of the purposes of this book.

The late twentieth century was the heyday of regional planning, with new towns rationally distributed throughout a metropolitan area. Ebenezer Howard's garden city was the original version of this utopian dream, but there were many others. They merged into a broad agenda for managing regional resettlement that took hold in all the countries of Europe as well as in the United States, along with their former colonial empires. This is what makes twentieth-century planning history such a conundrum. Rather than emphasizing centrality, it must focus on diffusion. The ideas, designs, and plans that thrilled urban visionaries entailed not just inner cities but also urban peripheries, suburbs, new towns, and regional planning. Planners spent their time tracking the spread of the "urban cancer" across the territory and wringing their hands over how to control it. They toiled over demographic statistics, maps, and graphs, and learned early computer programs that could accurately forecast movement and change across vast geographic areas.

The Second World War transformed the entire scale of planning discourse and propelled it into the era of postwar reconstruction. The war had been a monumental conflict defined by massive population shifts, fluctuating territorial boundaries, titanic planning of material resources and infrastructure, and unprecedented state control over economies and societies. This legacy is vital to understanding the extraordinary admiration for and acceptance of planning in the second half of the twentieth century, and the scale at which it was carried out. City building across regions became state policy, a mechanism for controlling assets and territory at an unprecedented scale. It made new towns an instrument of political and economic power. Urban practitioners articulated complex notions about how this metropolitan dispersion should take place, and theorized extensively about how settlements should be formed. They went far beyond the garden city in imagining the future. Dispersal, in their view, would be concentrated around specific nodes or towns that would reproduce the urbanity of big cities on a smaller, more humane scale. The towns would be interconnected in a web of communications and transportation as a complete territorial *system*.

This tension between dispersal and concentration is one of the most significant features of twentieth-century planning. It forces us to consider the nexus between spatial planning and social engineering from a broad geographic point of view. It also entails an examination of how new town visions and regional planning were enacted not only in the West but in the developing world, where their relationship to colonial and postcolonial politics is too obvious to ignore. Planning was from its inception a transnational endeavor; the new town concept traveled around the world. Planners shared an elevated vision of urban and regional development, and planning indeed became professionalized and internationalized in no small part due to new town development. It was a dynamic concept, and the experience of numerous towns and cities across the globe contributed to its philosophy and operation.

From Craft to Science: Planners and New Towns

Studying the movement to build new towns thus tells us much about the global transmission of what constituted ideal places and how this influenced planning culture. Planners and architects arrived at their desks with a set of preconceptions derived from complex political and ideological cultures as well as the continuous circulation and appropriation of ideas. For many of them, their approach to urban utopia was conditioned by years in wartime exile, and then by the immediacy of reconstruction. They began their work while the war's rubble was still smoldering. The passions of progressive urban reformers were fired by their desire to create a better world, to unite as "one world" in a global community, to fight social injustice and poverty, and to promote freedom. These were highminded goals that resonated for them. The artifacts of their ambitions were an extraordinary cache of plans, diagrams, and illustrations of perfected urban places. By the 1960s, these visions were generated increasingly by cybernetics and systems theory, which opened the prospect of networked "intelligent" cities. The two, along with the radical culture of the 1960s, produced an effervescence of utopian imaginaries that added to this mix. Urban reformers believed in the possibilities of the future and their ability to create it. They foresaw the colonization of outer space, and divined colossal megastructures, automobile utopias, fantastic webs of "plugged-in" cities, and far-flung universal settlements. They toiled on projects worldwide, cooking up ideas in a crusade to save the planet from imminent urban and ecological doom. This impassioned journeying of people, urban texts, and utopian imagery acted as a shared global corpus that could be manipulated, modified, and operated within local circumstance.

In a certain sense, urban and regional planning became a global production network with its own properties, actors, institutions, and organizations. As professional expertise, it ranged from artistic craft to technocratic science. We will meet a variety of actors time and again in this book, setting out new towns in England, Canada, and the United States, in Israel and Iran, in China and Africa. Practitioners such as Constantinos Doxiadis, Victor Gruen, and James Rouse became international celebrities. Other planners worked unheralded for the RAND Corporation, the Ford Foundation, and the United Nations. Many more made successful careers in government and in new town development corporations. These last have remained obscure in accounts of twentieth-century urban history. Architects, engineers, social scientists all contributed to the new town movement as an intellectual endeavor. The nature of their work ran the gamut from the flamboyant in the case of the architectural avant-garde to the mundane in the case of highway engineers. Altogether as planning elites, they succeeded in designing and building an extraordinary number of groundbreaking new towns, and they left a rich intellectual legacy of planning theory and utopian aspiration.

Planning new towns helped forge a new self-awareness and confidence in professional planning as a fully transportable, rational endeavor. These new town builders came from two crucial generations; my discussion of them rather neatly divides the book into two parts. The first generation of builders began their careers in the 1920s and 1930s under the influence of the Bauhaus and CIAM (Congrès internationaux d'architecture moderne), which forged early new town and regionalist thinking. They endured the Great Depression, wartime rupture, and exile as frustrating obstacles to their aspirations. Some suffered violent political repression and internment in concentration camps, whereas others managed to skirt the political minefields. They arrived in the postwar world eager to put their ideas into action, and produced the first great wave of new towns. Reconstruction gave them the platform to implement projects unimaginable in the early days of their careers. Their conceptual frame of reference came largely from the prewar world, especially the regionalist vision of garden cities with their neighborhood units. They were largely architects by training and came from a tradition of planning as art.

Chapter 1 provides the background of this generation's prewar inheritance. Chapters 2 and 3 examine the new towns they produced during the postwar reconstruction years of the late 1940s and 1950s. The analysis concentrates in particular on the application and export of the garden city ideal, which emanated from Great Britain and the United States and spread worldwide before experiencing its demise in the resource towns of the Middle East and Africa.

The second generation of new town builders came of professional age after the war, in the 1950s and 1960s. They started their careers ready to cut ties with the past and launch into the future of urban form. The result was an entirely new scale and complexity in the second great wave of new towns in the 1960s and 1970s, which is the subject of the last three chapters. Planners worked hand in hand with experts in the military and defense industry, with sociologists and economists, computer geeks and civil engineers. They formed a new postwar class of technocratic elites at the center of decision making. They traveled extensively, networked prodigiously, and shared an emergent body of knowledge tied to cybernetics and systems analysis. Their work was not steeped in any soppy sentimentality about garden cities. Instead, it was a full-bodied debate over the practices and spatiality of collective life in the age of mass society.

The influence of cybernetics and systems theory on this generation of urban practitioners was immense. Recognition of this intellectual impact is one of the objectives of this book. Information and communication theory, applied mathematics, and game theory were regarded as the answers to urban problems. The new town operated as an intelligent system, and the ability to grasp and control all the intertwined connections and feedback loops was dramatically enhanced by the computer. Cybernetics was also deeply bound to advances in aerospace and to the technological wizardry associated with the Space Age.

Chapter 4 focuses on the impact of cybernetics and of the Space Age more generally on the practice of utopia, and on the techniques and circuits of knowledge and international exchange. American and Eastern Bloc versions of cybernetic cities are evaluated in relation to the Cold War and the space race between the United States and the Soviet Union. In chapter 5, the impact of this cutting-edge technological culture of the 1960s is teased apart in terms of comprehensive metropolitan planning, and four new town case studies in Britain, France, India, and the United States are considered. Lastly, in chapter 6, the impact of Space Age advances is assessed from the vantage point of avant-garde architecture and the myriad visionary urban designs and fantastical cities of the far future it produced.

Altogether, an enormous intellectual debate took place in the mid- to late twentieth century about the theory and form of new towns. For some planners, a new town was a settlement started from scratch—it was newly built. Others claimed that it was a comprehensively planned, self-contained community based on precise formulas of size, population, and economic and employment characteristics. There also was an obsession with new town typologies and hierarchies. New town enthusiasts spent countless pages attempting to divine the physiognomies of freestanding new towns, satellite towns, and company towns, and the attributes of small versus medium-sized new towns. They pored over regional maps that depicted new towns as nodes in elaborate highway systems. They brainstormed with sociologists and psychologists to plan the precise qualities of neighborhood and community, of public space and centrality, and the correct architectural style and landscape design. They ruminated over the correct computer modeling techniques. For some, it was the "heart" of the new town that mattered; for others, it was sustaining life and the earth's environment. Few planners began with any knowledge of or experience with the vast lived territory in which they were planting their imaginaries. If anything, they viewed urban peripheries as a disorganized emptiness to be rationalized and modernized. This opened up the possibility of creating towns ex nihilo.

Along with the discourse went endless drawings and illustrations, statistical data, cartographic and photographic evidence, logic diagrams, and mathematical formulas that both figuratively and literally mapped out visions of urban utopia. They became a corpus of avant-garde visual art that deserves to be treated in its own right as a creative endeavor. The book attempts a deep reading of these designs—how they dramatized the urban landscape and gave it the appearance of spectacle, at times atmospheric and emotional.

The consequences of all this planning were enormous: cities and regions were viewed as complex systems with rational patterns of land use and transportation. Each new town was itself a unified system of carefully arranged buildings, spaces, and traffic arteries. Social stability would be assured by a perfected spatial realm. New towns put planning in high relief and elevated urban visionaries to a scientific avant-garde. The global dimension of this planning discourse was remarkable, though little has been written about its having produced some of the most influential texts of twentieth-century planning. These writings were shared internationally as recipes for designing and building the future. The result was a new town *movement* that stimulated an extraordinary intellectual engagement of diversity and consequence.

The Shadow of Dystopia

Of course, utopias are notoriously fickle enterprises. They are always *elsewhere*, just beyond the horizon. Even more pertinent to our discussion, lurking just below their mesmerizing vision of the future is the dark shadow of dystopia. Lewis Mumford, who had tremendous influence on the postwar new town movement, perhaps best captured the dialectic deeply rooted in the city as ideal aspiration. Writing in the 1960s, he described the historical genesis of the city as a symbolic representation of the universe. It was Tommaso Campanella's famous *City of the Sun* (1602), now brought down from heaven. But the price of this ideal was "isolation, stratification, fixation, regimentation, standardization, militarization." In the end, Mumford postulated, "utopia merges into the dystopia of the twentieth century; and one suddenly realizes that the distance between the positive ideal and the negative one was never so great as the advocates or admirers of utopia had professed."[16]

Just as quickly as they were hailed as utopias, new towns were pronounced complete fiascoes. At its core, after all, the concept of utopia implies a ruthless assessment of reality. For Ernst Bloch, the preeminent philosopher of utopia, "The essential function of utopia is a critique of what is present."[17] Despite the ideal conditions a new town supposedly embodied, blowing the whistle on the miseries of living in one was a way of venting these underlying tensions between ideal and debacle. New towns were pronounced inhuman, brutal, and psychologically stultifying. They were ugly from the start and fell into ruin almost immediately. They were sobering examples of state hubris—colossal megaprojects that ended in colossal failure. It was precisely because so much was expected of them as utopian apparatus that the reaction against them could be so violently critical. Feelings about new towns oscillated wildly between love and hate. This was just as much about the feverish role utopias play in the modern imagination as it was about the places themselves. The quest for perfection can slide toward collapse with amazing ease.

This is precisely why the new town movement in the second half of the twentieth century is worth studying. New towns can seem quite trivial in comparison with the other vast forces of urbanization and suburbanization that characterized this period. It is easy to be entirely skeptical about what new towns actually are, to say nothing of their out-of-the-way locations. Moreover, the Anglo-American colloquial meaning of *utopia* is generally pejorative, suggesting a project that is naïve and wildly impractical. In reality, whether or not these projects were actually successful depended not on their utopian lineage but on where they were located, how they were maintained, who lived in them, and a variety of homegrown political, social, and cultural exigencies.

All cities are produced through a complex, often opaque system of collective decision making by government and industry, architects, planners and builders, and residents who may cooperate or resist. Despite the myth that they were free of politics, large-scale development projects such as new towns were always fraught with convolutions, from local political wrangling to nation building and the Cold War–era jockeying for territorial power. For the most part, the evolution of individual new towns actually differed little from other development projects. They suffered all the usual problems of cost overruns, construction delays, and problems with housing and services. A few became stellar fantasies, part of the urbanistic lexicon of the twentieth century. Some were highly successful; others were dismal failures. Some became the equivalent of gated communities for privileged residents who enjoyed the exclusivity of a perfectly planned and carefully controlled suburban paradise. More humdrum new town projects evolved; these merged into the general sprawl of suburbia. Others simply died away. Although they may seem marginal, new towns actually played an outsized, fantastical, and, we might even say, perverse role in the urban imagination.

Framework of the Book

The first question I am usually asked about this book is the obvious one— what exactly is a new town? I have consciously resisted positing any kind of ultimate definition, any more than I would for the idea of "neighborhood" or "community," or even "city" for that matter. The meaning of these concepts is contingent on time and place. The task of the historian is to discover these meanings and investigate why the new town was such a powerful talisman of the future. This book does not attempt to describe individual new towns in detail or to cover their extraordinary numbers. Nor is it intended in any way as a survey of new towns, which would produce a monotonous litany of places. There are also myriad ways one could assess the new town *movement*: its success and failure, and the social consequences of living in utopia, have always been controversial. How planning affected the lives of people who actually lived in new towns, for example, deserves analysis in its own right. My study is not a physical description of these settlements as they were actually built or an assessment of their social impact. Instead, it offers an intellectual dissection of them as visionary dreamscapes.

To that end, I have selected as case studies new towns that range across geography and across the various intellectual crosscurrents that spawned them. Some were quite ordinary and quickly forgotten; others were the megastars of the new town movement. They are considered in terms of practice, culture, and discourse, rather than mechanics—how they were actually fashioned or what resulted. The focus is on the intellectual, cultural, and political expressions of new towns as a vibrant *movement*, and the ways in which these places represented an imagined utopia. They are proof of the extraordinary depth and extent of the mid- to late twentieth-century debate on the future of cities as well as the range of experiments that was tried and the fluidity of ideas and concepts. A host of planners, architects, urban reformers, and futurists engaged in an international dialogue about how to design and build an ideal town. The book's focus is on these individuals as much as on the new towns themselves. The artifacts left by their work are remarkable in their scope and diversity, and deserve our scholarly attention.

Writing a book about new towns requires wading through volumes of planning whimsies and wonkery. Rather than take a heavy-handed critical posture toward it, I have tried to let this new town crusade speak for itself. In studying documents and other materials, I have relied on the historian's skills and played the role of investigator, mythbuster, and skeptic as warranted. At the same time, I was conscious of our own twenty-first-century urban crises: megacities spreading across vast hinterlands with little logic; millions of people living in favelas, neglected urban outskirts, and refugee camps; environmental degradation; global warming and sea-level rise. From this vantage point, I became increasingly impatient with any contemporary fixations on the garden city as a stand-in for sustainability, any belief that the fashion for "systems of systems," intelligent cities, or smart cities can save the day. I was also aware of the current crusade to build new towns, and the explosion of new urban places popping up seemingly everywhere, especially in Asia and the Middle East. As the world's population surges and becomes ever more urban, building new towns is becoming a genuine fixation once again. It seemed an appropriate time to dig out the neglected history of the new town movement and get a better fix on what planners were trying to achieve and why. I have chosen new town projects that seem to answer these questions the best and exemplify the evolution of urban ideals.

No doubt there are many other new towns that would fit these criteria just as well: the sheer number of towns to study was dizzying. Researching a transnational history was also a daunting task. I found new towns in predictable as well as unpredictable places across the globe. The limitations of evidence, language, and understanding of local circumstance played a role in my selection of the towns studied. I have the utmost respect for the painstaking and extraordinary work done by scholars focusing on individual new town case studies. I have relied heavily on their scholarship along with archival and published sources.

There is far more research to be done on towns and cities—new or not—outside the West, particularly in Africa and the Middle East. This is also true of the impact of cybernetics and systems analysis on the history of architecture and urban planning. The chapters in this book that address the systems revolution are only surface treatments of a topic of extraordinary depth and complexity. Given these limitations and constraints, the evidence presented here treats the city—in this case the new town—as the legible artifact of an era. Some of the new towns in this book remained on paper as fantasies, while others were actually built: some of the towns were quite small, others were stunning megaproductions. I concentrate on their power as a utopian vision; how these projects invented and spread a modern Western ideal of the good life and urban form; and the twentieth-century context in which they operated.

My goal in these pages is to provide an interdisciplinary analysis of the new town movement that explores its intellectual and ideological foundations, what it suggests about the politics of modernization and urban planning, and how it has shaped our understanding of the urban world. My hope as well is to present a fresh perspective on the strengths and weaknesses of comprehensive urban policies, and provide a new basis on which to understand the challenges of contemporary urbanization. Ultimately, my position is that utopia is a fundamental aspect of humanity, and that utopian projects spur people into bringing about a better future. To quote Karl Mannheim, "The disappearance of utopia brings about a static state of affairs in which man himself becomes no more than a thing"; he would be left without ideals, and "would lose his will to shape history and therewith his ability to understand it."[18]

ONE

The Origins of the New Town Movement

From the very beginning, new towns shared a complicated birthright. In traditional planning narratives, they all traced their lineage back to Ebenezer Howard's garden city movement. But this pedigree was very selective: by the mid-twentieth century, Howard's radical socialist program and cooperative vision were already consigned to the curiosity cabinet. The garden city had become an Arcadian reverie that more than anything represented a longing for the past. Yet this "garden" imagery remained one of the taproots of the new town movement through the end of the twentieth century, making it a blend of progressive, yet conservative ideology that hindered thinking about cities as much as it helped. Claiming origins in the garden city also sidestepped the more problematic influences on urban and regional planning that had intervened in the years after Howard's groundbreaking proposal. In truth, planning new towns was deeply rooted in the experience of colonialism, war, and military policy, and the expansion of state authority. It was wrapped up in successive visions of regionalism that dramatically enlarged the scale and power of settlement strategies. Tracking the origins of the new town movement thus takes us initially to Howard's legendary garden city of Letchworth, England, but it quickly moves us in other directions. Along with the British garden city movement, German and Soviet reform initiatives were vital to the birth of the new town movement, as was the American experience. Accordingly, the examples drawn in this chapter focus on these histories and weave the struggle for urban reform directly into the crises of the mid-twentieth century.

Garden Cities of Tomorrow

In the original formulation of *Garden Cities of To-Morrow* (1902), Howard had suggested a radical transformation of the social and physical environments. The ghastly slums of the turn of the century would be left behind. The working classes would find new life in self-sufficient planned communities that balanced individual and community needs. These hopes underlay all utopian urban experiments and were the core ambitions of the entire new town movement. Howard's version suggested a population of thirty-two thousand people living on one thousand acres of land. The city would be a locally managed, limited-liability company that attracted light industry, jobs, and services. Land would be owned in common. In the pioneering spirit of homesteading, people would build their own homes and open their own small-scale shops.

Howard was influenced by an amalgam of ideas, from early utopian socialism and cooperative movements to anarchist Peter Kropotkin's belief that new technologies would allow the dispersal of populations away from the horrors of the industrial city. The garden city trusted the restorative qualities of nature and the ways it could be merged with vernacular culture to create a perfected living environment. It offered the benefits of urban living together with the advantages of country life, and ultimately both individual freedom and social cooperation. It was imagined as circular in form, about a kilometer in radius, so that people could easily walk between home and work. The town center would have a park, public buildings, and a shopping arcade. As soon as a settlement reached its maximum size, a new version would be established with an agricultural belt separating the two, until finally a cluster of towns set in the countryside would emerge connected by a rapid transit system. Howard called this polycentric vision Social City. It was a "commonsense socialism" visually depicted in his iconic diagram as the "Third Magnet" that married the merits of city and country, and was superior to both Victorian capitalism and centralized socialism.[1]

The reaction to Howard's sophisticated fusing of urban design with social reform was unrestrained enthusiasm. The campaign for garden cities merged seamlessly into the climate of internationalism already well established by the progressive social reform and social philanthropy movements. Soon after its publication, *Garden Cities of To-Morrow* was translated into a host of languages, and local garden city associations sprung up across Europe.

The garden city concept was introduced into Russia by 1908, and social reformers in Saint Petersburg published a Russian translation of Howard's text in 1911. A small party of Russians made the pilgrimage to Letchworth in 1909 with German garden city enthusiasts, and again in 1911 with a Danish and German delegation. The influential Russian architectural journal *Gorodskoe Delo* eagerly promoted garden city ventures, while architect Vladimir Semionov worked with architect and urban reformer Raymond Unwin in England and wrote extensively on garden city ideals. Based on Semionov's design, the Moscow-Kazan Railway Company began construction of Russia's first model garden city at the Prozorovskaia Station, forty kilometers east of Moscow.[2] It was such a success that the Russian Ministry of Transport began building similar settlements for railway employees. Garden city–style projects popped up in Siberia, where an All-Russian Garden Cities Society was founded.

Over the course of the twentieth century, Russia established more garden cities and new towns than any other nation. But experiments in garden cities, garden suburbs, and model villages were also cropping up everywhere in Europe and beyond in the United States, Canada, Australia, and Japan,[3] although many were no more than residential developments dressed in greenery. As in Russia, a good portion of these was developed by railway companies eager to provide suburban housing for middle-class commuters. The development of the railways introduced an entirely new spatial geography of metropolitan growth.

Exploiting new technologies was what made the garden city a locus for the dreams and desires of progressive reformers. Yet it was flavored with a curious blend of conservative values. Nostalgia for vernacular forms permeated the garden city fantasy, as did an elitist penchant for rural reveries and old-style rusticity. Reformers envisioned pleasant, untroubled towns and villages set amid foliage and trees. These sylvan hollows would temper the forces of modern life. Living in nature was associated with healthiness and purity of the soul, with getting "back to the land" and a simpler way of living. Garden city projects were often associated with quirky alternative communities of privileged middle-class reformers who espoused everything from vegetarianism to utopian communitarianism.

Essential to the international acclaim for the garden city ideal were Ebenezer Howard and Raymond Unwin. They were the great proselytizers, a veritable two-man, globe-trotting traveling show. The diagram of the Three Magnets and photographs of Letchworth became promotional icons for spreading the gospel. In addition, texts by renowned practitioners such as Unwin provided a conceptual canon for early planning professionals. Unwin's *Town Planning in Practice* (1909) and *Nothing Gained by Overcrowding* (1912) were perhaps even more influential than Howard's work, with a reach as far as the garden suburb experiment on the fringes of Adelaide in Australia. England's Garden City Association became a fine-tuned publicity machine, hosting conferences and overseas lecture tours. It welcomed foreign visitors for tours of Letchworth and Hampstead Garden Suburb, and, in Yorkshire, tours of Woodlands, laid out for the miners at the Brodsworth Colliery, and Hull Garden Suburb, for the workers at Reckitt's chemical works.

In the summer of 1912, while traveling to Cracow in Russian Poland, Howard announced the formation of the International Garden Cities and Town Planning Association. It was meant to capitalize on international activities, but it was also a strategy for maintaining some control over the rampant garden city projects taking place.[4] To the chagrin of the faithful, Unwin's arresting Arts and Crafts architectural imagery set amid pastoral gardens overwhelmed Howard's original social reforms. Howard never defined his garden city vision by its built environment, only by its social processes. But in the general eagerness to pursue utopia, the garden city was becoming an urban design strategy in the narrow sense of the word. The physical quality of these places was undeniable, and it was easy for urban reformers to evade the more challenging social communitarianism that was essential to Howard's formulation. The militant orthodoxy of the English garden city movement was also related to competition from the International Housing Congresses and the International Union of Towns, to take just two of the emerging rival organizations. Although the International Garden Cities and Town Planning Association was a welcome forum for cooperation and ideas, its members almost immediately felt strongarmed by its English kingpins.

However, the most formidable competition for leadership of the garden city movement came from Germany. The English-German relationship was the axis around which the international movement spun, and it was caught in the undertow of political suspicions between the two nations. While they blithely smiled together at garden city events, reformers from both countries surreptitiously looked for signs of "national vitality" in each other's planning practices. Yet historian Stephen Ward has made the point that despite these frictions, British reformers constructed Germany as an imaginative geography of innovation that served as a source of inspiration and ideas.[5] Some eighty garden city experiments appeared in that country. Beginning in 1912, the garden village of Margarethenhöhe on the outskirts of Essen was developed by the Krupp family of munition makers, based on the ideas of Camilo Sitte, Howard, and Unwin. Two of the best-known early efforts were Berlin's garden suburb of Falkenberg and the garden suburb of Hellerau near Dresden. Hellerau, sponsored by the Deutsche Werkstätten as a vehicle for fashioning applied-arts workers into a cohesive trade class, was not far from Ebenezer Howard's original social reform ideals. The Werkstätten and the garden city movement allied with each other in a common mission of national education and economic prosperity.[6]

By the eve of the First World War, the German Garden City Society already claimed over two thousand members. It appealed to urban reformers and municipal officials, industrialists, and a gamut of land reform enthusiasts. Some followers were immersed in nostalgia for the *völkisch* folkloric past, while others sought a progressive, modern lifestyle. For the most part, however, planning remained a "wealthy man's toy."[7] The German garden city movement was saturated in conservative desires for healthy village living that skirted the edges of racial consciousness. Many of the new recruits were admirers of Theodor Fritsch, author of the virulently anti-Semitic 1896 pamphlet *Die Stadt der Zukunft* (The City of the Future), which vaguely resembled Howard's garden city ideal. A periodical published by Fritsch, the *Hammer*, warned that housing conditions in large cities eroded the vitality of the German people. In addtion, it espoused a rigidly segregated social order in small towns and railed against Jewish land speculators.

As this brief survey suggests, by the early twentieth century the garden city movement was the main channel for the exchange of urban planning ideas and experiences throughout the world. The Garden City Association and the small garden suburbs and garden-style housing estates that sprang up in the early years of the century served, according to historian Thomas Sieverts, as "experimental fields of modernity," places where the interplay between social practices, site planning, and design could be tried out.[8] The career of pioneering urban planner Thomas Adams illustrates this transnational voyaging of people and texts. Adams was the first manager at Letchworth and secretary of the Garden City Association. Invited to become an adviser to the Canadian government, he developed a series of model plans for its nation's towns, including the garden suburb for the Richmond District of Halifax[9] and Temiskaming, constructed in 1919 along the Saint Lawrence Seaway by the Riordan Pulp and Paper Company. He then moved to New York, where he led the preparation and writing of the *Regional Plan of New York and Its Environs* (1929),[10] the first master plan for the New York metropolitan region.

The First World War and the New Regionalism

In their comparative study of the interwar years, sociologists Christian Topalov and Susanna Magri argue that reconstruction after the First World

War was a turning point in the history of urban planning.[11] The dream of Ebenezer Howard's garden city transmuted into a far broader vision of scientific management at the regional scale. The horrifying brutality of the war ignited a moving vision of reconstruction and a dream of soldiers returning home to a better life. The emotions behind these ideas translated new towns into places of transcendence, arising from the ashes. They gave the embryonic movement immediacy and ardor, a sense of obligation to bring humanity out of the darkness and into a new and better world. The early trials at garden cities turned into a duty to construct a new social order. Faced with the war's destruction in Belgium, in 1919 the International Garden City Congress, held in Ghent, made an impassioned plea to clear away the ruins and build a better, peaceful world. There was a sense of urgency, a need for comprehensive town and regional planning that dealt with everything from transportation to food supply, and especially housing.

The quaint cottage designs of the original garden cities had little to offer in the face of massive privation. In England, David Lloyd George won the 1918 election for prime minister with his Homes for Heroes campaign on behalf of troops returning from the western front. Frederic J. Osborn, one of the most influential evangelists for the garden city movement, urged the building of a hundred new towns on garden city principles in his *New Towns after the War* (1918). Osborn and a small band of media-savvy enthusiasts organized themselves into the New Towns Men to proselytize their message during the Homes for Heroes drive.

This connection to war and reconstruction is vital to understanding the visionary quality of new towns and the budding regionalist movement. James Scott has argued that the blossoming of modernist projects took place at particular moments of crises, "such as wars and economic depressions, and circumstances in which a state's power for relatively unimpeded planning is greatly enhanced."[12] It was precisely the calamities of the First World War through the Great Depression and especially the Second World War that created this kind of opportunity for massive, state-led public works projects such as new towns. They gave the new town movement a martial quality, a fervor and intensity that was intimately connected to violent upheaval and renewal. For instance, during the First World War, Raymond Unwin worked for the Ministry of Munitions, building the settlement of Gretna-Eastriggs in southern Scotland for workers in the armaments industry. He argued passionately for satellite cities laid out in a regional array. The idea of regional dispersal was directly related to protecting wartime industries from enemy attack. Details of Unwin's Gretna-Eastriggs plans were sent to the United States and used as the basis for garden city–style armaments settlements once that nation entered the war in 1917.[13] In addition, Unwin became involved with the London Society, a reformist group interested in thinking through a plan to restructure the London metropolitan area. In 1921, it published a small volume of essays entitled *London of the Future*.

Unwin continually refined his ideas on metropolitan decentralization, presenting them at the annual congresses of the International Garden City and Town Planning Federation.[14] As the most powerful voice of the garden city movement, his efforts to broaden Howard's original concepts into regional planning were instrumental. By 1927, he was acting as technical adviser to the Ministry of Health's first committee charged with hashing out a Greater London regional plan.

The need for regional dispersion was made even more urgent in Britain by the Great Depression and the suffering of regions caught in the spiral of economic decline. A hue and cry were raised for regional planning and the relocation of industries to redress geographic inequities and provide relief for areas entangled in the depression's snare. The government-sponsored *Marley Report* (1935) and the *Malcolm Stewart Report* (1936) outlined an ambitious program of industrial dispersion and the building of new towns on the garden city model. Also during the 1930s, the Town and Country Planning Association launched a whirlwind new town campaign. A cavalcade of urban reformers pounded on the urban problems of Britain and, step by step, pushed government officials into action. Arthur Trystan Edwards founded the Hundred New Towns Association in 1933 with the publication of his pamphlet *A Hundred New Towns for Britain*, written under the pseudonym "Ex-serviceman J 47485." His plan was for 5 million Britons to be rehoused in towns of fifty thousand people apiece.

On becoming Conservative prime minister in 1938, Neville Chamberlain appointed a Royal Commission under the chairmanship of Sir Anderson Montague-Barlow that for the first time raised new towns as official public policy. The *Barlow Report* was a watershed publication that carried out the long-overdue survey of British towns. According to historian Peter Hall, the report established the principles of urban containment and the dispersal of population and industry that underpinned all postwar new town legislation. And it endorsed the idea of a Ministry of Town and Country Planning to carry this policy out.

A wide consensus took shape among progressive reformers on both sides of the Atlantic. For them, the way forward entailed not just decent working-class housing but the rational provision of services and schools, clean air, water, and sanitation; the improvement of daily life; and a sense of social solidarity with the masses of people who had for too long endured the miseries of slums.[15] Reform was about creating new cities for a new life. The built environment and social life were realized to be interdependent. The planner's challenge was to coordinate and balance the relationship between the parts and the whole—central city, satellite towns, and metropolitan region. Technical studies and social science methods had to be called on. This scientific approach required surveys, data, charts, and schematic diagrams.

The problems associated with urbanization provoked a general rationalization of geography. In Britain, for example, planning pioneer Patrick Abercrombie worked on a host of regional plans for blighted areas. In Germany in 1920, a regional planning association titled Siedlungsverband Ruhrkohlenbezirk was organized for the Ruhr industrial region to guide future growth and dispersion into new towns, although it lacked any formal power. Regional plans also were developed for the cities of Berlin, Stuttgart, Hamburg, and Mainz. In the Netherlands, the vast engineering project to dam the Zuiderzee and reclaim the Ijsselmeer polders of South Holland triggered a regional plan that eventually became the Randstad.[16] French urban reformer Henri Sellier planned sixteen garden suburbs around Paris between 1916 and 1939. The first generation of French urban planners began lobbying for a regional vision of "Grand Paris." Planning luminary Henri Prost led the initial surveys of the metropolitan region at the new Institut d'urbanisme de Paris. The result was a comprehensive map of the entire area and a new understanding of metropolitan expansion. It became the basis for the original *Plan d'aménagement de la région parisienne*, legally enacted in 1939. In New York, the Regional Plan Association (founded in 1922 by business leaders and progressive reformers) was lobbying hard for implementing the landmark *Regional Plan of New York and Its Environs*.

These interwar years of experimentation with new towns and regionalism bore a variety of fruit. The modernist movement emerging in Weimar Germany also captured the imagination of urban reformers. Bruno Taut and Walter Gropius both worked on garden city housing estates, or *Siedlungen*, before the war: the former at Britz, the latter at Siemensstadt. They largely eschewed the picturesque vernacular for blocks of simple modernist flats, but the ideal of relocating the city's teeming masses amid gardens and greenery remained intact. More moderate in his ideas than Taut or Gropius, architect Ernst May worked for Raymond Unwin at Letchworth and at Hampstead Garden Suburb. He built on Unwin's theories as well as those for the Deutsche Werkstätten garden suburb at Hellerau, especially as head of the Silesian Rural Settlement Authority between 1919 and 1925, when he produced more than a dozen settlements.

May put his concepts to work most successfully in the New Frankfurt initiative. Frankfurt was a focal point in the productive merging of reform initiatives in the 1920s and early 1930s. The city's newly elected social democratic municipal government incorporated outlying towns and suburbs into an enlarged metropolitan area and appointed May its planning director. With May's help, mayor Ludwig Landmann conceived of a New Frankfurt that would embody a new era of modernity and social reform.[17] Brilliant marketing tacticians, the two men created a city logo, publicized their ideas in the slick *Das Neue Frankfurt* magazine, gave lecture tours, and went on the radio.

Ten percent of the city's population would be resettled in a plan that originally laid out twenty-four satellite towns. They were experiments in a new concept of living culture, or *Wohnkultur*, one of the key utopian concepts developed by modernists in their attempt to use architecture and planning as instruments for restructuring society. Only in new settlements could this communitarian culture be created comprehensively, without the intrusion of the old city's outmoded values. In New Frankfurt, housing was constructed using the latest technologies in prefabricated concrete panels, state-of-the-art kitchens and bathrooms, and rooms flooded with natural light. The communities were built up using simple geometric forms, with housing arranged in zigzags and linear rows set amid avenues and alleyways that assembled into neighborhoods. The design coherence was reinforced by promenades and pedestrian walkways, reflecting pools, and streets painted in contrasting colors.

May conceived of the settlements as a concentric ring of "daughter towns" connected to the mother city of Frankfurt, yet separated from it by a series of gardens and parks. Each new town would be self-sufficient, providing schools and day-care facilities, churches, community centers and recreation, shops and workplaces, and access to nature. Among them, the new town of Römerstadt was the most evocative of May's ideals. The town plan followed the contours of the landscape. Functional, affordable housing was set in rows and terraces, their facades painted in earth tones, in a highly successful combination of organic design principles inherited from the garden city tradition and the avant-gardism of modern architecture.[18]

New Frankfurt won international acclaim at the second meeting of CIAM, held in Frankfurt in 1929, and was widely praised by members of the Bauhaus. Like Letchworth, it became a pilgrimage site for reformers searching for models of metropolitan planning and modern town forms. American housing activist Catherine Bauer visited along with Lewis Mumford, who was traveling in Germany for research on his book *Technics and Civilization*. They struck up a friendship with May that continued through the war.

The next year, May took a staff of seventeen from Frankfurt to the Soviet Union with the hope of contributing to the workers' struggle and constructing entire cities for the new socialist paradise. The great potential of the moment also drew Bruno Taut, Hannes Meyer, and Le Corbusier. Even after the Bolshevik Revolution, garden cities had remained the operative framework of urban reform in Russia, although they were now baptized as "Red." It was not until the first Five-Year Plan begun in 1928 that the government started encouraging the formulation of new theories for building cities. A passionate debate ensued on the nature of the *sotsgorod*, or socialist city, as the Soviet Union hurtled into urban and industrial transformation. The battle was initially drawn between two camps: the *disurbanists*, who argued for decentralization mostly following the garden city ideal, and *urbanists*, who demanded an increased scale of urbanization and industrialization. The compromise was to build medium-sized cities and new industrial cities. Many architects contributed to the ideal *sotsgorod* and the fraught deliberations about the nature of urban reform.

In the context of shifting communist ideology, the urban designs were daring, but they could also be politically risky. Linear industrial towns were proposed by El Lissitzky and by Nikolai Miliutin, especially the latter as outlined in his seminal publication *The Problem of Building Socialist Cities* (1930).[19] The linear city had political appeal, because it seemed to abolish the division between city and country according to the principles outlined by Karl Marx. Miliutin produced such plans for the new industrial towns of Magnitogorsk in the Urals, Stalingrad on the Volga River, and Avtozavod, where an automobile plant was taking shape under the direction of Ford Motor Company. Parallel industrial and residential strips were separated by greenbelts and highways. The towns would be nodes along transportation routes in one continuous band of development. Miliutin's groundbreaking concepts were published in Ernst May's *Das Neue Frankfurt* and were featured in the Proletarian Building Display in Berlin in 1931.[20] The linear city ideal survived as one of the most viable alternatives to the concentric pattern of garden and satellite cities. The Association internationale des Cités Linéaires was founded in Paris in 1928 to promote the concept, and was endorsed by CIAM.

It was this whirlwind debate about socialist cities that "May's Brigade" stepped into. Working for the Standargoproject (a trust for standardized industrial cities), they arrived in a world where "flags, slogans, placards, memorials, and loudspeakers spreading political and anti-religious propaganda were everywhere. Here and there churches were being demolished. Station waiting rooms were graced by identical artificial palm trees. Most people were very poorly dressed, and smiling faces rare."[21] May and his team proposed a system of twenty-four satellite towns, or *gorod kollektiv*, for the Moscow metropolitan region. They won the competition for Magnitogorsk and designed plans for some twenty other new towns, including Orsk, Novokuznetsk, and Kemerovo, although none of the designs ever came to full fruition. The plan for Magnitogorsk echoed that for New Frankfurt with a series of satellite residential communities in rows around the town center. But the plan was drastically changed once May left the Soviet Union after three years of frustration.

These gyrations in imagining cities for the New Socialist Citizen masked the reality that planning in these tumultuous years was at best trial and error. Functional zoning, with the factory separated from the residential area by greenbelts, was eventually implemented after the Second World War in a variety of new towns across the Soviet Bloc, with Magnitogorsk as the ideal. But once Stalinism was imposed in the 1930s, any form of utopian planning was outlawed, and new industrial towns became pastiche landscapes of social realism or haphazard settlements with little planning or architectural composition at all. One reason for the hit-or-miss approach was the extraordinary pace and extent of urbanization in the Soviet Union. Between 1926 and 1939, the urban population there more than doubled, with the vast majority accounted for by collectivization and the mass migration from the countryside. During the first Five-Year Plan, sixty new cities were begun, and hundreds more added during the second and third plans. Although some were started from scratch, most joined together older, haphazard rural settlements into the *sotsgorod* of the future.

The American Influence

The American interpretation of garden cities is equally illustrative of the fusion of influences comprising both the new town birthright and the regional vision that framed it. In 1923, Clarence Stein and Henry Wright sailed for England to meet with Ebenezer Howard and Raymond Unwin and make the pilgrimage to Letchworth, Hampstead Garden Suburb, and the newly established Welwyn Garden City. The experience was clearly the inspiration for the "city planning atelier" that Stein organized to discuss how the garden city could fit the needs of America. He and Wright persuaded Unwin to help shape the group's theoretical framework, with the result that Unwin became actively involved in the American planning scene. But the garden city movement was not the only inspiration for Stein's ideas; others clearly were his education at the Ethical Culture Society in New York City, his work with the Hudson Guild Settlement House in the city's Chelsea neighborhood, and his years at the École des Beaux-Arts in Paris.

The "atelier" that eventually became the Regional Planning Association of America was a cauldron of progressive planning and social reform ideas expounded by Stein, Wright, Lewis Mumford, Clarence Perry, and Benton MacKaye. They were left leaning and cosmopolitan, and promoted a radical rethinking of the social needs and physical layout of the American urban experience. The New York meeting of the International Town Planning and Garden Cities Association in 1925 was the opportunity to publish their ideas in a special issue of the *Survey Graphic*. It laid out their vision of regionalism for a new age. The roots of their convictions lay in the socialist underpinnings of Howard's garden city movement, the economics of Thorstein Veblen, and the regionalist philosophy of Patrick Geddes.

Geddes was of course the great originator of regionalism and the regional survey. His idiosyncratic thinking melded biology with the French geography of Élisée Reclus and Paul Vidal de la Blache, the anarchism of Peter Kropotkin, and the global information systems of Paul Otlet. Geddes began with the natural geographic region, its topography and ecology, its culture and history as the basic units of his analysis, for he viewed these as the motivic forces of urban life and the seeds of transformation. Towns were living environments: they melded the physical, social, and spiritual. Town and regional planning and social welfare were forms of civic education that would stop the spread of amoeba-like conurbations and open the door to a democratic awakening. Geddes's call for civic action and the renovation of cities had immense appeal, especially for Stein and a young Mumford. Stein formalized the purpose of the new regionalism: "To improve living and working conditions through comprehensive planning of regions including urban and rural communities and particularly through decentralization of vast urban populations by the creation of garden cities."[22]

The regionalists gathering around Stein and Mumford fully believed in a "fourth migration"[23] out of the metropolis and into a regional pattern that would return settlement to the dispersed patterns of the past—before *tyrannopolis*, in Mumford's words, or the imperial metropolis, had created such havoc. A network of planned new towns set in open space would be linked by networks of highways and telephone and electrical power grids, and would bring advanced technology to every point in the region. Yet each new town would be of human scale, with its own local identity and small-town atmosphere. This was the new "middle ground" between the large-scale metropolis and the old rural world. It was a distinct locale associated with a specific stage of modernization. The spatial concept was taken from Geddes and developed by Mumford and MacKaye, who noted that "the basic geographic unit of organic human society is the single town of definite physical limits and integrity."[24] It was a living organism, a source of alternative values and communal responsibility, the meeting ground for a localized "spirit of place"; it invigorated civic life. The towns served as diversified and dispersed fields of experiment "in between" that would stabilize American life and hold back the deadly conformity of the big city.[25]

These ideas formed the basis of the Regional Planning Association of America's (RPAA's) report for the New York State Commission on Housing and Regional Planning. Its proposed plan was based on a settlement pattern of decentralized cities located in the belt stretching from New York City north up the Hudson River to the Mohawk valley, and then westward to Buffalo. Patrick Geddes visited New York in 1923 and then again in 1925, when he met with the RPAA. Although the meetings were less than successful, they no doubt influenced the conception and design of the plan, especially the Epoch III diagram for a future of "wholesome activity and good living."[26] The network of garden cities would be developed by limited-profit private organizations in alliance with regional planning boards. Each town would be organized around schools, community services, and shopping centers defined at the neighborhood scale.

The RPAA's understanding of regionalism focused on the community as the foundation of a balanced region, in harmony with nature.[27] The initial test cases were carried out at Sunnyside in Queens, New York City, and in Radburn, New Jersey, in 1924 and 1928. Radburn updated the garden city to the motor age by strictly separating pedestrians from cars. Its superblock layout featured cluster housing on residential cul-de-sacs protected from road traffic. The result was an insular, self-contained community space with housing set amid rustic gardens and pathways.

Social reformer Clarence Perry filled in this self-contained space with the "neighborhood unit." Initially working as a high school principal in Puerto Rico, Perry joined the Russell Sage Foundation, where he studied local community issues and school and recreation reforms that were offshoots of the settlement house movement. He understood the neighborhood as both an extension of individual personality and an organic entity that would strengthen social bonds. In his diagrams, the elementary school is placed in the village common at the heart of the neighborhood unit, and often is flanked by churches. In two early texts written for the Russell Sage Foundation, Perry explained the transformation of school buildings into multifunctional social and service centers for the community. The activities there elevated political life, stimulated civic spirit, and developed neighborhood consciousness and responsibility.[28] These institutions operated, along with small parks and playgrounds, local shops, and the surrounding residential environment, as a social ensemble.

Perry's neighborhood unit was one of the most powerful concepts in twentieth-century urban planning, and was embraced as fundamental to new town ideology. It was a universal ideal that was applied in town planning schemes across Europe, the United States, and beyond.[29] Integrated into the Radburn layout of superblocks, it was a seductive vision of family-oriented communities and a mechanism for social construction. The ideal assumed that great cities might be radically reshaped and telescoped into smaller places featuring all the virtues of village life.

By the late 1930s, constructing garden cities in a regional constellation was accepted by politically progressive and conservative circles alike on both sides of the Atlantic as a de rigueur solution to all that ailed urban existence. This ambition would be implemented as part of a broad housing and social reform agenda based on scientific planning principles in which government played a decisive role. Architects and planners formed part of a richly honed transnational "civil society"[30] that propagated these ideas and concepts. In Berlin, Clarence Stein met with members of CIAM as a representative of the RPAA at the Thirteenth Congress of the International Federation for Housing and Town Planning. Walter Gropius and José Luis Sert met with Lewis Mumford and were inspired by his work. The Radburn principle, projects such as Norris, Tennessee (part of the Tennessee Valley Authority), and the Greenbelt Towns were international models of urban reform.

On one of his visits to the United States, Raymond Unwin toured Greenbelt, Maryland, one of three garden city projects supported by the New Deal Resettlement Administration of the federal government. Rexford Tugwell, the program's administrator and confidant of both Franklin and Eleanor Roosevelt, modeled the projects on the English garden city and on Ernst May's New Frankfurt experiment. He envisaged eventually settling three thousand Greenbelt Towns as self-sufficient, cooperative communities. But the plan came under withering attack in the press as a sinister socialist plot. Americans remained highly suspicious of government interventionist programs. The Greenbelt scheme was stopped after only three prototypes were under way: Greenbelt in Maryland, Greenhills in Ohio, and Greendale in Wisconsin. Although their immediate impact was negligible, in the long run they were a powerful demonstration of what could be accomplished by a progressive-minded social welfare state.

Mumford was the master at articulating this vision of communitarian regionalism. In 1938, his *The Culture of Cities* was published. The book thrust Mumford into the international spotlight and onto the cover of *Time* magazine. It was the definitive statement of the RPAA's ideas, and among the most influential texts about twentieth-century planning. In the midst of the Great Depression, its dream of a better urban environment, integrated into the surrounding region and attuned to the rhythms of daily life, found a wide audience. The message was optimistic and inspirational, and rippled through the United States and across Europe. Mumford defined himself within the currents of left-wing progressive politics and cultural radicalism. Yet it was his sharp emotional intelligence and the eloquence of his ideas that stirred the urban reformers who pored over *The Culture of Cities* as the rumblings of war grew ever closer.

The book's ideals were visually portrayed in the short documentary film *The City*, produced for the 1939 New York World's Fair as part of its City of Tomorrow exhibit. Funded by the Carnegie Corporation of New York, it was Stein's brainchild, with narration written by Mumford and music composed by Aaron Copland. Stein enlisted members of the RPAA to promote the film, including New York architect Albert Mayer, his brother-in-law and longtime friend, as well as Tracy Augur, who was chief town planner for the Tennessee Valley Authority.[31] It begins with the lost Eden of the New England village, whose social world is in balance with nature and technology. But Coketown (actually Pittsburgh, where these scenes were filmed) dissolves this idyll. There, viewers are thrown into the molten fires of the industrial inferno. Then the megalopolis of New York destroys any humanity left with its mechanization, congestion, and daily stress.

Melodramatic and sententious, the film swings between entertainment, propaganda, and social commentary. Nearly half its footage is devoted to the garden city and neighborhood planning solution, and was shot in Greenbelt, Maryland. These idyllic scenes depict families and children experiencing domestic bliss. Clearly, the new town and its neighborhoods are places of social harmony and happiness. Although Mumford bemoaned his lack of influence on the American urban scene, both *The Culture of Cities* and the documentary had a profound effect internationally on the progressive imagination and were referred to repeatedly as sources for new town ideology.

Experiments in Fascist New Towns

Ironically, it was the fascist regimes of the 1920s and 1930s that carried out the ambitions of communitarian regionalism combined with centralized state planning. The resulting new town projects were ample evidence of the ways in which garden cities and regionalism could become dogmatic political practice, particularly when they reeked of anti-urban rhetoric and the nostalgic yearning for small-town life. Utopia has always been packed with moral overtones. Between 1932 and 1940 in Italy, Mussolini's regime constructed numerous ideal *città nuova*, including Mussolinia, Saubadia, Littoria, Pontinina, Guidonia, and Aprilia. Each was inaugurated with great fanfare: groundbreaking ceremonies were attended by Il Duce himself. Despite their rural pastiche and traditionalism, they were a means to increase economic productivity and create new loyalties to the fascist state. Their designs were generally grandiose exercises in architectural monumentality, with Fascist Party regalia as decor. Some, such as the plan for Saubadia, became models for a young generation of Italian urbanists. The most admired were those associated with the land reclamation project in the Pontine Marshes near Rome. Projects such as this were seen as a form of regionalism in which new towns acted as a purification and modernization of the countryside.[32] They reproduced fascist notions of hierarchy in their rational ordering by size and function. Like all new towns, the Italian variants were propagandistic spaces. They were symbols of state power and nationalistic aspiration, and a stage set for the emergence of the fascist New Man.

In Germany, Gottfried Feder, the Third Reich's Settlement Commissar, emerged as the National Socialist Party's main ideologue on regional planning and new town schemes. His *Die neue Stadt* (1939) promoted a heavily racist version of the garden city ideal. All urban growth was to stop, and the old-style, working-class *Mietskaserne* housing blocks abolished. Cities would be reduced to populations of no more than one hundred thousand inhabitants, with the remainder dispersed to small communities in rural settings. Examples used in Feder's text included designs for the Italian new towns of Littoria and Sabaudia and the American Greenbelt Towns.

Like many garden city concepts, Feder's was steeped in the vocabulary of the "organic city," a term that set urban life within a bucolic vision of nature in which the town could grow as a plant in native soil. It was also wrapped up in a discourse of cultural and ethnic homogeneity, and in a fantasy of the authentic in the urban landscape and environment. Feder's new towns followed the traditional garden city design principles, with radial-concentric street patterns and neighborhood units bounded by major roads and centered on community facilities. The optimum size of each new town was twenty thousand "settlers." The focal point for each neighborhood of thirty-five hundred residents was the *Volksschule* (elementary–middle school), around which retail establishments and service centers were positioned. Feder's concept was a fascist version of Clarence Perry's ideal. The new town would provide its regional hinterland with services, while the surrounding farms and rural villages assured a steady food supply.[33]

Feder was eventually removed from his position, but his replacement, Walther Darré, was equally anti-urban. The National Socialist *Siedlungsprogramme* succeeded in laying out hundreds of new villages. They were meant to facilitate German movement eastward in the Nazis' *Generalplan Ost* for the colonization of *Lebensraum*. The new towns, in other words, would expedite the expansion of German living space along the eastern frontiers. The spatial scheme was grounded in the work of Carl Culemann and Walter Christaller, who were commissioned by the Nazi government to create a "standard city," or generic *Normalstadt*, that could be mass-produced to redistribute the German population.

First published in 1933, Christaller's thesis on central place theory demonstrated that at the root of any successful economic system was a form of settlement that rationalized the exploitation of resources. Settlement patterns were not accidental but depended on a specific territorial structure and an ordered hierarchy of urban centers. Christaller's work provided the scientific basis for a polycentric web of small and medium-sized urban places. The coming of the Third Reich provided the opportunity to implement his ideal regional geography around a reinvigorated concept of national community.

In 1940, Christaller joined the staff of well-known regional planner and SS member Konrad Meyer, who was head of the Soil and Planning Department under Heinrich Himmler's Reich Commission for the Strengthening of Germandom, and charged with formulating *Lebensraum* policy and the *Generalplan Ost*.[34] Christaller argued for comprehensive regional planning and an orderly system of communities covering the entire German Empire. His hierarchy of central places would descend from the national capital of Berlin to provincial capitals, then to towns, farming villages, and finally to concentration camps, all of which were planned to the smallest detail and would be constructed by forced labor. The map of urban-centered regions spread across Germany, France, and the whole of eastern Europe. New towns would be created from scratch anywhere they were needed in order for territory to conform to Christaller's theory. Each was standardized with a National Socialist Party ceremonial hall, a central parade ground, and the visual trappings of Nazi society. For example, Christaller planned thirty-six new, medium-sized market towns (*Hauptdorf*), or "settlement pearls," as they were called, for the German-annexed areas of western Poland. They were to be located at rail and highway junctions and surrounded by a network of ideal German villages.

As part of the *Siedlungsprogramme*, two large industrial cities were planned in the Braunschweig region of northern Germany as part of the Kraft durch Freude (Strength through Joy) movement promoted by the German Labor Front. The first new town was associated with the KdF-Wagen, or Volkswagen, factory about 175 miles west of Berlin. Founded in 1938 as the City of the Strength-through-Joy Car, it was renamed Wolfsburg in 1945. Architect Peter Koller designed a garden city for ninety thousand people that was laid out with a ring-shaped system of streets surrounding Klieversberg Castle.

The second new town was the Stadt des Hermann-Göring-Werke, or Salzgitter, founded in 1942 as part of a huge industrial complex of foundries and chemical and electrical facilities constructed under the Nazis' four-year economic plan. Salzgitter was one of the great industrial monoliths built in preparation for war—Magnitogorsk in the Soviet Union being the other of this size, also built by slave labor. Architect Herbert Rimpl, who was a great admirer of Ebenezer Howard, designed three large residential sections for 130,000 people that were separated by green spaces. The intersection between each was the site for the town center, featuring the monumental square, stadium, and assembly hall that were fascist standard issue.[35] In the hands of fascists, Howard's original concept of integrating the best of the urban and rural worlds could resonate with a treacherous political timbre.

Wartime Utopias

As evidenced by the example of Germany, rearmament and the Second World War dramatically intensified the debate on cities and regions, and put into high relief the relationship between militarism and spatial planning. It immediately became obvious that cities were prime targets for military bombardment from the air. From the first days of the war, some of Europe's most venerable urban places were wiped out, and the rest were under constant threat. The destruction and death toll were horrifying. But protecting war-related industries was even more of an immediate concern. It meant scattering them to isolated areas away from urban bombing targets. The scale of wartime production thus created a vast new industrial geography. In anticipation of an invasion by the West, Stalin disassembled and evacuated whole factories from the western regions of the Soviet Union, moving them deep into the eastern territories. The most famous example was of course Magnitogorsk, built in the 1930s in the Ural Mountains to protect Stalin's nascent steel industry from assault.

Country after country shifted its critical industries away from its cities and into its protected heartland. Germany moved production sites toward the center and eastern parts of the country. The British government's shadow factory program scattered vital industries and especially aircraft factories to the west and northwest of England and to Scotland. Once London was bombed, there was a general dispersal of industries, even of engineering firms. Welwyn Garden City became an armaments factory site. The "self-contained and protected community unit" of Speke outside Liverpool was hastily built with the arrival of wartime aircraft factories in the area.[36]

Industrial dispersion was often treated as an opportunity to construct an ideal city from scratch—that is, to carry out the plans that had proved so frustratingly impossible before the war. Housing for the workers toiling in the relocated plants was the first priority, but locating them in connection with vital transportation arteries was also the chance to experiment in utopian designs. This was evident in Britain, where the Civil Defense Regions created as a wartime protective web became the administrative geography of the postwar Ministry of Town and Country Planning.

The claim that people and industry should be dispersed for their protection found ready listeners among stalwart urban reformers. In notes for an unfinished book, Clarence Stein contended that the threat of air strikes demanded a national policy for scattering industry beyond existing population and manufacturing centers. As a further safeguard, the residential areas in these war industry towns could be separated from factories by open greenbelts.[37] In the United States, war production sites were shifted from the East Coast to the Midwest and the West Coast. The US Public Housing Authority, the Department of Defense, and the Kaiser Steel Corporation together built among the most complete new town projects of the Second World War outside Vancouver, Washington (McLoughlin Heights), for aircraft production and near Portland, Oregon (Kaiserville), for wartime shipyards. These included affordable housing, energy-efficient infrastructure, public transportation and the separation of pedestrians from automobile traffic, schools and hospitals, recreation and day-care centers, and shopping districts planned around a central community square. They were a triumph in new town utopian typology. They also were a clear signal to elated urban reformers that states could intervene on a massive scale not only in industrial production but in constructing infrastructure, housing, and even entirely new cities founded on progressive planning principles.

Further, wartime strategic operations were dispersed to Oak Ridge, Tennessee, Richland, Washington, and in the case of the atomic program, Los Alamos, New Mexico. At the Willow Run automobile plant in Michigan, where Ford Motor Company produced B-24 bombers, a team including Eero Saarinen, George Howe, Louis Kahn, and Oscar Stonorov proposed a design for a "Bomber City" that, although defeated, included housing, public transportation, and neighborhood amenities such as schools, childcare centers, and health clinics. *Architectural Forum* called it "the best guide for postwar planning we have yet produced."[38] The US War Production Board began plans for satellite industrial towns that would surround large cities in a metropolitan regional configuration. Sites for the aircraft industry and future airports were the essential pieces in this puzzle.

Private aircraft companies also began searching out new locations for their factory towns. Douglas Aircraft created the town of Daggett, thirteen miles from Barstow, California, in the Mojave Desert. Nine miles southeast of Oklahoma City, enterprising community builder W. P. Atkinson secured a 330-acre tract opposite a new Douglas cargo plane plant. Working with the Army Air Service Command and the Federal Housing Administration, he created Midwest City as an "air industry city." In 1944, the Urban Land Institute featured an aerial view of the town on the cover of its *Urban Land* magazine, under the heading "Model Community."[39] All told, during the war the US Army built more than 1,000 air bases, while the Navy built over 60 major airfields. At the war's end, over 500 surplus military airfields were transferred to municipal authorities. They became prime areas for urban development.

The war also altered the scale of urban and regional geography. It dramatically expanded the use of aerial vision that had already become an accepted part of urban and regional planning discourse in the years before the war. From Patrick Geddes and Le Corbusier[40] to French urban reformer Henri Chombart de Lauwe, the enlarged visual field of flight was a cognitive, synoptic shift. Architectural historian Mitchell Schwarzer refers to this as "aerial globalism" that both enlightened and estranged the realms of landscape and architectural perception.[41] Flying over miles of territory, viewing the contours of the landscape, and imagining development from a bird's-eye view became the framework for visualizing cities and urban regions. It was a visual spectacle, readily adaptable to large-scale land-use planning. Geography was magnified and enlarged: it crossed political borders and illuminated broad regional patterns, and flows of information and communication. An aerial perch abstracted regional geography as a flat surface, completely accessible and available for exploitation. The airplane's mastery of the regional landscape not only produced a new way of seeing but lent itself to utopian visions of development. Planning could be totalized across vast areas. Geography could be brought into economic production and used to solve the unbridled chaos of urban expansion.

While architects like Le Corbusier might have only dreamt of this kind of capability, the war made it a reality. The aerial photographs taken during reconnaissance missions were nothing short of extraordinary. Thousands upon thousands of images at high and low altitudes, by day and night, were taken worldwide across the fields of battle, showing military installations, cities and towns, key infrastructure and transportation routes, and industrial sites. They provided a new reading of landscape and a vital instrument of wartime planning. By 1945, the British Air Ministry's Central Interpretation Unit had a daily intake averaging 25,000 negatives and 60,000 prints. By V-E Day, its library had amassed some 5 million prints of sites from around the globe. Geography had been dehumanized and militarized. Land mass, infrastructural terrain, and cities were all viewed as targets by military aircraft and aerial reconnaissance missions, their pilots peering down through bomb scopes and cameras.

Aerial photographs, both vertical and oblique views, were translated as objective tools that could provide the evidence, the detailed survey and factual data, Geddes had called for. The wartime prints were used consistently in the *Greater London Plan of 1944* to evidence urban chaos and suburban sprawl, and to provide a visual mapping of spatial planning.[42] In the United States, planner Melville Branch argued in *Aerial Photography in Urban Planning and Research* (1948) that wartime aerial photographic techniques were vital to a more scientific approach to planning and to the analysis of the urban environment. Private firms began marketing aerial photographic surveys to states and municipalities as a tool to assess land for taxing, delineate zoning districts, and plan in the regional context.[43]

In 1945, Lewis Mumford prophesized about the postwar era, a time when, it was imagined, the airplane would be the magic solution for the future: "At the beginning of the twentieth century, two great inventions took form before our eyes: the airplane and the Garden City, both harbingers of a new age." Mumford believed that the new transportation technology would transform urban regions into garden cities with wide belts of open land.[44] The Los Angeles Regional Planning Commission spoke directly to this vision of satellite communities connected by airstrips in a new kind of metropolitan region. Wartime Los Angeles was the boomtown of aviation, with nearly 228,000 workers engaged in the aeronautics industry centered on the Douglas, Lockheed, North American, and Northrop factories.[45] The Douglas Aircraft Company's Bomber Plant was in Long Beach, and its footprint in the economy and geography of the Los Angeles basin was enormous. In 1945, it was Douglas that, under contract with the US Air Force, established a think tank for air power and aerial reconnaissance which became the acclaimed RAND Corporation, with its base in Santa Monica.

At the war's end, maintaining the aircraft industry in the region's economy was essential. Along with the Los Angeles Board of Public Works, the Regional Planning Commission devised its *Master Plan of Airports* (1940, revised 1945), which identified fifty airstrips located across the area in a crystalline diagram with satellite communities. In Long Beach, developers created the planned community of Lakewood to provide housing for workers from the Douglas aircraft plant. Lakewood became a model new town, the "city of the future."[46] Chicago, too, began to plan for its future with flight. Its Planning Commission along with its local Regional Planning Association issued a report in 1941 that envisioned concentric rings of some thirty airports surrounding the city. The plan covered fifteen counties, with new communities built around air travel.

Planners believed that new airports could invigorate the economies of older US industrial cities, and architects as celebrated as Frank Lloyd Wright saw decentralized airports spread across the country as the future. In 1945, industrial designer Norman Bel Geddes built a model of a "Future City" for Toledo, Ohio, based on five different airports that could accommodate everything from private planes to the largest jet-propulsion aircraft. Air travel as vital to metropolitan regional planning continued to be one of the most consistent features of postwar urban utopias. Illustrations of future cities inevitably dramatized helicopters and svelte flying machines whizzing between hypermodern skyscrapers outfitted with landing pads. This imagery would only intensify as the Space Age took hold of the cultural imagination.

The Call to Arms

The war also made the urban crisis an immediate and passionate call to arms. The clamor for an urban revolution came from every quarter. Many of the texts defining the way cities were imagined in the postwar years had been written either at the height of the Great Depression or when the war was casting its long shadow. Most of the architects associated with the Neues Bauen (New Building) formalism and the Bauhaus had gone into exile: some fled to the Soviet Union; others, such as Walter Gropius, Marcel Breuer, and Lazło Moholy-Nagy, fled to the United States. Ernst May went to Kenya and Rhodesia. Richard Paulick worked in China. CIAM members shifted their activities to England and especially the United States, where a CIAM Chapter for Relief and Postwar Planning was set up in New York. It was there that José Luis Sert and his CIAM colleagues wrote *Can Our Cities Survive?*

In 1942, when the book was published, the question was not a rhetorical one. Sert's analysis set out CIAM's modernist vision and its concern with the entirety of the city, especially since architects would be faced "with the problem of reconstruction and the development of new regions demanding the creation of new communities,"[47] postulates that would be reiterated in the Athens Charter, published in 1943. CIAM's was one call among many. The city became a complex utopian project—an elaborate projection of divergent, at times contradictory ideals.

There was a sense of pathos in the way urban visionaries tried to light the way through the dark hours of war. In 1943, architect Eliel Saarinen published *The City: Its Growth, Its Decay, Its Future*, in which he hoped, "This war—once over—might help to realize the obligation of the post-war period to make good the indifference of the pre-war period. Indeed, this is an essential obligation which, in fact, must be a war in itself. . . . It must be a war against slums and urban decay." Saarinen argued for decentralization "around the nucleus of the original compactness, an organic grouping of new or reformed communities of adequate functional order according to the best principles of forward town-building."[48] In Sweden, Sven Markelius was using his position as head of Stockholm's Urban Planning Office to lay out a regional plan that would carry through the social reforms he and Alva Myrdal had begun in the 1930s. By 1944, his office was delivering the preliminary version of the *Stockholm in the Future* report that became the source for regional satellite towns around the capital.

Jean Gottmann, who had fled France for England ahead of the advancing Nazi army, was writing furiously on the nature of regional geography. Although his vision of the metropolitan region differed radically from that of Lewis Mumford's, he nevertheless called for abandoning old conceptions about urban, rural, and suburban geographies. Instead, Gottmann advocated new practices of cooperation and coordination across metropolitan regions. His understanding of geographic space was based on land-use planning as an applied science. It offered a new rational and relational milieu that could assure the harmony and equilibrium of the entire regional geographic system.[49]

All these theorists were instrumental in promoting the vision of regional planning and new towns as an alternative to what Mumford called the intolerable "necropolis" of the old cities. The sound rejection of uncontrolled growth dictated by the market and its greed led to championing a settlement system organized equitably and independently. This philosophy had permeated Ebenezer Howard's garden city ideal and remained fundamental to the way urban reformers imagined future cities. But the war heightened the sense of social justice and moral righteousness, and the steadfast demand for redressing so many long-standing wrongs. Theirs was a deeply altruistic cause.

The immediate crisis was the war's extraordinary destruction. The worldwide devastation and landscape of ruin was mind boggling. Entire cities had been flattened by aerial bombardment. In London's East End, the Docklands had been destroyed in the Blitz along with acres of working-class housing. Sections of urban England had been reduced to rubble. Other cities had been destroyed by fierce ground combat and reduced to rubble-strewn landscapes. The number of homeless was staggering: 25 million in the Soviet Union, another 20 million in Germany. A report written by the US Housing Administration in 1951 estimated that the war had reduced 25 million to 30 million dwellings worldwide to ruins, and damaged another 10 million to 15 million too severely to repair. All this added to "20 years of gross neglect of . . . housing,"[50] creating an emergency of global proportions. Statistics on worldwide housing shortages were rolled out in the popular media in a relentless torrent.

The housing crisis became one of the most hotly debated and stormiest political issues of the twentieth century. In the aftermath of war, it was second only to the desperate search for food. The number of mass evacuations, deportations, and ethnic cleansings, along with political refugees and the displaced and expelled, the wounded and dying, was in the millions. Wave after wave of peoples were on the move, attempting either under official auspices or on their own to find safety and shelter. The human catastrophe was made worse by political breakdown, shifting territorial boundaries, the painful steps toward decolonization in some places, and the establishment of new forms of colonial authority in others. Added to this, some 80 million enlisted men from sixty nations were being demobilized in some manner or another, and were expecting to find a place to live and continue with their lives.

Faced with this overwhelming reality, there was a convergence across the reformist spectrum on the need for a new urbanism and a new regionalism. It was voiced as a chorus of revulsion against the grim pathologies of old cities and the chaotic sprawl that surrounded them: big cities offered nothing but misery and paralysis. These angry accusations against old cities and past failures were a product of the war and its aftermath. In the British government's *Target for Tomorrow* booklet series (1945) outlining reconstruction, Sir William Beverage waved an accusatory finger at the appalling "over-crowding, dirt, noise, darkening of the air by smoke." Cities "breed and multiply disease, and cut at the roots of family life." In the same publication, author John Madge complained: "On the one hand, we have urban confusion, overcrowding and slums: on the other, we have rural poverty and backwardness, the result of generations of neglect."[51]

This vehemence against old cities saturated reformist thinking. Utopian aspiration has always required this display and disparagement of the present, which will fall away to reveal a new temporal age. Writing in 1945 for the Re-building Britain pamphlet series, Lewis Mumford argued that "the demolition that is taking place through the war has not gone far enough . . . the bulk of our building no longer corresponds to the needs and possibilities of human life. We must therefore continue to do, in a more deliberate and rational fashion, what the bombs have done by brutal hit-or-miss."[52] It was but one in the long litany of curses against the failure of cities. The verbal and visual depictions of disease and pollution, congestion and overcrowding, repulsive blight, loss of any human dignity, reverberated across the world. Not only Western cities such as Glasgow, London, or Paris that had long come under critical ire, but now the teeming masses huddled in the slums of Beirut, Karachi, and Mumbai (formerly known as Bombay) left urban crusaders aghast.

Reformers were tormented about population density, social disorder, and political volatility in wild, out-of-control cities. Perceptions of urban life were defined negatively as problems. Although the loudest condemnations were directed against big cities, reformers also pointed to the myriad disheveled informal settlements around towns or industrial sites that pocked the countryside. Composed of ramshackle self-made housing and worker shanties, the lack of infrastructure, services, or governmental authority in these forgotten places was appalling. In poor countries, the endemic poverty was explained as the by-product of archaic traditions, ethnic and religious strife, and the lack of assimilation into the modern world. These problems were denounced endlessly in a broad-based campaign to save civilization with rational planning. Solutions could only be found from a regional, even national geographic viewpoint that balanced growth across territory. What gave this crusade immediate cogency was the Second World War, and it was only in the context of the war and the reconstruction years afterward that this kind of regional planning was actually realized.

The "population problem" drove in the icy spike of fear. It was a phrase that plagued mid-twentieth-century urban reform discourse. Both Mumford and Jean Gottmann pointed to population growth as *the* urban problem." Their writings were filled with statistics evidencing the unsustainable levels of population growth, replete with unnerving prophesies of future catastrophe. This "population problem" was formulated in good part by the expert conviction that population change was both intelligible and predictable. Population and housing surveys became a normative part of municipal planning by the 1930s. Professional demographers incorporated census data and statistical surveys of birthrates and death rates, age and gender, fertility rates, migration patterns, and number of households into their methodologies. Mathematical formulas, matrices, and modeling techniques were used to predict population growth scenarios. Many of the major cities of Europe and the United States carried out their first large-scale demographic studies and population estimates either just before or immediately after the war. This invention of population as a category of scientific knowledge was also a powerful political instrument. Evidence of the "baby boom" and the dramatic decline in death rates after the Second World War were regarded as predictors of future havoc. The densely populated urban regions of India and Asia were a doomsayer's spectacle.

Apprehensions were increasingly raised in both intellectual debate and policy making, spurred on by alarmist popular tracts such as eugenicists Guy Burch and Elmer Pendell's *Population Roads to Peace or War* (1945) and the once-forgotten best sellers of 1948, Fairfield Osborn's *Our Plundered Planet* and William Vogt's *Road to Survival*. Hugh Everett Moore's twenty-two-page pamphlet *The Population Bomb!*, first published in 1954, introduced the expression and went through thirteen editions by the early 1960s.[53] The population explosion became an international cause célèbre, and made the cover of *Time* in 1960. The population bomb, teeming cities on the edge of chaos, the looming catastrophe of starvation, were etched into the public mind and used as weapons in the battle for influence between different ideals of planning and reform. Unrelenting population growth made cities into giant fulcrums for masses of people, "disorganized congestion, decline, dilapidation, blighted areas, and then, slums," in the words of Eliel Saarinen.[54]

The war and its aftermath also transformed the imaginary of "populations" into grand, sweeping gestures. Millions of soldiers and millions on the home front had been mobilized. The war regimented and militarized masses of people. The vast migratory movements caused by the wartime political settlements heightened the worldwide human spectacle. This spatial vision, the mosaic patterns of migration and land use, the synoptic view of settlement and infrastructure, fashioned far-reaching perspectives on reform.

Experts peering down on such a totalizing landscape talked of "worldwide planning" on a heroic scale. The atmosphere of crisis demanded a new way forward. The expectations of millions of people would not be satisfied with piecemeal urban renewal projects diminished by political haggling. Responding to their hopes meant brand-new, modern places to live. As the war ended, planners and architects from every country embarked on study tours and attended international conferences with a steadfast resolve to seize the moment. They took up positions in the new United Nations with a sense of passion and moral responsibility. They huddled over Clarence Stein's and Clarence Perry's garden city and neighborhood unit ideals. Lewis Mumford's *The Culture of Cities* was rapidly translated and devoured by urban reformers, as were the first copies of Abercrombie and Forshaw's plan for the London County Council that appeared in 1943. For Mumford and for many urban theorists surveying the wreckage left in the war's aftermath,

> Utopia can no longer be an unknown land on the other side of the globe: it is, rather, the land one knows best, reapportioned, reshaped, and recultivated for permanent human occupation. This conclusion makes imperative in our age what was still only an ideal possibility at the time when Ebenezer Howard outlined his project for the first Garden City: the internal recolonization of every country.[55]

TWO

The Futurology of the Ordinary

Whatever their individual proclivities or politics may have been, by the Second World War urban reformers were agreed on one thing: only the construction of entirely new planned communities could solve the urban crisis. Town building and regional planning were the mantras of the mid-twentieth century, and reformers believed that such projects would require leadership and massive public investment. The war had made clear the capacity for national mobilization and state control over resources and key infrastructure. Reformers were intent on harnessing this muscle to comprehensive master planning. They were determined to make a fresh start. They were driven in part by fears of social and political volatility, but also by a passionate resolve to construct a new world order from the ashes of the past.

The fervor of the new town movement is evident in architect H. Peter Oberlander's article "New Towns: An Approach to Urban Reconstruction" (1947). Oberlander was Canada's first professor of urban and regional planning, and he later became secretary of its Ministry for Urban Affairs. "Our 'megalopoli' from coast to coast are sick organisms, sick as a whole and as such must submit to a surgeon, no mere physician, who will carve out the fested city flesh by the square mile," he declared. "We must rebuild totally. . . . The great challenge of our era is to build anew, to create New Towns, complete in themselves as places to live, work, shop, study and relax."[1]

Planning culture invented a new vocabulary and spatial syntax for assembling cohesive regional geographies and coordinating modernization on a regional scale. These were among the primary impulses of national governments, the UN, and philanthropic institutions such as the Ford Foundation and Rockefeller Foundation. The desire was for *balance, equilib-*

rium, harmony. Built from scratch, picture-perfect new towns were the leading edge in this crusade; they were the megastars of postwar reconstruction and proof of the promised world to come. Even though a majority of new town sites were already occupied by people going about their lives, these were treated entirely as empty spaces that could be filled up with the promises of the future. The fact that modern towns were imagined as miraculously appearing out of nowhere made the state and its planning apparatus seem magical. There was a benign paternalism to this alchemy in the early postwar years. The state was a reassuring hand on the shoulder. The new town program was a moral expression, part of the progressive social contract between the welfare state and its citizens. The values were sincere, confident, uncynical.

This promised land was captured in the endearing British cartoon film *Charley in New Town,* produced in 1948 by the government's Central Office of Information and directed by the trailblazing animators Joy Batchelor and John Halas. It was one in a series of animated films in which the Everyman character Charley introduced the British public to new government programs. In *Charley in New Town,* Charley leaves behind the daily grind and congestion, the drab houses and ugly streets of the big city (most certainly London), where there are no places for kids to play, for the ideal town in the countryside: "Our town is a good place to work in and a grand place to live in."[2] The film then sets out all the planning principles reformers had been fighting for, and all the arguments for new towns: functional zoning and greenbelts, five-minute commutes from home to work, "up-do-date factories with light and air," decent houses and neighborhoods, parks and playing fields, schools and shopping centers . . . and plenty of pubs. These were the expectations of reconstruction and the hopes for a new age under the guidance of the state—whether it took the form of a welfare state, a communist state, or a new state emerging from colonialism. These also were dreams shared by millions of people.

The ultimate signatures of this ideology were new capital cities, such as Le Corbusier's acclaimed design for Chandigarh in the Punjab state of newly independent India. By the mid-1950s, the master plan for the city of Baghdad in Iraq and the new capital of Islamabad in Pakistan (the latter designed by Constantinos Doxiadis) signaled the arrival of modern nations in the Middle East. In Brazil, the high-modernist capital of Brasilia was conceived by Oscar Niemeyer and Lúcio Costa in the late 1950s. Canberra in Australia was developed as the new seat of government, surrounded by satellite towns.

These new capitals, designed by the most illustrious architects of the day, received the most fanfare. However, town building went far beyond these high-profile projects. The priorities of reconstruction were "the people and their needs." States assumed the responsibility for rebuilding a strong and secure postwar society for Charley, for Everyman. It was not the towering spectacle of capital cities that captured the postwar imagination so much as a futurology of the ordinary. Everyday life was the arena of reform, and in new towns it was a display of aspiration and anticipation. These desires reflected Ernst Bloch's concept of utopianism as a "principle of hope" situated at the level of everyday life.[3] The mesmerizing dream of the new town was about decent homes, parks and playgrounds, schools and clubs, as well as the neighborliness and social harmony that would make life better for everyone. It was a social imaginary of extraordinary power.

This optimistic humanism, the social justice quality of modernization discourse in the postwar years, is perhaps one of its most cogent yet problematic features. "Let our New Towns return to the human scale of life; they should become an environment where man is the scale of values once again," Oberlander remarked in his essay on reconstruction. "Cities must reflect this spirit of erect citizens, proud to be free and aware of their rights and responsibilities towards themselves and their neighbors."[4] Echoing this sentiment, Finnish social reformer Heikki von Hertzen described his new town of Tapiola as "a whole town for everyone."[5] The new town formed the lexicon of modern social improvement and quality of life. It was the deus ex machina of the welfare state and set the parameters of individual and collective expectations. At the same time, however, the outcomes were framed by notions of societies as singularly bounded and internally integrated realms in which people were more or less the same. The accent was on young nuclear families that acted as the stabilizing structure of the nation. The whole design and built environment of the new town were made for them, and the everyday world it promised was geared to their needs as modern people.

The backdrop to all this was that new towns were conceived as a textbook strategy for acquiring and assembling investment capital and infrastructural assets. They were massive public works programs. Nations rebuilt through public spending on large-scale development projects based on the Keynesian economic model. These would stimulate the economy with new material resources, provide full employment, expand output and productivity, and consolidate territorial dominion. New towns were tied to land reclamation, the drive for industrial development, and the extraction of raw materials. They were deployed for the resettlement of people and production facilities. Constellations of new towns connected to one another by transportation webs defined regions and ultimately the nation as a whole. New town ideology was thus deeply embedded in an expansive and modernized national geography and in a rational hierarchy of neighborhoods, cities, regions, and productive life. There was a social fairness sentiment to these planning policies, a sense of public duty and a deliberate intent to equalize economic opportunity and standards of living.

Planning was presented as above politics and entirely justified after the tragedy of war. The quest for a rationally planned territory and society was shared by both sides of the Cold War abyss: American-style capitalism in the West and Soviet-led communism in the Eastern Bloc. Solidifying and expanding territorial control, hardening borders, stabilizing populations, and maximizing wealth and productivity were fixations of the state. In this respect, new towns were integral to nationalistic, patriotic drives for consolidation and modernization.

This chapter begins with an assessment of very ordinary places—new towns that are mostly unfamiliar, but at war's end represented the struggle for reconstruction and the conversion of industry to peacetime production. Across both the East-West divide and the gulf between north and south, heavy industry, mining, and hydroelectric energy powered the great postwar "economic miracle." New industrial and resource towns sparked the initial movement for urban reform.[6] They made up an entire geography of production tied to natural wealth, and set this phase of the new town movement squarely in the crusade for output and productivity of the early postwar years. It is this tie to specific industrial locations that brings these new towns together as a functional group and establishes their utopian qualities. They also symbolized the social goals of the welfare state. For young families, luggage in hand, who moved into these perfect worlds, these new towns were first and foremost about jobs, houses, and schools. They were creations of reconstruction idealism and society's dream of a happy future. Focusing on these often overlooked industrial towns underscores their impact on the late 1940s and early 1950s, their importance to utopian typologies, and the ways their ordinariness seemed so remarkable.

The second part of the chapter examines the drive to remake large capital cities by dispersing their populations to a constellation of new towns within a metropolitan region. Here, the urban territory is more familiar, and covers the celebrated regional plans for London, Stockholm, and Helsinki. By the late 1950s into the early 1960s, these plans spawned a remarkable series of new towns that became models of the future and achieved mythic status in the planning world—places such as Stevenage in Britain and especially Vällingby in Sweden and Tapiola in Finland. Here, regional planning was about emptying out the big city and transforming its downtown into a magnet for commerce and consumerism. Once-miserable families stuck in inner districts would find happiness in the new towns set amid greenery that encircled the capital. These offered the best in housing, along with every service and recreational advantage. They were embellished with public transportation, parks, and public spaces that were indeed astonishing achievements, especially given the continued scarcity of resources after the war.

In all these early new towns, the quixotic imagery of the garden city and neighborhood unit was an immensely appealing orthodoxy. Although Clarence Perry's and Clarence Stein's work had some currency in the 1930s, their theories reached an international zenith during the reconstruction years and afterward. Their reforms became part of the return to order, with the family as the bedrock of social life. Nothing seemed more necessary after years of brutal war and violent upheaval. The antidotes to these disasters were daily community practices, the ordinary banalities of homes, schools, playgrounds. The neighborhood was an arena of normalcy, the incubator of citizenship and civic virtues. It was the natural setting for young families eager to begin their lives and raise children. Family was understood as parents and children, but it also registered the familiarity and human association of the everyday world.

The neighborhood was also a mechanism for coordinating state investments around a specific set of goals: family policy, educational reforms, functional housing, modernization, and consumerism. New towns were the prototypes for the arrangement of these investments, and "neighborhood" was the socially engineered space in which they would take form. Traditional planning histories are largely oriented around the *physical* planning of the neighborhood unit, especially the superblock and the separation of pedestrians from vehicular traffic, as exemplified in Clarence Stein and Henry Wright's plan for Radburn, New Jersey. No doubt these were significant aspects of neighborhood development. However, in this first wave of reconstruction planning, it was the state's *social welfare* institutions that provided legibility to the neighborhood unit ideal. Crucial to the model of social construction were nursery schools, elementary and high schools, community centers, and health clinics. These became a typology of requisite object-symbols or metonymies for a planned community. They were derived from the work of Clarence Perry and his design for a neighborhood fabric woven around the school and the community center at its heart (fig. 2.1). Again and again, schools appeared as the formal points of reference in postwar designs for neighborhood units. Housing was placed in

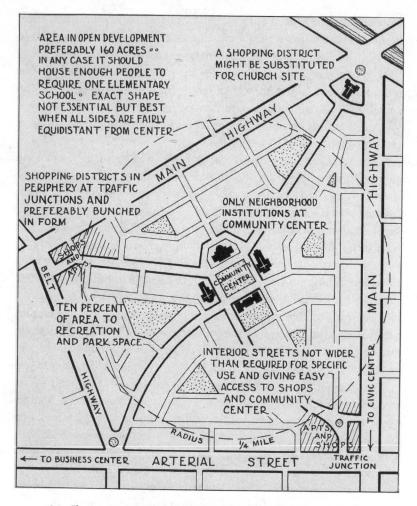

Text within the image:

AREA IN OPEN DEVELOPMENT PREFERABLY 160 ACRES °° IN ANY CASE IT SHOULD HOUSE ENOUGH PEOPLE TO REQUIRE ONE ELEMENTARY SCHOOL ° EXACT SHAPE NOT ESSENTIAL BUT BEST WHEN ALL SIDES ARE FAIRLY EQUIDISTANT FROM CENTER

A SHOPPING DISTRICT MIGHT BE SUBSTITUTED FOR CHURCH SITE

SHOPPING DISTRICTS IN PERIPHERY AT TRAFFIC JUNCTIONS AND PREFERABLY BUNCHED IN FORM

ONLY NEIGHBORHOOD INSTITUTIONS AT COMMUNITY CENTER

SHOPS AND APTS

BELT

COMMUNITY CENTER

TEN PERCENT OF AREA TO RECREATION AND PARK SPACE

INTERIOR STREETS NOT WIDER THAN REQUIRED FOR SPECIFIC USE AND GIVING EASY ACCESS TO SHOPS AND COMMUNITY CENTER

HIGHWAY

HIGHWAY

MAIN

HIGHWAY

MAIN

TO CIVIC CENTER

RADIUS   ¼ MILE

APTS AND SHOPS

← TO BUSINESS CENTER   ARTERIAL   STREET   TRAFFIC JUNCTION

2.1. Clarence Perry, "Neighborhood Unit" illustration from the Russell Sage Foundation publication *Regional Survey of New York and Its Environs* (1929). © Russell Sage Foundation. Courtesy of Rockefeller Archive Center and the Division of Rare and Manuscript Collections, Cornell University.

relation to them, and in walking distance from them. The relation between state and citizen was articulated through a pedagogical framework.

The neighborhood unit flashed around the world as the basic building block of ideal towns across Europe, the United States, and beyond in the Middle East and Asia. Perry's celebrated 1929 diagram of a "model neighborhood unit" became an omnipresent feature of planning texts in the early postwar era. More than just a symbol or representation, the diagram became an actor on the historical stage. It was endowed with legendary status and echoed across the globe, outdistancing, literally and figuratively, Ebenezer Howard's Three Magnets illustration. The reproduction of visual and textual discourse invented the neighborhood unit as ritual performance and self-evident category. The imaginaries of neighborhood and community allowed the state to navigate into everyday life and create the social reality it desired. In a classic utopian posture, the neighborhood was a land of enchantment.

"A Piece of Utopia": Industrial and Resource Towns

Although new towns were the brightest stars in a variety of utopian constellations, their actual territorial footprint in the immediate postwar years was usually tied to heavy industry and natural resource extraction. Fueled first by the war and then by reconstruction, industrial productivity reached an all-time high, together with full employment. Huge investments were poured into material assets: coal, iron and steel, aluminum, chemicals, crude oil, copper and magnesium, uranium. These resources drove the new consumer industries and were vital to the military buildup and atomic weaponry of the Cold War. The new town movement as part of these operations has largely been lost in the scholarly emphasis on housing, architecture, and urban design. Yet new towns formed a pattern of localities connected to the military-industrial complex. New towns were steel towns: they were the sites of aircraft factories. They sprung up to support the global oil industry and to extract mineral ore deposits and the precious uranium used in atomic weapons. They sat next to huge hydroelectric and petrochemical complexes and massive ports that loaded precious resources for transshipment. They were planted in "backward" or aging industrial areas to help retool production. Their locations were explicit to resource exploitation and territorial hegemony.

This bond with the postwar industrial boom differentiates new towns from the more widespread, prosaic housing schemes that were also part of the reconstruction years. The state allied with large-scale corporations in converting the economic engine to national rebuilding and modernization, providing comprehensive urban infrastructure and a stable, contented workforce. Power over economic geography shifted to national governments that often organized Keynesian policies into Five-Year Plans. Military installations and nationalized industries could be easily shifted to peacetime production and new locations. Private industries could be enticed with generous tax breaks, cheap land, and the promise of infrastructure. Construction ex novo would create a harmonious world of citizen-workers and render visible the nation's industrial power.

These economic motivations were fused with a welfare state philosophy that tied together production and consumption, family and social services in a broad spatial context. This integrated vision lifted new towns into the category of utopianism. The plunge by government into new town construction was driven by emotion as much as by hardheaded economic strategy as officials surveyed the wreckage of war and the genuine human suffering. It aimed to do far more than rebuild what had been destroyed. Reformers held a passionate determination to solve the tragic conditions in poor regions, the squalid housing pocking a landscape marred by the belching smokestacks of factories, the poverty and unemployment that stalked the depression years. The enthusiasm and sense of realizing radical ideas was enormous. Families would live well and enjoy jobs and pensions, health care, and decent public services. The pent-up desires after the deprivation and sacrifices of the Great Depression and the Second World War stoked this fantasy of abundance. It was a grandiose gesture, a product of the optimism and sense of state commitment that underlay the reconstruction years.

A revolution was afoot. In the case of Britain, the postwar world would mean not only reconstruction but a national renaissance. An entire series of legislative acts handed the national government extraordinary power to buy land, control planning, and construct housing in ways that were unthinkable before the war. The initial step had been the 1940 *Barlow Report* that recommended a national policy for how and where factories and the industrial workforce would be located, something which had already taken place with the defensive scattering of critical wartime industries. Ultimately, however, the establishment of the Ministry of Town and Country Planning in 1943 was the major breakthrough. Lewis Silkin was appointed its first leader by the new Labour government in 1945, and indicated his intention to tackle new town legislation before local authorities reverted to their perennial squabbling. In October of that year, he appointed a New Towns Committee under the chairmanship of Sir John Reith. With great haste, the Reith Committee pushed through recommendations that became the New Towns Act of 1946. The first generation of fourteen Mark I new towns was started immediately, between 1946 and 1950. Most were meant to welcome families from the slums of London and Glasgow as set out by the two great regional plans of the postwar era: the *Greater London Plan of 1944* and the *Clyde Valley Plan of 1946*, both by Patrick Abercrombie. Not only were these plans the most significant land-use visions for Britain, but their international influence was profound. They laid out a rational model of spatiality and settlement that guided planning doctrine throughout the postwar era.

Abercrombie is a legend in the history of planning. His career began at the University of Liverpool, where he was founding editor of *Town Planning Review*. In 1935, he became professor of town planning at University College London. For Abercrombie, planning's moral mission was the reconstruction of a lost state of harmony, which required a top-down ordering.[7] He embarked on town planning as a professional expert reliant on historical and spatial analysis and civic and regional surveys. Like Patrick Geddes before him, Abercrombie was alive to the particularity and distinctiveness of place. The region was a great economic and geographic unit, a "body" that absorbed historic experience. There was an incongruity between this historical and environmental hyperawareness and the formulaic tactics Abercrombie reiterated in the dozens of regional plans he helped produce both before and after the war, in Britain and elsewhere. Yet he was a master at intertwining the anatomy of place with planning as a science that took in the full extent of socioeconomic development.

Abercrombie's *Greater London Plan of 1944* and *Clyde Valley Plan of 1946* recommended sweeping controls on industry in London and Glasgow, and their dispersal to satellite towns and regional centers within a national planning framework. The plans were not just about housing. They were about building up factories and opportunities for technological innovation. They were an attempt to retrofit aging industry and revitalize depressed economies. These policies would provide jobs and create a more equitable and productive economic geography. For the new town of Crawley outside London, for example, Abercrombie recommended a wide spectrum of light industry centered on electrical, printing, and pharmaceutical products. At Stevenage, also outside London, the aim was to build up the industrial workforce from three thousand to nearly thirty thousand, concentrated mainly in engineering. In the *Clyde Valley Plan*, Abercrombie advocated replacing Glasgow's shapeless and outmoded physical plant with modern industry in new towns. This was the best prospect for attracting new companies and diversifying the city's industrial base. The first generation of Mark I new towns have been criticized in hindsight for lacking a cohesive economic strategy, but a remarkably high proportion of their economies was based in manufacturing, heavy industry, and resource extraction.

Designation as a new town meant the possibility of revitalizing aging coal, steel, and shipbuilding industries. This was the case for the Scottish new town of Glenrothes, which replaced the squalid slums near the Rothes coal mine as part of the *Forth Valley Plan*[8] for central and southeast Scotland. The plan recommended four new towns to revitalize coal mining. The population at Glenrothes would work in the mines, but also in paper mills and light industry that would help diversify the town's economy. The new town of Cwmbran in South Wales would renew the ironworks and offer new jobs in nylon and glass factories. In the case of Peterlee in northeast England, the coal miners themselves demanded that a new town be built to replace the soot-blackened colliery villages. The designation of Newtown Aycliffe in the North of England as a new town meant the conversion of its munitions factories to peacetime use. The new towns were economic stimulus packages to modernize sagging economies and attract new industries. But they were more than a government handout. Every town was eventually expected to stand on its own feet and pay its way.

Yet more often than not, this reshuffling of factories and people ran into a thicket of difficulties. Local municipal authorities, already saddled with overcrowded tenements and desperate pleas for housing, had little patience for what naysayers saw as dreamscapes. The Glasgow city council, for example, fought the *Clyde Valley Plan of 1946* tooth and nail. The city's press sensationalized conditions in the notorious slum districts: they were rat infested, disease ridden, controlled by street gangs and immigrants. Added to the urban blight, Glasgow had suffered severe damage during the wartime Blitz. Beleaguered city officials wanted the rubble cleared, the slums demolished, and better housing erected in the city, not promises to ship people off to some industrial Oz.

The Mark I towns were also criticized by their opponents as more a matter of political expediency than genuine regional economic appraisal. In some areas, local opposition to designation as a new town was intense. Stormy public meetings and initial challenges in the courts were not uncommon. There were the inevitable political horse-trading and compromises. Industrialists were doubtful about building factories in forgotten backwaters far from London. There were false starts. Labor was scarce. Vital materials, from brick to steel, were still being rationed. The sheer bravado of the new town program in the face of such material shortages marked it as a courageous new beginning.

The new town of Corby in the English Midlands is a more detailed example of how designation as a new town played out. Corby was located near England's largest supply of accessible ironstone and also the coalfields in Yorkshire and Durham. Mining rights and steel production were owned by the Stewarts and Lloyds Company. By the mid-1930s, the town was budding into a premier steelworks and steel-tube-making center. Thousands poured into the sleepy village of fifteen hundred souls, especially young

Scottish workers from the Clyde Valley. Ten years later, in 1945, Corby was a boomtown of twelve thousand people, its steel mills churning out war materiel, including the steel tubes used in the cross-channel pipeline supplying fuel to Allied forces on the European Continent. The makeover also made Corby into a layered landscape of progressive social reform experiments. Stewarts and Lloyds constructed housing, shops, cinemas, and a football pitch. It hired architects to design a road system and town plan for Corby's future, replete with a town square and central avenue. But it did little to resolve frequent skirmishes between the new immigrants and the local population. Townspeople were divided along the sociological frontier of the railroad tracks, with the old village to the south and the "company town" spreading north around the steel smelting plant. A hodgepodge of council housing dotted the landscape. Corby gained a reputation as a hard-knuckle steel town of "broken bottles and ragged weans . . . prodigious feats of drinking," and frequent violence.[9]

When Corby was designated a new town in 1950, its population stood at eighteen thousand, and its newest immigrants were workers from Eastern Europe. The official title came at the same moment the British steel industry was nationalized, putting it directly under government control. This raised the expectation that Corby's steel production would be critical to the nation's recovery, as would the workers themselves. New town status "would last 100 years,"[10] and the population was forecast to reach forty thousand.

New town status meant the chance to consolidate a scattered industrial workforce into "an integrated and socially balanced community" with housing, infrastructure, and services.[11] The master plan laid out a shopping, entertainment, and social center for Corby, with seven low-density residential neighborhoods to welcome the anticipated growth. Each neighborhood unit would contain all the facilities typical of a small community. Corby suddenly transformed into a vast construction site with new housing estates, new shops and pubs, and a new bus station. The design was a model of state-sponsored industrial modernization and social reform. As the Manchester *Guardian* effused, Corby "breathed steel . . . steel is the meaning of Corby."

It resolved to establish a new kind of industrial society. All the old squalors of industrial living would be swept away. All is gentle, delicate . . . thoroughly nice. Hardly anything is ugly. Hardly anything is whole-hog. The Corby Development Corporation has set out to create a sensible, decent, healthy, neighbourly kind of Elysium, and has succeeded absolutely.[12]

It was the new England.

These new towns were quite ordinary, yet paradoxically exemplary spaces in the huge project of national modernization, ones that would forge new ideals and new citizens. In the early postwar years, the international reputation of British town planning was at its zenith, as was the fame of Patrick Abercrombie, who traveled widely to promote his ideas and carry out projects in Haifa, Karachi, Ceylon, Addis Ababa, and Hong Kong.[13] Countries clamored for British planning exhibitions and for lecture tours by British planning luminaries. Abercrombie's *Clyde Valley* and *Greater London* plans with portraits of the earliest new towns were staged worldwide in posters, photographs, and film. They were accompanied by Abercrombie's students, who found jobs applying British new town strategies in the Commonwealth countries.

In Australia, the wartime emergency and the need to pool national resources initiated its first steps toward regionalism. New South Wales set up a "regional survey of the resources of the State with a view to facilitating planned development, notably in connection with housing and public works, but also to provide a basis for investigation . . . of the employment structure, decentralization, and primary and secondary industry."[14] Other states followed suit, so that by the war's end, the map of the Australian Continent was filled with regional districts ready for postwar reconstruction. In 1951, a "First Six New Cities" movement[15] was proposed by Australian urban planners as part of the "prosperity through expansion" policy of the conservative federal government. Modernization enthusiasts envisioned wide-open expansion across the continent, with a doubling of the country's population in twenty years as well as extensive mineral exploitation and industrial development, especially in the western territories. Exploration for energy and natural resources became the focus of national patriotism.

At the same time, there was general agreement that Australia's major east coast cities were too large and overcrowded. They were described as pits of moral depravity, disease, and poverty. The solution was to disperse the urban population into the rest of the country. By the early 1950s, defense reasons were also given for dispersal: fears of atomic attack along the east coast fed into calls for resettlement away from Brisbane, Melbourne, and Sydney. New towns built around industries in government-designated territories were the most logical approach, and would encourage foreign investment. They would fulfill the dream of national integration by domesticating the country's vast reaches and making them beneficiaries of progress. They also were exercises in territorial domination and sovereignty: the quest for a new and better order inevitably involved the removal of local populations deemed nonessential. But public officials and the planning experts who supported these policies did not see this as colonizing: it was simply a part of modernity. Such a gargantuan territorial scheme evoked the abstracted geography of the aerial view in that indigenous peoples disappeared from sight or were shunted away as subjects of anthropological study. It was a modernization logic that made the new town a symbol of extraordinary magnitude in development aspirations while ignoring the odious consequences of colonialism.

Beginning in 1951, the new town of Kwinana was constructed as part of a massive petrochemical complex and deepwater port on Cockburn Sound south of Perth. It included the largest oil refinery on the Australian Continent run by the Anglo-Iranian (later British Petroleum) Oil Company, a steel-rolling mill operated by the BHP Company, a fertilizer plant, and cement works controlled by the British firm Rugby Portland Cement. Eventually, an Alcoa aluminum plant and a BHP pig-iron blast furnace were built, as well as a naval base and host of other industries. Driven by modernization doctrine, the Labour government of A. R. G. Hawke (1953–59) poured unprecedented investments into land acquisition, industrial infrastructure, and comprehensive town building for a population of forty thousand. Together with the Anglo-Iranian Oil Company, the State Housing Commission saw the development of Kwinana as vital to the industrial development of the state of Western Australia. The town became Australia's prototype for the future.[16]

As people swarmed into the Perth complex in search of jobs, Australian urban planner Margaret Feilman (fig. 2.2) was hired to design the new settlement. Feilman was the only woman working in Western Australia's Public Works Department. In 1948, she won a scholarship to the University of Durham in Great Britain, where she studied the new industrial towns of Peterlee and Ayecliffe and toured European cities undergoing reconstruction. She returned to Australia convinced of the need for urban planning, and quickly adapted the blueprint of the British new town to Kwinana. Then she began the site analysis and local surveys.

Feilman's sensitivity to the local setting was evident in a garden city plan that took advantage of landscape and topography. She articulated the city as a network of four neighborhood units surrounding Freemantle Port and the industrial complex: Medina, Calista, Orelia, and Parmelia (fig. 2.3). Traffic circulation was a primary factor in her design: the four settlements were connected by Gilmore Parkway. Each was planned for one thousand timber-framed and brick homes situated around a central spine avenue, with narrower roadways providing access through each neighborhood unit.

2.2. Margaret Feilman, 1953. © *The West Australian.*

To avoid monotonous rows of identical housing, some twenty different home designs were available to residents. A primary school marked the center of each neighborhood, along with shops, a cinema, a community center, and open-air gardens. In another new feature for Western Australia, parking lots and service stations were integrated directly into the plan. The residential zones were separated from Kwinana's industrial complex by parkland and open spaces. The neighborhood of Calista was regarded as the urban heart of Kwinana, with banks and businesses, a city hall, a library, a concert hall, and a museum grouped around a formal square. Feilman's ideal was to create "an Australian town in an attractive landscape." She reported in 1955, "A social life with its own organizations is already functioning and has given a very real community identity to the town, which is essentially a social experiment."[17]

Photographs of Kwinana from the mid-1950s depict an outback world of single-story buildings in the cleared bushland alongside the massive oil refinery and port. The first residents to move into their new home were a pipefitter and his family from Scotland, pictured in the local newspaper happily washing up after their first meal in their modern kitchen.[18] The photograph was a celebratory portrait of the new working class in alliance with industry and the state.

Kwinana was one of the most comprehensive applications of the British new town model in Australia, and was heralded as a triumph in conception and design. The Cockburn Sound complex and the Kwinana neighborhoods displayed the deep bond between industrial modernization and the welfare state in the postwar decade. Massive public spending on large-scale infrastructure projects such as these would modernize the economy, provide full employment, and consolidate territorial dominion. The social investment in Kwinana would create a stable, happy workforce and increase capital gain.

As in Australia, nationwide population resettlement was official policy in Canada. The nation's entire territorial space was understood as *terra nullius*, land belonging to no one. It was a colonizer's dream. By the early twentieth century, there was already a landscape of single-industry resource

2.3. Aerial Photograph of Kwinana, Australia, date unknown.
© Kwinana Council. Courtesy of the City of Kwinana.

towns that had accompanied mining and forestry into the frontier territories. Corner Brook in Newfoundland was a pulp and paper town, Thompson in Manitoba a nickel mining town. A 1953 survey indicated that Canada had more than 166 single-company settlements, with a combined population of 189,000.[19] Their number increased rapidly as Canadian companies hunted for minerals, especially uranium—the most coveted element of the atomic age. Elliot Lake was a boomtown near the uranium mines along Lake Ontario, where planners looked to acclaimed new towns such as Britain's Harlow and Sweden's Vällingby for inspiration.[20] Uranium City was built in Saskatchewan near the Beaverlodge uranium mines. Even before these, the planned community of Deep River in Ontario was constructed as part of the Manhattan Project. The site was chosen for its proximity to the uranium mines along the Chalk River, where an atomic energy facility was established (fig. 2.4). Deep River, along with the Manhattan Project sites in the United States and Britain, was among the West's secret science cities.[21]

2.4. Chalk River Laboratories at the new town of Deep River, Ontario, date unknown. Courtesy of the National Research Council Canada Archives.

They were strictly regulated and access was possible only with special permission. This covert network expanded further with the Cold War. Frobisher Bay in the remote northern reaches of Canada was constructed in the late 1950s as the hub station at the Dew Line, the air defense warning system against enemy missile attack in the NORAD radar web. Originally an Arctic Inuit settlement, it was reimagined by the Canadian government in a bold futuristic design. In stylized drawings, thirty-six cylindrical high rise apartment towers encircle a town center enclosed in a protective dome heated by atomic power. Although never built, Frobisher Bay was a triumphant example of Space-Age utopian fantasies in steel, concrete, aluminum, and plastic meant to defy the harsh northern environment.[22]

Frobisher Bay and Deep River were talismans of the atomic future. But as paper dreamscape in the case of the former, or actual site in the latter, the whimsical ideals behind their plans hid reality. The irony was that living in Deep River could be bromidic. The atmosphere was famously captured by Australian journalist Peter Newman, in his satire of that picture-perfect town for *Maclean's Magazine* in 1958. "Deep River is a Utopian attempt to create a happy environment where all is ordered for the best," he wrote. "There have been no major crimes in Deep River, it has no slums, no stoplights, no unemployment, no cemetery, no beer parlors, and few mothers-in-law." But the highly skilled atomic workers who lived in scientific paradise found its perfection tiresome. To chase their boredom, they organized sixty-eight clubs and associations for a range of interests, from glassblowing to square dancing. One physicist vented his feelings in a poem for the community paper that expressed the enigma of new towns everywhere: "Although the town is trim and neat / With cozy houses on every street / Though saying so is indiscreet / I hate it."[23]

All this development paled in comparison to what was known as Canada's Big Triangle, a gigantic industrial complex four hundred miles north of Vancouver that was inaugurated in the early 1950s to exploit the natural wealth of British Columbia. Its plans included a seven-hundred-mile oil pipeline backed by six major oil companies; massive oil and gas exploration projects; the Canyon Dam on the Nechako River, with huge water tunnels carved through the mountains; the Alaska-Canada Highway; electric transmission lines and a subterranean powerhouse described as "the largest power-generating plant ever built"; a giant pulp mill; and the "world's biggest aluminum smelter."[24] Raw materials for aluminum production were brought in by ship at a new industrial port. In a deterministic vision of the environment, the entire wild, mountainous terrain was surveyed by engineers using helicopters and airplanes, and then mapped out for development from a bird's-eye view. This was modernization's mastery of the natural world.

To house the workers streaming into the region, the Aluminum Company of Canada (Alcan) developed the new town of Kitimat alongside the smelting facility on the fjord coastline. Originally a Haisla Indian fishing village, it was remade into a major port town for fifty thousand residents. To create the settlement, Alcan hired none other than Clarence Stein, the premier American advocate of the garden city and neighborhood unit ideals. Working with New York architects Albert Mayer and Julian Whittlesey, he assembled a multidisciplinary team to carry out the design for the new community.

Mayer was himself a well-known advocate of garden cities and new towns since the 1930s, and he worked with Lewis Mumford and Henry Wright as an active campaigner for progressive housing reform. At the same time, he was also working in India on the plans for New Delhi and his pilot projects in the state of Uttar Pradesh (see chapter 3). He was heavily influenced by Gandhi's concept of village life, and used comparisons with Indian cities in formulating the plan for the Canadian wilderness settlement.

For Stein and Mayer, these international itineraries were crucial creative sources for their new town concepts. Most of all, however, Kitimat was the culmination of Stein's long career and his vision for "Cities to Come," the title of his unfinished manuscript. Stein's notion of neighborhood at Kitimat remained the small-scale settlement nestled around its elementary school, community center, general store for daily needs, parks and playgrounds, and traffic-free pedestrian paths. It was "the focus of community life, a symbol of neighborhood unity."[25] The ordinary life of families within this sentimentalized setting was the utopian bonus of pleasure and good living.

But it was the enthusiasm of the first residents of Kitimat and their confidence in the future that most demonstrated new town magnetism. They were young and multiethnic: they came from around the world. As recounted in the town's first official history,

There is a man who came to Kitimat from central Europe five years ago. He and his wife have been steadily employed since that time. They own their own home which commands a view of the Douglas Channel, breathtaking in its wonder. Their standard of living would be as high in any North American city. They have improved their house in their own time by the work of their own hands. They drive a car. And they have made a fine garden.[26]

These were humble desires, but Kitimat embodied the fantasy of an idyllic ordinary life made possible by a massive industrial development scheme. In 1958, the United Nations Film Services arrived there to film *Power among Men*. Kitimat was selected as one of four on-the-scene encounters with worthy development projects to be directed by renowned filmmakers Thorold Dickinson and J. C. Sheers. Its townspeople were featured in the cinematic tribute to the human capacity for transformation.[27]

There was more than a hint of paternalism on Alcan's part in this new town endeavor (fig. 2.5). That a private industrial corporation would be at the forefront of new town planning was part of the social compromise of the postwar period. The more than two thousand men and women working at the Alcan plant demanded not just better factory conditions but decent housing and social services and well-planned communities. New towns were a stage for good management of labor relations and attracted the trained personnel needed for industry. The orderly, family-oriented environment enhanced production and profits. Alcan's vice president echoed these sentiments: "No modern large-scale business can be successful without a loyal, competent and happy work force. . . . We want the very best

2.5. Proposed Kitimat City Center commercial building for Kitimat, British Columbia, by Alcan, 1955. © Rio Tinto Alcan. Courtesy of Northern Sentinel Press Collection, Kitimat Museum and Archives.

possible living and working conditions at Kitimat."[28] Planned towns also echoed the rule-governed, predictable atmosphere of large-scale corporations. Indeed, these new towns were generally managed by development *corporations*, which tended to be public entities but styled on the private corporative model. By the late 1950s, there were eleven churches and over sixty cultural, recreational, ethnic, and fraternal organizations in Kitimat: "busy, busy" free time made for a content labor force.[29]

Steel Towns as Socialist Ideal

The decisive alliance between government, industry, and labor defined postwar modernization on both sides of the Cold War. New towns heralded state control of material assets and capital investments, and the building of a new, more equitable society. Just as in Corby, Kwinana, and Kitimat, so the ideal socialist city, built from scratch, was the experimental arena for a new society, one in which harmony and happiness would reign.[30] These idealistic places were reproduced across the global resource landscape.

About one thousand "new towns" appeared in the Soviet Union, and sixty additional town projects were built in other Eastern Bloc countries. For the most part, they have been written off by scholars as worker dormitories at steel plants and oil refinery sites. Indeed, these new towns were the flagships of the Five-Year Plans. Steel still forged the power of nations, and it made these settlements into far more than just base camps. Their ideological content was enormous. Each country in the Eastern Bloc had its socialist showcase: Nowa Huta and Nowa Tychy in Poland, Eisenhütten-stadt in East Germany, Sztálinváros in Hungary, Dimitrovgrad in Bulgaria, Havirov and Vorosilov in Czechoslovakia, Titograd and Velenje in Yugoslavia. As political symbols, they were almost as important as the Red Flag. They were imagined as "splendid living environments, economically and culturally, that would promote the collective life of mankind."[31]

This strapping fantasy was conveyed by news articles about the building operations, and by films and novels, poems and paintings, and popular music. New towns were enchanted places. Ordinary people would have good jobs, access to education, modern housing, parks and playgrounds, health care, and a myriad of social and cultural entitlements. These hopes were the prerequisites for the New Socialist Man, the archetypical citizen who would emerge with the socialist revolution. The ideal socialist city that emanated from the Soviet Union was clearly the primary source of inspiration. The design and construction of new towns were intimately associated with politics, as all utopias are. The socialist new towns, however, were imagined as "new" and different because they were planned, thereby avoiding the incoherency and bourgeois cosmopolitanism of the capitalist city. They were also new because they gave shape to the "socialist way of life" that would give workers the right to the city, along with the right to work, housing, culture, and recreation. They embodied the next stage in the spatial history of modernity. The new towns would be free of conflict and beautiful. In them, a new socialist generation would mature in peace and happiness. Socialism had the power to bring this urban future to pass immediately.

Yet underneath this political swagger, ideal socialist cities were not fundamentally different from the new town dreamscapes in the capitalist world. Both sides of the Cold War divide shared deeply in the regenerative, utopian aspirations of the reconstruction years. Rebuilding gave rise to enormous hopes for the perfectibility of the urban realm. Socialist new towns shared the traits of modernism and modernization, interpreting them within the context of the Soviet Bloc.[32] Planners and architects in the East held a wide assortment of aesthetic influences, professional relationships, and urban theories in common with their Western counterparts. These roots were found as much in prerevolutionary Eastern Europe as in the ideological blueprint of socialism—which itself was subject to change.

The architects and urban planners who designed and built Eastern Europe's new towns were members of a key midcentury generation of students who came of professional age during the 1920s and 1930s. For many of them, their years of exile during National Socialism and the war enhanced their international networks and the transfer of planning concepts. Their knowledge and professional experience allowed them to weather the shifting ideological tides of socialist planning and architectural doctrine.[33] These transnational influences were as significant to the socialist urban ideal as the received wisdom of Soviet planning policy—and perhaps they were longer lasting. Planning expertise was accepted and promoted by the regime as the framework for imagining an ideal socialist world. Both the Eastern and the Western Blocs focused on reconstruction, economic development, and social ordering, and on providing their citizens with material and social well-being, collectively and individually.

Consequently, the neighborhood unit became the official modus operandi for Eastern Bloc new towns, although it was termed *microrayon* in socialist parlance. The term was chosen to distinguish neighborhood planning as a specifically socialist concept. It was originally used in 1930 by noted economist, demographer, and Party theoretician Stanislav Strumilin,[34] who adapted the neighborhood unit formula of Clarence Perry to

Soviet planning. What initially differentiated the socialist version of the ideal from its capitalist correlative was its rhetorical and design emphasis on the collective nature of urban life. Collectivity was evidenced by the schools, day care and medical facilities, and community clubhouses that each neighborhood featured. The *microrayon* would overcome class divisions. The uniquely socialist space of everyday collectivity would prevent the alienation and social segregation of the capitalist city.

Yet the differences in interpretation of *community* and *collectivity* were largely exaggerated by the ideological conflicts of the Cold War era. Social equity, peace and harmony, the desire for ordinary happiness, were dreams shared in both the Eastern and the Western Blocs during the reconstruction years. The neighborhood ideal had been thoroughly absorbed into planning discourse. By the 1964 United Nations Symposium on New Towns held in Moscow, which featured a procession of international new town experts from both sides of the Iron Curtain, it was agreed that "the neighborhood unit and the residential district should be the basic scales of new town planning." To underscore the point, N. V. Baranov, deputy chairman of the Soviet State Committee for Engineering and Architecture, added, "The neighborhood, as the basic structural component, retains its significance and is organized on the same lines whatever the size of the town."[35]

In the case of Poland, the immensity of wartime devastation meant a public crusade for reconstruction with only the scarcest of resources to rely on. The country lay in ruins. At the Yalta Conference, Polish frontiers were summarily shifted westward with massive, antagonistic removal and transfers of population. After the war, the nation's productive capacity was at a standstill, and its cities were barely functioning. One-sixth of the population had died. Polish-educated elites had been decimated. Initially, a Ministry of Reconstruction was set up that consolidated the remaining skilled professionals and university staffs around an ambitious program of physical planning at the national, regional, and local levels. By 1949, with some stable governance in place, the responsibilities for reconstruction were divided among an array of agencies. The Polish Academy of Sciences was established with a network of research institutes. Extensive regional surveys and cartographic analyses were carried out.[36]

The national census of 1950 provided an up-to-date demographic portrait of the country. Studies of settlement networks and the structure of cities were undertaken based on Walter Christaller's central place theory (see chapter 1) as well as on the British Reith Report about new towns. Planners mulled over the plan taking shape for a cluster of new towns around Stockholm.[37] Mapping out and appraising the national territory were preparations for rationalizing space and pooling scant resources in a countrywide plan. Given the dreadful postwar conditions in Poland, it was a heroic undertaking.

Poland initiated a number of new towns projects. Among them, Nowa Huta and Nowa Tychy were the most significant, and have been extensively studied. Nowa Huta (New Steelworks) on the Vistula River in Upper Silesia was celebrated as the biggest project of the country's Six-Year Plan, which was in place by 1953. When completed, the V. I. Lenin Steelworks there would produce, according to propaganda, as much steel as all the prewar Polish steel mills put together. The new town was linked by road and rail lines extending eastward into the Soviet Union, which invested heavily in the project. Technical assistance, rolling mills, and raw materials also came from the United States.[38]

Nowa Huta's location was also politically motivated. Silesia was one of the most contested and politically volatile areas of Europe. The bulk of the territory had been transferred to Polish jurisdiction after the Second World War. The proposed land-use map for Upper Silesia displayed an ideal schema of transportation arteries interlacing old and new towns in a strict pecking order. Large-scale development projects and new towns established territorial control and solidified the new border with Germany. Nowa Huta was also a deliberate replacement for Catholic Cracow, which was one of the centers of resistance against the Soviet-backed communist government. The rival towns were only seven miles apart. Cracow was described in official media as bourgeois, reactionary, old-fashioned, and weak. Nowa Huta was modern, young, healthy, and strong.

The reality facing Polish planners was that Silesia was a dense, aging industrial area, where 80 percent of the population worked in mining and heavy industry. And it was a social powder keg. The 1920s and 1930s had been marked by two full-blown uprisings by ethnic Poles and Germans, occupation by French troops during the border plebiscite, ongoing strikes by miners, and perennial food shortages. Clean water was a scarce resource. Exhausted mines and outmoded industrial plants processed zinc, lead, iron ore, and steel in a demoralizing landscape strewn with crumbling slums made worse by wartime bombardment. Industrial waste and pollution made living conditions insufferable. Dank, overcrowded tenements jutted up against zinc foundries and blast furnaces.[39] It was precisely these dreadful conditions that made planning imperative for Poland's future. New towns functioned as utopian ideals, but within specific historical conditions—in this case the gruesome residue of the past.

Nowa Huta was originally planned for a population of one hundred thousand, although it eventually grew to over twice that size. The planning team led by architect-engineer Tadeusz Ptaszycki made the new town a script in socialist realism with a design grounded in Renaissance theory. Historic tradition became the hallmark of Stalin's campaign against the modernist influence of the West as capitalist and formalist, and for a heroic nationalism. Socialist architecture and urban design were to be based on national heritage as an expression of the political life and national consciousness of the people. Nowa Huta as well as the early postwar new towns of Nová Dubnica in Slovakia, Nova Gorica in Slovenia, Sztálinváros in Hungary, Stalinstadt (Eisenhüttenstadt) in East Germany, and Dimitrovgrad in Bulgaria were all representatives of socialist realist style.

Nowa Huta's layout was a scrupulous geometric composition. Three boulevards radiated from the central square (fig. 2.6) selected for its views of the Vistula River and surrounding valley. The central axis was marked by two monumental buildings in classical style: the theater and the city hall. This last copied one of the most famous buildings of the Polish renaissance, the town hall of Zamość. The entire urban configuration was set in relation to the steel plant to the east (figs. 2.7, 2.8). The entrance gateway to the Lenin Steelworks flashed a modernist insignia, while its administrative offices were decorated with architectural flourishes reminiscent of

2.6. Central Square, Nowa Huta; Cracow, Poland, ca. 1960. Photograph by Henryk Makarewicz. Courtesy of Imago Mundi Foundation.

2.7. Engineer Jan Anioła, the first general director of the Lenin Steel Mill, with volunteers from the Służba Polsce (Service for Poland) work brigade at the building site of blast furnace no. 1, Nowa Huta; Cracow, Poland, May 16, 1952. Unidentified photographer. © Jan Anioła Family Private Collection, Historical Museum of Krakow.

2.8. Celebration of the opening of the blast furnace at the Lenin Steel Mill in Nowa Huta; Cracow, Poland, July 21, 1954. Photograph by Sovfoto / UIG. © Getty Images.

Polish palaces. From there, collective housing estates were meticulously laid out along with sports fields, parks, and greenbelts. There was an all-embracing narrative quality to the model settlement.

Of all the new towns of Eastern Europe, Nowa Huta was the biggest venture. As such, it occupied an outsized role, not only in official propaganda but also in literature, films, and music.[40] It was built as performative urban theater—a city of wonder. For author Marian Brandys in his novel *The Beginning of a Story* (1952) and composer Witold Lutosławski in his *Song of Nowa Huta* (1952), Nowa Huta was "the pride of the nation," "the forge of our prosperity," and "a work of Polish-Soviet friendship." Wisława Szymborska published poems to the city in the Nowa Huta weekly newsletter *Budujemy Socjalizm* (We Are Building Socialism). Photographer Wiktor Pental, who lived and worked in Nowa Huta in the 1950s, captured its life in an extraordinary series of images of the perfect city. Outside Warsaw, Nowa Huta was the subject of more films and media coverage than any other city in Poland. Some three hundred films were produced, along with nearly three hundred episodes of the *Nowa Huta Chronicles* detailing the city's life that were shown weekly in cinemas and on television.

The town name became so pregnant with meaning that everyone intuitively grasped its symbolism: the eight letters of Nowa Huta, carried aloft on monumental placards at marches through Warsaw, were sufficient in themselves. Moreover, bulletins kept the entire country apprised of the town's construction. Nowa Huta's street names were Six-Year Plan Avenue, Lenin Avenue, Avenue of the Shock Workers, Soviet Army Street, Great Proletariat. Most important, the town represented Poland's socialist future: its steel plant was a temple to labor, and its very existence and form were read as political ideology.

This ideal city was built by the people themselves, organized into the youthful work brigades that were an omnipresent feature of the reconstruction years. On radio, in print media and posters, the young people of Poland were called to Nowa Huta to find a new life. Fresh-faced young men organized into worker battalions were captured in official documentaries such as *Destination Nowa Huta!* by Polish filmmaker Andrzej Munk, *Birth of the City* by Jan Łomnicki, and *I Was Building the City* by Bohdan Kosiński. Although they initially lived in austere construction camps and waited in food lines, propaganda hailed these new arrivals as the heroes of Nowa Huta. The best known was Piotr Ożański, who was awarded the coveted title of Stakhanovite for his bricklaying talents and marched at the head of the May Day Parade.

Nowa Huta and its people were a massive productivity machine. Its citizens were the vanguard of the new socialist world, and as such would be showered with vocational training, health services, and the promise of modern housing. In reality, however, much like Corby in England, Nowa Huta was a raw place, and its hard-drinking population became increasingly troublesome and dissident. The young brigade members could easily morph into hooligans with a hard political edge.[41] In addition, Poland's Romani population was displaced to Nowa Huta, and the city became a safe haven for Ukrainian and Greek refugees fleeing political violence and for former prisoners released from Soviet labor camps. The lags in construction, the dearth of services, and the broken utopian promises provided plenty to grouse about. Alongside the flow of official propaganda, there was surprising criticism of the project both in planning documents and in the press. The frontier atmosphere was captured in an excerpt from Adam Ważyk's *A Poem for Adults*,[42] published in August 1955 in *Nowa Kultura*:

> From villages, from little towns, they go in wagons
> To build a foundry, to conjure up a town
> To dig out a new Eldorado.
> A pioneer's army, a gathered mob
> With a storehouse of oaths, with a little feather pillow
> Bestial with vodka, boasting of tarts
> A migrating mass, this inhuman Poland
> Howling with boredom in December evenings
> Fed on great empty words, lives wildly from day to day . . .

This hard-boiled impression was eventually portrayed in Andrzej Wajda's 1976 film *Man of Marble* with its fictional Nowa Huta bricklayer Mateusz Birkut. In the film, Birkut gains fame as the propagandistic overachieving worker, but eventually falls from official favor for his involvement in demonstrations. He dies in the 1970 clashes at the Gdansk shipyards—a portent of things to come.

Nonetheless, newspapers were packed with articles about the utopian world rising from the mud. Nowa Huta was Poland's first experiment in prefabrication, concrete construction material, and modern building methods.[43] The plasticity and practicality of concrete were associated with progress, liberation, and social justice. The conviction was that a complete modern town would produce new social behavior and a new society. The city's residential areas were arranged into neighborhood units of 5,000 to 6,000 residents. There, young families would enjoy domestic bliss. They lived an ordinary, everyday modernity that was extraordinary for the times.

By 1952, Nowa Huta already had a population of some 57,000, and by 1956 the number of residents exceeded 100,000. Long waits for apartments ended in flats with modern kitchens, parquet floors, and central heating. These dwellings stood for the revolution in social relations, the esteem and quality of life that the working class would enjoy. The town was slowly outfitted with health centers and schools, libraries, theaters and cinemas, and shops. Hundreds of trees were planted to create a city of greenery. This socialist urban archetype would produce a sophisticated modern citizen. It was predicted that Nowa Huta would become a first-class European tourist attraction.[44]

This same mesmerizing discourse surrounded the new towns of the German Democratic Republic (GDR). Four official new towns were built there: Stalinstadt, Schwedt, Hoyerswerda, and Halle-Neustadt. Stalinstadt was founded in 1950 on the Polish border near Frankfurt/Oder and was heralded as the first socialist city in Germany (the city's official name was changed to Eisenhüttenstadt in 1961). It was a steel town of the same mythic status as Magnitogorsk, the Soviet showcase beyond the Ural Mountains.[45] For East German architects, Stalinstadt was "a piece of utopia," a "social model." Early propaganda pamphlets gushed over a place where "the future bids you good morning." The city would gather accretions of myth that had less to do with the place than with the way the communist GDR saw itself and its emerging identity. Well over half its industrial infrastructure had been destroyed during the war, and much of what was left had been dismantled and carted off to the Soviet Union as reparations or lost to West Germany or Poland. In this landscape of desolation, the founding of Stalinstadt was interpreted as a momentous sign of national reemergence.

Stalinstadt was promoted as the counterpart to Nowa Huta along a new German-Polish border of peace. The former enemies would be reconciled in their humming steel production within a new economic geography overseen by COMECON, the Eastern Bloc's Council for Mutual Economic Assistance. Under its multilateral trade provisions, coke and coal would be shipped from Poland, iron ore from the Soviet Union. The GDR encouraged the industrialization of its underdeveloped northeast by major production and engineering projects. Stalinstadt was thus founded as a workers' foothold in the rural Uckermark area. Poor regions such as these, "inherited from capitalism, would be eliminated forever." The socialist city would loosen the grip of the reactionary rural and petit bourgeois classes.[46] The region was also far from the US airbases in West Germany. Even more significantly, Stalinstadt was meant to absorb the refugees expelled from the east by the new German-Polish border, and provide them with the stabilizing influences of jobs and homes in the new socialist system.

The town was famously based on the GDR's new principles of socialist urban development, aimed at the harmonious satisfaction of the human demand for work, dwelling, culture, and recreation. In the spring of 1950, a delegation of high-level planners and architects from East Germany made a study tour to Moscow, Leningrad, Stalingrad, and Kiev. The result was the promulgation of the legally binding Sixteen Principles of Urban Design, based on the General Plan for Moscow. The document has been interpreted by architectural historians as a political reaction against CIAM's Athens Charter, which along with the Bauhaus had been condemned by the socialist regime as functionalist and cosmopolitanism. Questions about aesthetics and urban design were wracked with political overtones. The extraordinary attention to urban planning at Stalinstadt was meant to reflect the GDR's social revolution in the landscape. Plans for new towns were supervised closely by Party leaders. Decisions were made at the highest levels at the Ministry of Reconstruction and by GDR president Walter Ulbricht himself.

After a review of a bevy of early experimental designs, architect Kurt Leucht's urban typology for a new town of thirty thousand inhabitants was officially approved for Stalinstadt in 1951 (fig. 2.9). The town was compact and featured a formal unity in the tradition of Renaissance classicism. A ceremonial *magistrale* named Lenin Avenue led to the main square.[47] The urban center was an absolute priority in socialist urban planning. The Communist Party (Sozialistische Einheitspartei Deutschlands, or SED) expounded on the primary role of the center at the Third Party Congress in 1950. Rather than a buzzing downtown with Western-style traffic and commerce, the center of the town was to be a measure of political mankind. It would be a site of "grandeur and beauty" at which everyone would feel welcome.[48] Leucht planned its all-important House of Culture and the city hall flanked by the grand entranceway into the steel mill in a formal geometric ensemble. These three institutions were the mechanisms for social transformation. Their monumental staging served as theatrical backdrop for political demonstrations, parades, and popular celebrations; at these the workers would claim their right to the city, which should be "national, beautiful, generous." Rounding out Leucht's classical design tribute to socialist urban utopia were a library, a theater, and a cinema.

Some 50 percent of Stalinstadt's steel plant workers were refugees from the East,[49] newly settled in the city of the future. Others were farmworkers and young people looking for jobs in industry. They found homes in

2.9. The design of Stalinstadt, East Germany, by Kurt W. Leucht, 1952. Photographer unknown. Courtesy of the Leibniz Institute for Regional Development and Structural Planning, Scientific Collections, A12 (Eisenhüttenstadt).

four residential communities arranged in a tight radial around the town center. The complexes featured four-story buildings in classical style with balconies and arcades (fig. 2.10). The buildings were divided into spacious modern flats and surrounded by gardens, playgrounds, and pedestrian pathways. Each neighborhood was a self-contained wedge with a communal day-care center and school, clubhouse, health center, and social services calculated by the number of inhabitants. Beyond them lay allotment gardens, sports fields, and parks. Local topography was integrated into the urban design to provide a particularity of place. Historian Ruth May argues that the uniqueness of Leucht's approach was "the attempt to construct an ideal congruence between traditional urban features and the new characteristics of a socialist industrial town." Work and life became reconciled in a new urban ideal.[50] The city would provide workers with the opportunity for unlimited potential. It was a collectivist dream.

Stalinstadt appropriated and expressed various strains of European urban theory. In preparation for his design, Leucht had traveled to the Soviet Union and studied Magnitogorsk as well as the Georgian industrial town of Roustavi, which was founded in 1948. Nowa Huta just across the border was also a model. Both Nowa Huta and Stalinstadt were grounded in the hallmark classicism of socialist realism. The layouts have a formal geometry, with a central square and axis marked by monumental buildings and radial boulevards. In an irony famously lost on Communist Party officials, classicism was also the signature of National Socialist design as

well as Le Corbusier's renderings for the reconstruction of towns such as
Saint-Dié or even his Radiant City. Nevertheless, socialist realism design
idioms provided communism with aesthetic legitimacy and meaning.[51] It
was the antithesis of the disorder and urban misery of the past. Reconstruction provided the opportunity to rectify these evils. The return to classical
ordering was also in juxtaposition to Western-style functionalism as well as
to the bucolic meanderings of the garden city.

However, this socialist realism phase was short-lived and riddled with
contradiction. Despite the official political posturing and hostility toward
modernism, Bauhaus alumni were everywhere in Central and Eastern
Europe.[52] They either skirted or dove through a minefield of official pro-
nouncements about national architectural heritage and urban aesthetics.
Below the surface rendering of Stalinist style, East German architects remained fascinated with prefabrication and standardization. They shared
tremendous interest in the British and Swedish new towns, and in construction techniques from France and the Netherlands.[53] Stalinstadt's functional
zoning, the attention to neighborhood, schools, and recreation and park
areas, followed in step with modernist creed and even with CIAM's Athens Charter. Like the garden city movement and CIAM, socialist planners

2.10. Socialist realism at Stalinstadt, East Germany. Residential block on Friedrich-Engels-
Straße with shops on the ground floor, ca. late 1950s. © Stadtarchiv Eisenhüttenstadt.

envisioned sunlit dwellings amid greenery and parks. By the 1950s, this was a shared ethos across the globe, much of it steeped in a postwar picturesque of the healing benefits of nature. It was a deeply poignant search for a better life. The design of neighborhood units containing everyday social services was also emblematic of the transnational character of mid-twentieth-century urban planning and its profound engagement with the ordinary.

Both Stalinstadt and Nowa Huta[54] attained allegorical status in the so cialist imagination. Both places shared the heroic characteristics of new town archetypes across Eastern Europe during this early reconstruction period. As each country implemented urban reform measures and desperately needed housing programs, new towns became the experimental gateways into the socialist future. As ideological and propagandist strategies, they gave hope to a postwar generation embarking on the great experiment in socialism. Like Nowa Huta or the new towns of Sztálinváros in Hungary and Kunčice in the Czech Republic,[55] Stalinstadt was depicted as a "city of youth." Propagandistic photographs of Stalinstadt typically staged young workers on an elevation, surveying the rise of utopia on the plain along the Oder-Neisse frontier (fig. 2.11). They are a self-actualizing people building a new world.

2.11. East Germans surveying Stalinstadt from a nearby hill, 1954.
© Bundesarchiv. Bild 183-26012-0001 / Photograph by Horst Sturm.

This totalizing gaze, pure and unencumbered, was found in such socialist realist novels as Karl Mundstock's *Helle Nächte* (1952), in which the female protagonist dreams of standing atop a crane and looking down on what will become Stalinstadt. The revered construction site transmutes into an idyllic urban setting. In official photographs, young families with children, and young men and women are captured in the streets of Stalinstadt, in the schools, playgrounds, and health centers. They gather in the clubhouse for community dances and festivities. They are visitors to the future: their everyday lives have been transformed. In the extraordinary collection of photographs in *Stalinstadt, Neues Leben-Neue Menschen* (1958), the great iron and steel combine dominates the city, with workers tending to their tasks among the fiery furnaces and machinery. The individual portraits, the valiant faces, the joyful children and families, produce a heroic working people, their lives "freed forever from anxiety about daily existence. . . . Stalinstadt is the beginning."[56]

Building a Regional Metropolis

Although they have been largely overlooked in postwar urban history, industrial and resource towns such as Corby, Kwinana, Kitimat, and Stalinstadt were the leading edge of postwar reconstruction and formed a decisive economic geography. By the mid- to late 1950s, a second set of new towns began taking shape around the great cities of Europe. The *Greater London Plan*, Helsinki's Seven Towns Plan, Copenhagen's Finger Plan, and the *General Plan for Stockholm* became international models of metropolitan planning. These were future visions for prominent capital cities with large populations, and their complexity and regional extent were of an entirely different scale.

The plans involved three crucial elements that made each of them into a cohesive schema. First was large-scale urban renewal that refashioned aging city centers into modern commercial and cultural zones. This is a topic in its own right that focuses on historic urban core districts. The purpose here is to instead spotlight the second element: the dispersal of the city's population to a network of satellite new towns arranged in a metropolitan region. The locations for these new settlements were coupled with transportation corridors in a new geography entirely invented by planning. Last was the framework of social welfare that made these new towns into a utopian landscape of social democratic reform. Indeed, the most immediate and successful metropolitan plans were carried out by the European countries (especially Scandinavia) most committed to progressive social welfare.[57]

What distinguished these places was far less the factories and steel mills, the jobs that symbolized production, than the social services that represented state magnanimity and the perfected life that could be enjoyed. These spanking-new towns popped up on a regional landscape as privileged courtiers surrounding the capital city, glittering stars circling the sun.

It was the plan for metropolitan London that initially captured the imagination of planners in the early postwar years. It became the most imitated model worldwide, embraced and adapted by a dizzying array of offspring. The principle of containing London's growth with separate garden cities in a regional nexus was of course a long-standing British tradition derived directly from Ebenezer Howard (see chapter 1). But it was the immense damage suffered during the Nazi Blitz that transformed the early garden city experiments into a far-reaching scheme for metropolitan planning. At the height of Nazi bombing in 1941, Lord John Reith, Minister of Works and Buildings, asked the City of London and the London County Council to start preparing plans for postwar reconstruction. His was one among many bids for the future of London that appeared while the city was still ablaze. Another was the linear plan proposed by MARS (the British branch of CIAM) for a high-density metropolis stretching along the river Thames.[58] But there was little doubt that the breathtaking regional vision proposed by Patrick Abercrombie and F. J. Forshaw would carry the day.

Abercrombie was introduced earlier in the discussion of resource towns, because his role in conceptualizing industrial revitalization throughout Britain was so significant. However, the County of London Plan (1943) and the Greater London Plan (1944) were his most illustrious achievements. They called for a radical dispersal of population and industry from the overcrowded central districts to the city's periphery. London's growth would be contained within a series of concentric rings that established an orderly regional pattern with defined boundaries (fig. 2.12). Moving outward from central London to the County of London, and then the Greater London metropolitan area, each ring would be less urbanized until the outer edge was a vast greenbelt. Beyond it, some 1 million people would be shifted out to ten new towns arranged in a wide orbit around the capital. Only eight were actually built: Stevenage, Hemel Hempstead, Hatfield, and Welwyn Garden City in the county of Hertfordshire; Crawley in Sussex; Harlow and Basildon in Essex; and Bracknell in Berkshire. Each town would be designed around superblocks outlined by the road system, then filled in with neighborhood units. This was an audacious vision, a trailblazing new scale to planning. It was a long-term strategy that set the parameters of the future.

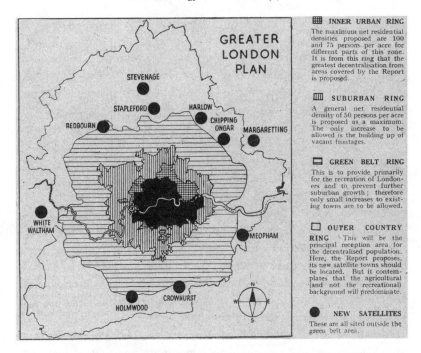

GREATER LONDON PLAN

STEVENAGE
STAPLEFORD
REDBOURN
HARLOW
CHIPPING ONGAR
MARGARETTING
WHITE WALTHAM
MEOPHAM
HOLMWOOD
CROWHURST

⊞ INNER URBAN RING
The maximum net residential densities proposed are 100 and 75 persons per acre for different parts of this zone. It is from this ring that the greatest decentralisation from areas covered by the Report is proposed.

▥ SUBURBAN RING
A general net residential density of 50 persons per acre is proposed as a maximum. The only increase to be allowed is the building up of vacant frontages.

▢ GREEN BELT RING
This is to provide primarily for the recreation of Londoners and to prevent further suburban growth ; therefore only small increases to existing towns are to be allowed.

▢ OUTER COUNTRY RING This will be the principal reception area for the decentralised population. Here, the Report proposes, its new satellite towns should be located. But it contemplates that the agricultural (and not the recreational) background will predominate.

● NEW SATELLITES
These are all sited outside the green belt area.

2.12. The *Greater London Plan of 1944*'s four zonal rings, with new satellite towns and villages indicated. © RIBA Library Photographs Collection.

Given the austere conditions in Britain in 1944, the *County of London Plan* and *Greater London Plan* were lavish productions including full-color plates of planning and survey maps. They were a mesmerizing visual spectacle of the future, and a visual aesthetics of power and propaganda. The diagrams take an aerial viewpoint, looking down on the London area from an exalted position and then transforming it into a color-coated plotting of communities and towns. The strategy was rooted in formal composition and the equilibrium of part and whole set into a unified scheme. As geography became an object of investigation and improvement, it was interpreted as designing and engineering abstract "space." There was a fixation with classifying populations in a rational hierarchy across this conceptual field. Precisely 1,033,000 inhabitants would be dispersed: 383,250 would be moved to new satellite towns, 125,000 to quasi satellites, and 261,000 to already existing towns. Diagnostics was imagined as transferable across both territory and time, as were planning solutions.

The metropolitan plan viewed from above became the dominant medium of planning language. It played a key role in Abercrombie's understanding of urban development, and how urban regions could and should be organized in a normative manner.

The *County of London Plan* imagined the creation of communities divided into smaller neighborhood units of between 5,000 and 10,000 people. As Forshaw and Abercrombie noted, "It is the intention of our proposals that children living in these units should not have to cross a main road from home to school. . . . The open spaces, apart from the regularly distributed playgrounds, are designed as far as is possible to surround the whole community, forming a natural cut-off between it and its neighbors." Community grouping, they continued, "helps in no small measure toward the inculcation of local pride, it facilitates control and organization."[59] The plan also contained a theoretical neighborhood study based on Clarence Stein's and Clarence Perry's concepts. Elegant full-color plates dramatized step-by-step construction of a London community neatly divided into neighborhood units that surrounded their schools.

The *Greater London Plan* carried this step further with an experimental model of how Stein and Henry Wright's system for the new town of Radburn, New Jersey, might be adapted to the proposed new town of Ongar northeast of London. Each of the six neighborhoods of ten thousand inhabitants was planned around its community center, two elementary schools, and four nursery schools. Each had its own shopping center, services, and light industries. Every two neighborhoods shared high schools, one for boys and one for girls. A variety of housing options were scattered through the site, from detached houses to flats. The neighborhood units were then separated from one another by greenbelts and arterial road systems. There is little porosity to this spatiality: neighborhood units were designed internally as a unified community experience. The full-color plates of this imaginary Ongar[60] were a sophisticated rendering of the neighborhood unit model superimposed over a blank spatial canvas. The sense of aesthetic satisfaction in these planning artifacts is palpable. They created an urbanistic sublime in which the future of Britain was assured.

Beyond espousing the belief in rational planning as fundamental to postwar recovery, both plans were publicity machines. The London County Council hired a public relations firm and released popular print versions of the county plan, held exhibitions and town hall meetings, and dispatched the leading lights of new town ideology for appearances on radio and in film.[61] The public relations campaign reached a crescendo at the Festival of Britain, held in London and throughout the United Kingdom in the summer of 1951. Organized by the Labour government, it was publicized as a "tonic for the nation," a celebration of Britain. It was a populist display of social welfare reforms, and the first new towns were trumpeted with fanfare. This was the occasion for planners and architects to stage their urban fantasies and shape the modern imagery for ordinary British people. Festival exhibits were organized in each of the new towns just moving into construction, among which was Stevenage. Located thirty miles north of London, it was the first new town to actually be built in Britain.

Stevenage was considered one of the most noteworthy urban planning models of the immediate postwar period. Its design and pedestrian spaces were dubbed "Festival Style." Stein acted as consultant for the Stevenage New Town Corporation. He spent several months in England in 1950 developing a plan for Stevenage that copied the principle that he and Wright had implemented for Radburn: the superblock, with neighborhood units separated by pristine green spaces and arterial roadways. Six complete, socially balanced neighborhood units were set out, each accommodating a population of ten thousand. This was a much larger number of residents than had been planned in the American version, and it came under substantial criticism. British planners also replaced Radburn's single-family homes with terrace housing.[62] This creative adaptation also leaped to futuristic landing pads for four-seater and twenty-seater commuter helicopters.

The Stevenage Town Center was an entirely car-free pedestrian shopping area. Its buildings formed a coherent architectural ensemble in precast concrete clad in steel, glass, and stone, connected by canopies and open malls with fountains. Advertising and lighting were incorporated directly into the design. In visual images from the 1950s,[63] the Town Center bustles with shoppers. Stores overflow with the consumer goods associated with the coming age of abundance while the co-op supermarket offers novel self-service shopping. Alongside the market are the latest in recreation and entertainment, from dance halls and bowling alleys to art centers and public gardens.

In addition to showcasing new towns such as Stevenage, the Festival of Britain's "live architecture" exhibit was an opportunity to experiment with neighborhood configurations. It was located on the Isle of Dogs in London's decimated East End, where over a million houses had been destroyed and forty-three thousand people killed during the Nazi Blitz. The Stepney-Poplar area had been divided into eleven neighborhood units in preparation for reconstruction, of which the Lansbury Estate was one. The reconstruction of Lansbury was taken on by the festival's organizers as a living demonstration of the future.[64] This presentation was intended not as a model for London but as a glimpse into the scientifically planned "new towns of tomorrow" then taking shape. The exhibit's Town Planning Pavilion, a steel-framed canvas tent, housed presentations of the principles of town planning and the urgent need for new towns. It depicted the transformation in everyday living that would accompany this move to paradise. Overall, the festival exhibit exuded the showy, eye-catching quality of the ordinary. Visitors were confronted by a massive aerial photograph of the London metropolitan region showing roads leading out to the new town dreamscapes set out by the *Greater London Plan*. Next to it was a panoramic mural of an idyllic village setting in the countryside—the pastoral urbanism of the garden city movement transposed into the future.

Architect Jaqueline Tyrwhitt, an active member of MARS, was the live-architecture exhibit scriptwriter, and she also prepared the planning design for the Lansbury presentation. She was engrossed in defining community models and typologies, and was essentially testing out her ideas at the Lansbury experiment.[65] She made the Ricardo Street Nursery and Primary School the heart of the new community. Around it were set small-scale public housing units (designed by Frederick Gibberd, who was also working on the master plan for the new town of Harlow) in a leafy setting with gardens and green space. The site simulated the English pastoral village displayed in the vast photographic mural inside the Town Planning Pavilion tent. Both the mural's Arcadian scenery and the reconstructed Lansbury neighborhood were meant to appeal to a broad British public optimistic about the future yet yearning for a romanticized past.

Literary critic Northrop Frye once argued that utopia is a kind of contract myth. On the one hand, it is the desire for the restoration of what society has lost or forfeited; on the other, it is an anticipatory illumination. It stands on the horizon of a new reality. The Lansbury exhibit fell into these kinds of utopian temporalities. It was filled with what philosopher Ernst Bloch called "a surplus of meaning," an overshooting of ideology and a radical imaginary of new kinds of spaces and social worlds. It and the other exhibits at the Festival of Britain were part of an opulent promotional machinery and captured the populist nature of the new town movement as a fantasy of the ordinary. But that campaign did not last long. Just months after the festival, the Labour Party was swept from power. The Conservative Party that won the election in 1951 had little interest in lavish and very costly new town fantasies. Except for a few notable exceptions (Cumbernauld in Scotland was designated a new town in 1955), the program went into hiatus.

The Swedish Model

There were, however, other urban dreamscapes taking shape on the horizon. The Scandinavian new towns were acclaimed as a spellbinding vision of modern living. Over the long run, they also achieved more success than their British counterparts. Images of Vällingby and Tapiola were reproduced endlessly and became towering tributes to rational planning methods. Both new towns followed the same metropolitan regional logic as that used in London: center-city urban renewal and the dispersion of the population to a constellation of idyllic towns in which the apparatus of the social welfare state could be ordered and enjoyed. And as in London and elsewhere, that logic was applied by the midcentury pioneers who launched the campaign for modern housing and urban reform in the years before the war.

Sven Markelius was instrumental in bringing modern architecture and urban planning to Sweden. He was one of the founding members of CIAM. During the prewar years he was involved in designing cooperative housing as part of the progressive social democratic reform agenda. Markelius helped organize the functionalist housing display at the Stockholm Exhibition of 1930, and was one of the authors, along with architect Uno Ahrén, of the manifesto *Acceptera!*, published the following year—one of the most noteworthy appeals for functionalism and mass production as a set of cultural values. This text was later integrated into the social democratic notion of *folkhemmet*, or people's home.[66] The concept sought a middle way between capitalism and socialism, and underpinned the entire Swedish welfare-state project.

In 1942, Ahrén published *Arkitektur och Demokrati* and initiated an interdisciplinary discussion between architects, planners, and social reformers that included Gunnar and Alva Myrdal, both of whom spearheaded the *folkhemmet* ideal. Ahrén and Gunnar Myrdal understood the socially aware technician as a society builder and social engineer. Alva Myrdal was a leader in advocating communal housing as a way of emancipating working women from the burdens of housekeeping and child rearing. Her work with Markelius resulted in the fifty-seven-unit Collective House for communal living in the center of Stockholm, featuring child-care facilities and shared kitchen and social space. It was inspired in part by pioneering Russian social housing, such as the Narkomfin Building in Moscow. Communal living was part of a broad debate about the nature of neighborhood and community that took place as reformers sought solutions to the dilapidated slums and social crisis of the metropolis.

Step by step, Swedish urban planners moved from collective housing projects such as these groundbreaking examples, to neighborhood and municipal planning, and then ultimately to regional planning that incorporated all these layers. They were helped by the fact that the Stockholm municipality held vast open acreage outside the old historic core and also owned its public transportation system. Moreover, in 1943, Swedish legislation set up the financial machinery for city-directed suburban development. It was followed by the 1947 Building Law, which made the provision of adequate housing and master planning a municipal responsibility. Cities, rather than private developers, had exclusive authority over how urban expansion would take place. The Building Law represented a decisive shift in power to leading politicians and technical experts in public service. These measures signaled the arrival of the Swedish welfare state and the massive public investment that would accompany it. And lastly, in what would become one of the most admired formulas for new town planning, in 1944 Markelius was appointed director of city planning for the City of Stockholm, a post he held for the next ten years.

Longtime politician and city planning commissioner Yngve Larsson, who selected Markelius, was one of the leading figures in an integrated approach to transportation, housing, and urban planning that distinguished Swedish reform ideals. In May 1944, Larsson proposed development of a new master plan for Stockholm that was published in a popular version as *Stockholm in the Future: Principles of the Outline Plan for Stockholm.*[67] Optimism and belief in progress were its bedrock. It projected that Stockholm's population, already 1 million, would reach 2 million by the end of the twentieth century. The plan, created by a team led by Markelius and Göran Sidenbladh, slated almost the entire central area of old Stockholm for clearance to make way for a modern downtown replete with high-rise towers, public spaces, and sunken pedestrian zones. The number of city residents would shrink by some 40 percent. People would start over in the suburbs and find happiness in a series of satellite towns, each welcoming around ten thousand to fifteen thousand people. The towns would be strung like pearls along new regional train lines and separated by protected countryside: Farsta and Högdalen southeast of Stockholm; Skärholmen to the southwest; Hansta to the northwest; and Vällingby eight miles to the west.

Both Markelius and Sidenbladh were instrumental in making the vital link between transportation planning and regional settlement patterns. Markelius, for instance, produced the 1949 report *Town-Planning in Stockholm: Housing and Traffic.* Under Sidenbladh's direction, the City Planning Office carried out and published a *Traffic Forecast for Stockholm* in the early

1950s. The idea was that the new towns and train system would be co-produced. Rather than building dormitory suburbs, the plan would lure industry and offices to the satellite towns through tax incentives. The new settlements would offer local employment opportunities for half the residents as well as community services and cultural and recreational activities that would make each a Stockholm in miniature.

Stockholm in the Future laid out its vision of the welfare state: education and child care, health care, job training, libraries and community centers, sports facilities, playgrounds. Ordinary life would be revolutionized. Apartment buildings near the core of each new town were to be surrounded by rings of lower-density terraced and detached housing and parkland. Urban and scenic beauty was described as the result of conscious planning. The goal was to create a rational, well-ordered regional metropolis that signified progress and prosperity.

Yngve Larsson, Sven Markelius, and the team of planners who created Vällingby and the other new towns around Stockholm drew from an expanding corpus of new town methodologies. Their inspiration shifted from the traditional German idea of *Siedlungen* (settlements) that had been so influential in prewar Sweden to the Anglo-American concepts that dominated the international scene by the 1940s and 1950s. Lewis Mumford's *The Culture of Cities* was translated into Swedish in 1942 and became a manifesto for the planning community. The 1952 *General Plan for Stockholm* was peppered with quotations from Mumford's book, which was cited as primary to the plan's objectives.[68] The British inspiration, whether it was garden cities or new towns, was considered by many as the origin of Stockholm's satellite towns. Patrick Abercrombie and F. J. Forshaw's *County of London Plan* was widely read, and Larsson was well versed in Abercrombie's regional principles. There were substantial contacts with the New Towns Committee in England, and in 1946 Markelius and his team attended the International Congress for Housing and Town Planning in London. They also visited the British new towns of Stevenage and Cumbernauld.

Yet Markelius downplayed all this influence, arguing that the British new towns were only one source for ideas. He freely admitted the sway of American concepts exemplified in Clarence Stein's plans for Radburn, especially its separation of pedestrian from vehicular traffic.[69] Stein was well known in Stockholm, friendly with the city's administration and with Markelius. Göran Sidenbladh had met with Stein in the United States in 1938, and along with Larsson and Markelius, went to see Radburn for himself. In exchange, Stein traveled to Stockholm in 1949, and then again in 1952 during an extended stay to visit Vällingby.

Members of the Vällingby planning team were also heavily influenced by the well-known regional "finger plan" for Copenhagen, which was published in a number of volumes in 1947. For their part, Danish planners were similarly aware of British concepts, and the 1942 Swedish translation of Mumford's *The Culture of Cities* was avidly read in Denmark. This evolving midcentury knowledge base was built up from key shared texts— regional plans, new town designs and visual imagery, and theoretical works such as Mumford's. It made metropolitan planning into a cohesive movement that spanned the immediate postwar years, and by the late 1950s and 1960s had become the model of utopian aspiration. For those in this midcentury generation, it was an unprecedented opportunity to actually see their ideals come to fruition in modern towns planned from scratch.

Few places are more closely associated with these utopian fantasies than Vällingby (fig. 2.13). It immediately became the poster child, the "holy temple," as Peter Hall has called it,[70] of city planning: thousands made the pilgrimage to see this urban paradise. Some one hundred thousand people came to the town's official opening in November 1954, which included official speeches and an enormous fireworks display. Vällingby instantly became a worldwide sensation. The speed and efficiency of construction were as much admired as the town itself—some fifteen thousand new town pioneers already lived there when the crowds gathered for the ribbon cutting.

Vällingby was Sweden's first ABC suburb, the acronym for Arbete-Bostad-Centrum (Work-Housing-Center), and represented the fullest realization of Markelius's planning concepts. Its ultramodern town center (fig. 2.14), easy train connections, and white apartment houses set in a garden landscape were a visual semiotics of the Swedish welfare state. The town was built on state-owned public land where the government had exclusive control over what could be built. Vällingby was a dream planning project that required few compromises; the design coherence and the quality of the built environment were exceptional. Everything about the town was fascinating. Comparing it to the new town of Harlow outside London, the *Observer* reported that "the pavement patterns, the commercial signs, the lettering, the lamp-posts, the colours" of Vällingby "[have] been the subject of popular interest and experiment."[71]

What distinguished the town was its sense of equitableness and "right to the city" from the midcentury social reformist point of view. It had been imagined as a cityscape of equality and freedom. Its allure was social harmony and domestic bliss, ease of access, and urban effectiveness. The hope was that if the correct design was applied, over time the residents would develop a "we-consciousness" and become a social community. This feeling,

2.13. Cover of Vällingby, Sweden, marketing brochure commissioned by the Stockholm City Property Board, 1952. Courtesy of Svenska Bostäder and Stockholms Stadsmuseum.

according to the 1952 *General Plan for Stockholm*, is not natural to big cities, where less than half the households are complete families. Big cities were too dense, and there was too much mobility. On the other hand, suburban areas offered the opportunity for the community feeling and sociability necessary to form a neighborhood unit.[72]

2.14. Aerial view of central Vällingby, Sweden, ca. 1954–55.
Photograph by Oscar Bladh. © Stockholms Stadsmuseum.

The Stockholm plan included a detailed model for a neighborhood unit that was applied to Vällingby. It was

> represented by a number of houses grouped around a children's playground, a sanded playground for younger children and a place for baby carriages— all near the houses and surveyable from their windows. Of such groups of homes is formed a greater unit, of size sufficient to support a shopping center, nursery, kindergarten and large playground with apparatus. This greater unit is laid out in a way that keeps all pedestrian traffic well away from roads and especially from highways.[73]

This ideal encapsulated the normative vision of the nuclear family that underlay postwar modernization and the welfare state.

Vällingby was a portrait of what the practices of everyday urban life could be. There was scant evidence of architectural monumentality, no grand insignia or symbol that proclaimed the city. Nor did it wallow in bucolic reverie. Instead, it offered modern living in the countryside to ordinary people. The town center of Vällingby was planned around a spacious concrete pedestrian deck covering the main train station (fig. 2.15). It was a civic square, with adjacent buildings celebrating the aesthetics of concrete and experimentation in red brick. Alongside was Sweden's first shopping mall. Built of steel, aluminum, and glass with transparent overhanging roof and bold neon signage, it was the epitome of modern styling. Crowds strolled through the Kvickly and Tempo American-style department stores and some forty shops. Another thirty shops and restaurants, a community center, and a public library were in place the next year. According to a marketing brochure, "shoppers could gaze at and enjoy without fear of traffic." Altogether, the town center epitomized the "intimate, comfortable atmosphere"[74] of modern everyday life. A web of pedestrian and bicycle paths led from it to the residential neighborhoods.

Vällingby's residential areas were composed of three linear districts of about ten thousand to fifteen thousand inhabitants apiece: Blackeberg,

2.15. Vällingby Central Plaza, ca. 1954–59. Photograph by
Lennart af Petersens. © Stockholms Stadsmuseum.

Råcksta, and Loviselund, plotted along train lines. Markelius and his team imagined each neighborhood district as a functional unit with a core area and a clear boundary. Each district was clustered around its *tunnelbana* station and sported a community center, social services, and schools. The residents were to carry out their everyday routines in their new town and make the trip by train to central Stockholm only for cultural events and special occasions. Everyone would have a daily commute of less than thirty minutes to a central-district job.

A variety of housing options and functionalist styles were constructed by the publicly owned building corporation Svenska Bostäder. All had up-to-date kitchens and bathrooms and were surrounded by playgrounds and park areas. As the population predictions for Greater Stockholm grew, planners relied increasingly on prefabricated three-story slab "lamella" blocks and six- to ten-story "point houses"[75] for the densest possible development around each district's core area.

Vällingby was a place of ordinary, everyday time and space made magical by its utopian setting. A marketing campaign launched at its opening helped to ensure the triumph of the Swedish new town model. Exhibits, brochures, and official ceremonies tempted potential employers and residents to succumb to the ideal. American architectural historian George Everard Kidder Smith's *Sweden Builds* (1950), with its depiction of Vällingby as offering "more lessons . . . than any other development yet built,"[76] was avidly read in the United States. Architects, planners, and municipal engineers flocked to Vällingby from around the world to study the future. The throngs of visitors were so large that the town provided "Miss Vällingby" tour guides who were proficient in multiple languages.

Images and descriptions of Vällingby, its housing, its public plaza landscaped with fountains, became part of an international planning lexicon, reproduced endlessly as a model aesthetic combining social reform with physical design. This visual narration and figurative language obfuscated the fact that Vällingby had been built by a social democratic government fully committed to a wide range of social welfare programming in the *folkhemmet* ideal. Instead, the town became a reproducible model by the almost exclusive focus on the characteristics of its built environment.

The Urban Future: Tapiola

Like Vällingby, by the 1960s the new town of Tapiola in Finland was one of the most powerful visions of the urban future. It was the superstar of the Seven Towns Plan to transform the area around the capital of Helsinki into a modern metropolitan region. The urban renewal plan mirrored those for London and Stockholm, calling for a population decrease in Helsinki's central area and then its redevelopment into a modern commercial center. Fifty percent of its inhabitants would be dispersed to the adjacent Uusimaa coast in seven new towns plotted along rail and road corridors. By siphoning Helsinki's population into discretely planned new towns, some 90 percent of the region would be preserved for agriculture, open coastline, and forest. The most beautiful and historically valuable places would be saved.

Tapiola was reported about in the architectural press of the 1960s with unrestrained enthusiasm. American architectural critic Ada Louise Huxtable waxed eloquent on Tapiola as a "beautifully clear, handsome and unequivocal demonstration of the virtues of the planned community" and "the antithesis of current practices in the United States." Architectural critic Wolf Von Eckardt effusively pronounced it an "urban Shangri-la . . . a new town that comes closest to meeting the ideal."[77] These kinds of accolades were not accidental. Heikki von Hertzen was the visionary behind Tapiola, and he was also its grandest salesman. Von Hertzen launched a high-octane media campaign that had few rivals in the new town movement.[78] Marketing was controlled through selectively produced aerial and architectural photographs, films, postcards, and scale models. Press kits were delivered to the Finnish press and architectural journals throughout western Europe. Tapiola was discovered in the United States in 1963 through urban essayist Frederick Gutheim, who visited the town and then organized a Tapiola Show in New York City at the Architecture League and in Washington, DC, at the National Housing Foundation. Afterward, the presentation toured the United States for three years.

Interest in Tapiola peaked in 1965 with the visit to the town by the University of Pennsylvania's New Towns Seminar. The group comprised a coterie of American planning personalities, including officials from the federal government's New Communities program, the president of the Regional Planning Association of New York, and developer James Rouse, who was busy developing his own new town of Columbia, Maryland (see chapter 5). Once again, von Hertzen acted as impresario and heavily promoted the visit by the American delegation to the Finnish press.[79] The unprecedented coverage coincided with the arrival of the Tapiola Show exhibit that had toured the United States, transported to Finland as "Tapiola and the Province of Uusimaa, 1965," and shown with displays of the Seven Towns Plan. Dissatisfied with even this amount of hype, von Hertzen promoted his urban Shangri-la throughout the world, and welcomed enthralled devotees

2.16. The shah of Iran, Mohammed Reza Pahlavi (*left*), visiting Tapiola, Finland, along with Finnish president Urho Kekkonen (*center*), Asuntosäätiö representative Yrjö Riikonen (*right*), and a Tapiola guide, June 23, 1970. © Espoo City Museum.

that included the shah of Iran (fig. 2.16). The town's international stardom tipped the scales in favor of local acceptance of the Seven Towns Plan as a metropolitan regional framework for Helsinki.

The acclaimed Finnish new town program originated in architect Eliel Saarinen's prizewinning entries for a general plan for greater Helsinki: the Munkkiniemi-Haaga Plan of 1910–15 and the Pro Helsingfors Plan of 1918. Both had been written just after a bloody civil war and Finland's independence from Russia. The planning competitions for the capital of the newly independent nation were sponsored by the M. G. Stenius Company. Stenius was heavily involved in real estate speculation, and instigating a regional plan opened the door to Helsinki's expansion and to highly profitable land deals.

Saarinen's winning entries were inevitably steeped in Ebenezer Howard's garden city concept as well as Raymond Unwin and Barry Parker's versions of the English garden city. Architect Gustaf Strengell, who helped Saarinen with the planning texts, visited Hampstead Garden Suburb in 1910 and wrote an account of his trip in the Finnish magazine *Arkitekten*.

The Munkkiniemi-Haaga Plan also included several pictures from Unwin's book *Town Planning in Practice.*

The Pro Helsingfors Plan proposed a series of satellite communities around the old city of Helsinki that were separated by open space and connected by rail lines and a streetcar system. The size of each satellite town was estimated at around 10,000 to 12,500 people. For each, Saarinen sketched out a dense center with a multilevel main avenue, surrounded by row house blocks and villas. The Tapiola site, six miles from the center of Helsinki on the Gulf of Finland, appears as one of the satellite communities in his plan. The area was designated Hagalund (Woodland Meadow), the name it retained until Tapiola was substituted in the 1950s.

Despite these early initiatives, Finland was still a rural country dominated by agricultural production and open countryside. Only 20 percent of the population actually lived in towns. What dramatically changed the situation was the country's defeat in the 1939–40 Winter War with the Soviet Union and the loss of some 11 percent of its territory. These setbacks were devastating. Over four hundred thousand refugees from the lost border areas of the Karelian Isthmus and the Salla region flooded back into Finland. They joined the retreat of demoralized combatants at the war's end in March 1940. There was a halfhearted attempt to distribute the evacuees in the countryside, but it was Helsinki and its expanding urban fringe that they turned to. Suddenly, Helsinki was mired in a population crisis. Then the profound effect of fighting in the Second World War and the defeat of Germany (Finland's ally) brought another wave of refugees: some five hundred thousand people escaping the destruction and the surrendered Finnish territories streamed into Helsinki and its surrounding areas along with demobilized troops. The housing conditions became deplorable. It may have been rooted in the garden city legacy, but as elsewhere, the fervor and urgency of the new town movement in Finland was forged from the brutality of war.

Facing these national calamities, Finland once again turned to the Anglo-Saxon world for inspiration. Lewis Mumford's *The Culture of Cities* was published in Finnish in 1946, along with Patrick Abercrombie's plans for the London County Council and Greater London. The neighborhood unit concept, known in Finland as the residential neighborhood theory, was introduced in the 1947 publication of Otto-Iivari Meurman's *Asemakaavaoppi* (Town Planning Theory). Meurman was the first professor of city planning in the country, and his ideas formed the intellectual backbone of Tapiola. The overpopulated cities of the past, in Meurman's view, were deadening environments that turned people into packs of animals. He championed Saarinen's original plans for the Helsinki region as well the ideas of decentralization and living in nature as curatives for catastrophe.

However, the main influence in Finnish planning was architect Alvar Aalto, who offered a schematic planning prototype in his pamphlet *An American Town in Finland* (1940), his project to build an experimental town for Finns left homeless by bombing and evacuations. Aalto developed a strong friendship with Sven Markelius during his frequent trips to Sweden during the war years. Both were active in CIAM and leaders of modern functionalism in their respective countries.

Aalto drew up an early regional plan for the Kokemäenjoki (Kumo) River Valley (1942–43) and a master plan for the new town of Imatra. These developments fell neatly into the category of resource towns. Both were financed by private industrialists—in the case of Imatra by the massive Enso Gutzeit Corporation, which was developing hydroelectric power on the Vuoksi River for its lumber and brick mills. Imatra was founded along the river in 1948, and comprised the three settlements left to Finland after the loss of nearby Karelia. These were little more than dreadful refugee camps lacking even basic water and sewer infrastructure. Founding a new town transformed living conditions, stabilized the workforce, and also fortified the new border with the Soviet Union. Idealistic in tone, Aalto's plans called for a series of rural villages and individual communities established around a modern town center. These and his plan for Lappland written in the mid-1950s argued for a regional approach that distributed urban activities over extensive geographic areas in tightly concentrated villages and towns.[80]

Together, Saarinen, Aalto, Meurman, and Heikki von Hertzen led regional planning in Finland, with the goal of dispersing populations from large cities to a network of new towns. Von Hertzen's urban concepts were derived from his long-standing social activism on behalf of families and housing reform. He condemned the miserable conditions in Helsinki in his 1946 manifesto *Koti vaiko kasarmi lapsillemme?* (Homes or Barracks for Our Children?). Arguing that planning was a tool to create a better society, he drew his models from the American progressive movement: the garden cities of Radburn, New Jersey, and Greenbelt, Maryland, and the moving social vision of Mumford's 1939 film *The City*, which he had viewed in New York. His planning of Tapiola was thus a labor in social idealism.

The Family Welfare League, which von Hertzen directed, acquired the 660-acre Hagalund site, as it was known, and then in 1951 brought together five welfare and trade organizations to form Asuntosäätiö[81] for developing the project as a model community. This goal was inscribed in the town's founding declaration, buried with the foundation stone under the future shopping center with great ceremony: Tapiola was to be created as a "modern urban environment" and an "average" community that offered a vision of ideal housing and a standard of living for the entire nation.[82] The town was as much rhetorical gesture as real place.

Tapiola also shaped the emergence of environmentalism as a code of planning conduct. The entire community was to be self-sufficient and immersed in nature. Urban life bowed to a constructed landscape of extensive greenbelts, well-kept forests, and traditional farms. This production of nature was an integral part of the modernizing process. In a 1956 article in the Finnish journal *Arkkitehti-Arkitekten*, von Hertzen responded to the immediate comparison of Tapiola with Vällingby, which in his opinion was a distressing mass of buildings and people. Tapiola, in contrast, was characterized by the ideal of *waldstat*, or "living with nature."[83] The aesthetic values of the natural environment and the contours of the landscape determined Tapiola's urban pattern. Building sites followed the natural terrain rather than street lines. Swedish landscape architect Nils Orénto led the crusade to create gardens and green areas that inspired a kinship with soil and sky. This new sense of environmental design began to supersede the older concept of the garden city.

Yet despite von Hertzen's goal of "a beautiful town for everyman" where "businessmen, university professors, and everyone could live side-by-side," social hierarchies were embedded in this ideal. The private view of the grander houses was westward, to capture Finland's winter evening sun. They enjoyed broad vistas of uninterrupted natural scenery. The view of the more modest dwellings was toward the town.[84] Such was the reputation of Tapiola as arcadia for the well-off that it eventually inspired a polemical book by Ossi Hiisiö entitled *Tapiola—the Village of Better People* (1970). This stigma had followed the garden city from its very beginnings at Letchworth.

In the 1950s and 1960s, Asuntosäätiö retained Finland's best-known architects to design various sections of Tapiola, its residential housing and civic buildings. The new town became a permanent exhibition of avant-garde functionalism set in the woods along the gulf coast. Prefabrication and system-built construction techniques were first introduced in Finland at Tapiola. The greater portion of the housing stock comprised three- and four-story walkup apartment buildings. Their interiors featured open floor plans, abundant natural light, and state-of-the-art electric kitchens that had space for a refrigerator (one of the most sought-after postwar consumer

2.17. Aerial view of Tapiola, Finland, with Aarne Ervi's town center design, ca. 1969–71. Photograph by Atte Matilainen. © Asuntosäätiö / Espoo City Museum.

items). The 1954 competition to design the town center (fig. 2.17) was won by architect Aarne Ervi. His proposal earned the envy of every new town planner of the time. He was one of the few architects in Finland (or perhaps anywhere) who managed to successfully design CIAM-inspired suburban towns.[85]

Tapiola's town center was entirely pedestrianized and linked to the Tapionraitti, a 1.5-kilometer-long pedestrian and bicycle route. Roads were moved out to a peripheral ring. Perhaps the most daring element of the design was Ervi's conversion of an ugly gravel pit into a serene reflecting pool that graced the town center. It was a brilliant decision that, along with an adjacent high-rise office tower, transmuted into Tapiola's signature landmark. The tallest buildings were placed on the highest part of the terrain to accent this imagery. The elegant design was reiterated endlessly in photographs worldwide. It became the touchstone of the new town movement.

Finland's first shopping center (on which Victor Gruen consulted; see chapter 6) was adjacent. Essentially a protected shopping street, it featured a department store and shops, meeting halls, restaurants, and a discotheque. The youth center with swimming pool, bowling alley, and gymnasium provided all the leisure amenities of modern life.

Despite its reputation for "living in nature," Tapiola pioneered a modern lifestyle devoted to recreation and consumerism. In a 1960s promotional film produced by Asuntosäätiö and written and narrated by Heikki von Hertzen, nature is presented for its recreational values. Families ski out onto the winter terrain from their doorstep: they hit the beaches and water-ski on the Gulf of Finland in summer.[86] The film is a teaching tool of modern lifestyles and the leisure-based advocacy of von Hertzen's environmentalism and nature preservation. Tapiola was a fantasy machine, an idyllic setting that merged seamlessly with the state's progressive social welfare policies.

The People and Their Needs

The garden city archetype and the neighborhood unit reigned supreme as utopian principles for the new towns of the reconstruction era, whether they were hard-core industrial towns or the stylish suburban towns arranged around capital cities. They were the bulwark against social chaos. Neighborhood and community offered an urban fantasy on which to encode social solidarity and stability. The visual picture evoked a quality of *place* for the "natural" goodness of man to reemerge. The bucolic garden city landscape was an immutable, trusted form. After a devastating war, young families would be reconstituted in a verdant setting supported by state services, ready to welcome a new generation. Yet in large part, community and neighborhood were also instruments of national mobilization and modernist grandeur. They were closely tied to the drive for industrial productivity and to the rationalization of space and territory. The idyllic atmosphere made this broader agenda not only palatable but even applauded in the context of reconstruction.

The urban typology, size classifications, and population numbers built into this dreamscape were worked out with extraordinary exactitude. Planners were absorbed in calculations that mapped science to the social realm and created carefully managed urban communities. Elaborate charts and tables gauged the needs for each level of population and each category in the urban hierarchy. The traditional neighborhood fabric along city blocks, the "street life" that typified the unplanned spontaneity of working-class communities, was nullified. Individuals no longer encountered the city personally, but did so through the mediation of a collective community space. It was an "end-state" approach to design. Theoretically, expansion would take place by attaching new neighborhood units to the new town. But there was little consideration for the inevitability of change. Instead, planners focused on the neighborhood as a scientifically derived environment that was invariable and fixed. The perspective underlined the belief in the perfectibility of the social body.

Such was the confidence in this vision during the reconstruction era that there were few voices of opposition. The best known was that of philosopher Karl Popper, who argued against technocratic social science and utopian social engineering as unachievable and ending in unintended consequences. Their purpose, he maintained, was to extend brute power and control by the state. Attempting to wipe the slate clean and redraw society from scratch based on visionary blueprints for a "great society" would end in disaster—one Popper associated with the totalitarian regimes of the twentieth century.[87]

Indeed, the nobility of the garden city and its neighborhood units began to dim as shortcomings became more glaring.[88] Sociologists challenged neighborhood planning as consciously or unconsciously reinforcing class and social divisions, though it was based on the presumption of social homogeneity. The mental picture of a contained neighborhood with garden-style cottages and apartments clustered around a school was also disparaged as unscientific and unworkable. There were few early studies of planned neighborhoods, and little concrete evidence that they actually facilitated any notion of harmonious well-being. Perhaps the most devastating critique came from American architect and urban theorist Christopher Alexander, whose classic article "A City Is Not a Tree" (1965)[89] laid to waste the idea that the neighborhood unit represented a viable urban form. It was a neatly ordered category in the minds of planners that in no way corresponded to social realities. Work, play, and social life took place in a thousand places across everyday life. The dreamland of neighborhood actually crippled conceptions of the city.

On a more practical level, garden cities with individualized neighborhood units were exorbitantly expensive to build and maintain. Their green ambiance assumed a control and manipulation of the natural environment that could rarely be sustained, either by local residents or by the state. In the pages of the American *Harper's Magazine*, Wolf Von Eckardt verbally shook his head at the result: "With visions, no doubt, of deep woods, clear brooks, birds and bees, they [planners] paint green blotches on their plans. But, more often than not, they don't have the slightest idea how to make or keep them green on the actual landscape or even what people might do with them. Often they end up as nothing but a big weed patch."[90] There was little appreciation of landscape as a dynamic entity with a life history of its own, or that a deep ecological approach was a more efficient design methodology. The broad expanses of greenbelts and park areas easily led to scattered pockets of residential isolation without a cohesive town center. Any sense of urban legibility dissolved or faded into delusion.

Although the concepts drawn from the reconstruction years remained an indelible part of planning vocabulary and were indeed heavily promoted by international institutions such as the UN, they mutated and opened up. Moreover, by the 1960s the circumscribed world of neighborhood could feel downright stifling for young people. The automobile offered an escape and the opportunity to roam free. The city beckoned, as did the vision of new worlds. It was this cultural shift that more than any other factor blew apart the neighborhood unit as utopian ideal.

THREE

Exporting Utopia

Of all the influences on town planning practices in developing countries, the belief that the solutions to urban problems lay in starting anew was the most radical and perhaps the most controversial in the long run. In Europe, new towns were a form of redemption. They were a glimmer of hope for happiness and progress. In countries moving toward independence, the ideal of the tabula rasa, the virgin territory on which dreams unfold, was even more tantalizing. Starting from scratch was a way to break through the ruins of the colonial past, break away from the myriad problems of old cities, and chart a fresh course forward in the landscape. It fed into the hopes of newly independent nations for a better future. It led to the expensive construction of new capital cities such as Islamabad in West Pakistan, Dhaka in East Pakistan, and Chandigarh in India's Punjab state. It guided a kaleidoscope of experiments in neighborhood units, garden cities, and new towns—some of which were successful and others wasteful failures. Though they were meant to steer new nations into the future, these urban ambitions were ensnared in the present. They were instruments in ethnic conflict, national rivalries, the race for resources, and the jockeying for geopolitical position among the superpowers in the Cold War.

New towns embodied dreams of a better life, but they were also embedded in state control of economic assets and territory. They were massive public works projects and catalysts for modernization. Applied to colonial or former colonial areas, the new town strategy had even more cogency in this regard. Its value was substantiated by Walter Rostow's acclaimed "take-off" theory on modern economic growth, which was outlined in his *The Stages of Economic Growth: A Non-Communist Manifesto* (1960). It became Cold War philosophy for US political elites. Rostow argued that underdeveloped nations threatened by the temptations of communism could be redirected through progress and American-style modernization. As a strategy for development, modernization made the claim of benevolence and universal applicability. It would create a better, more humane world.

Rostow's theory consistently emphasized the social and cultural factors that imbued nations with the capacity for modern change. According to policy makers who avidly followed his ideas, progress was hampered in the so-called Third World by indigenous obstacles such as traditionalism, endemic poverty, a lack of skills and education, and the absence of infrastructure. These comprised the backwardness to be rooted out as a prerequisite for a nation's independence. Anthropologist Arturo Escobar makes the point that poverty was modernized after the Second World War. Poverty had long been accepted as the status quo in colonial possessions. But once they became self-determining nations, reformers regarded poverty and its conditions as social problems requiring immediate intervention.[1] Moreover, modernization enthusiasts imagined that Western-style development would solve the "population problem" in the teeming slums of the Third World's cities. Developmentalism became an article of faith in US foreign policy and the keynote for a resurgent imperialism. It emerged as a global managerial mandate increasingly couched in social theory, and it was grounded in the belief that progress should follow the path trod by Europe and North America.

New towns were widely adopted throughout the Middle East and Asia as part of this broad agenda, which was produced and managed by Western experts in alliance with local political and planning elites. They were powerful Possible Worlds derived from this specific postwar historical juncture. Although their sizes varied from city to village, new towns were constructed as spaces where development and modernization would be performed in a new kind of colonial urbanism. As infrastructure projects, they offered a host of advantages. They were a strategy for capital investment and job creation, managing valuable assets, and rectifying disparities in labor demand and supply. An integrated settlement system based on new towns conquered and consolidated national space and incorporated it into the modern world economy. Such actions could embrace overt militarization and forced population movements. Nevertheless, new towns were seen as a solution to some of the most implacable problems facing newly emerging nations. This was particularly the case in those faced with political unrest and ethnic violence, and with floods of desperate refugees and rural migrants making their way to already overcrowded cities. Town building was a mechanism for providing these people and the deprived poor living in slums and squatter settlements with the amenities of modern life.

The examples in this chapter from Asia, Africa, and the Middle East highlight new towns as this virtuosic enactment of developmental modernism. They also reflect the extraordinary *number and extent* of new towns as a strategy for modernization, and their accommodation to nationalist predilections. New towns became the modern spatial utopia associated with the formation of an independent national consciousness.

The postcolonial elites who assumed power in the early postwar years had little identification with the memories and histories of the traditional urban fabric. They embraced the ideology of developmentalism as the path toward a thoroughly modern, secular nation. Building entirely new settlements seemed an inspiring solution and a blueprint for the future. The first generation of planners, architects, and engineers in emerging independent nations were heavily influenced by the West, and internalized Western colonial practices as well as Western ideals of modernism in architecture and planning. The sway of the West came not only from the colonial past but from a host of international development agencies led by the Ford Foundation, the International Bank for Reconstruction and Development (commonly known as the World Bank), and overseas aid such as the US Point Four program. They sent out legions of foreign experts who were instrumental in articulating and transmitting developmental discourse. American and European sociologists, anthropologists, architects, and planners rushed to study and evaluate non-Western societies.

Yet people made themselves modern rather than being made modern. Newly minted architects and planners grappling with town planning for the first time fused Western utopian ideals with their own perceptions about vernacular urban design and built form. New towns offered a modern vision without overtly testing the traditional divisions of society inherited from either the precolonial or the colonial past. They were founded on traditional social segregation and structures of power. Formally sanctioned social hierarchies were embedded in new town housing, in neighborhood units, and in urban design. But these realities were smoothed over by enthusiasm for the future.

The wide range of new town programs featured in this chapter demonstrates how newly independent nations appropriated developmental modernism and made the discourse a complex hybrid of imported ideas and local realities.[2] The examples also dramatize the thicket of difficulties and duplicities in applying garden city and neighborhood unit models outside the West. Lastly and perhaps even more significantly for understanding the give-and-take of planning knowledge and discourse, Western architects and planners were themselves influenced by their projects in emerging nations, and transferred these experiences to new town ideology and the metropole.

The new towns discussed first were those meant to relocate refugees and displaced persons at the end of the Second World War, the numbers of which are incalculable. Millions upon millions of people had to be repatriated to their homelands or resettled somewhere else. The United Nations Relief and Rehabilitation Administration was set up in 1943 to provide immediate humanitarian aid, and was then replaced in 1947 by the International Refugee Organization and then in 1950 by the United Nations High Commissioner for Refugees. But the refugee crisis continued long after the guns had been silenced, evolving into the longer-term challenge of providing housing and infrastructure for people to reconstruct their lives. In response, the UN set up a Department of Social Affairs, headed by renowned Swedish social reformer Alva Myrdal (see chapter 2). Its Economic Commission, with branches in Europe, Latin America, Asia, and the Far East, went to work on town and country planning and the housing crisis. A host of distinguished architects and planners took up positions in its various units. They were impassioned by the hopes of reconstruction and ready to work.

The commission's Town and Country Planning unit was headed first by Dutch sociologist Ann Van der Goot and then by Yugoslavian architect Ernest Weissmann, both members of CIAM (Congrès internationaux d'architecture moderne). Weissmann in particular was one of the guiding lights of New York CIAM's Chapter for Relief and Postwar Planning. Under his mandate, the Town and Country Planning unit worked closely with CIAM members as well as with the International Federation of Housing and Town Planning and the International Union of Architects. It sent out hundreds of development missions featuring the most renowned names in social and housing reform.

Solving the heartbreaking misfortunes of refugees was imagined as part of the longer-term undertaking of Western-led modernization. A host of institutions provided the conduit for Western reform elites to investigate conditions and experiment with their theories in the Third World. UNESCO (United Nations Educational, Scientific, and Cultural Organization) was founded in 1946 and led by Charles S. Ascher, who was instrumental in the creation of the American new towns of Radburn, New Jersey, and Sunnyside in Queens, New York City. The Carnegie, Rockefeller, and Russell Sage Foundations all sponsored field research and social-scientific studies in distant places.

Above all, the Ford Foundation, the world's largest private philanthropy, became the most powerful private institution carrying out developmental modernism throughout the developing world. From its headquarters in New York and a myriad of international field offices, it carried out its mission of advancing human welfare with a zeal backed by millions of dollars in investments. Ford worked hand in hand with the US government and the CIA in implementing US Cold War modernization policies. Paul Hoffmann, who had been the administrator of the Marshall Plan, headed the foundation in the early 1950s. Don Price, who was in charge of the foundation's overseas investments, explained the foundation's work: "Something like a quarter of the world's population has become independent since World War Two, and anything we can do to help it get settled on a basis of free government and economic progress will be a good thing. The belt of new countries we are working with would have tremendous problems even if there was no such thing as Communism."[3]

The Ford Foundation's Urban and Metropolitan Development wing established an international affairs program that was instrumental in supporting urban research and the construction of new towns as the theater of development. In the 1950s, the foundation pumped $32 million into the Middle East for development projects and maintained full-time offices in Cairo, Beirut, Tehran, Ankara, and Tunis to conduct its far-flung operations. These projects plus the foundation's programs in India and Southeast Asia in the 1950s totaled $90 million.[4] Applied social science techniques were the framework for this philanthropic mission to modernize the world. Money was poured into social science research and what was called "social overhead capital," a term that included not just basic infrastructure but also housing, training, and education programs and support for the emergence of civic life.

The Refugee Crisis

By far, the most immediate postwar concerns lay in regions with newly established independent states where communist insurgencies, sectarian violence, and refugee populations living in misery threatened to tear apart any hope of national consolidation. In the Middle East and the Indian subcontinent as well as in Southeast Asia, the various postwar settlements, mandates, and independence movements created an arc of unstable nations open to the threat of communism.

The degree to which new town development was implicated in this Cold War landscape and the politics of decolonization is vividly revealed in Malaysia. Japan's withdrawal at the war's end left what was then called Ma-

laya torn apart and in chaos. Thousands had fled the countryside to escape Japanese atrocities and starvation. They found refuge in the jungle,
where they joined communist military hideouts and congregated in vast
squatter camps anywhere food was available. Rampant unemployment and
crippling inflation fueled violence and unrest, which were brutally sup-
pressed by the occupying British military forces. Sir Gerald Templer, the
British high commissioner of Malaya, led a massive military campaign to
defeat the guerrilla rebels of the Malayan Communist Party, whose numbers swelled as the catastrophe wore on.

The British developed an entire repertoire of counterinsurgency strategies, one of which was the forced relocation of some six hundred thousand Malayans, mainly ethnic Chinese, from the squatter camps on the
fringes of the jungle to resettlement compounds. These compounds were
decorously called New Villages, one of which was Petaling Jaya; it had been
constructed on a rubber plantation in the Klang Valley about seven miles
outside the capital of Kuala Lumpur, on the new federal highway leading
to Port Swettenham on the coast. Although surrounded by barbed wire and
under constant police supervision, the refugees were provided with modest homes, schools, and health services, and were given ownership of the
land. It was a soft campaign to win the "hearts and minds of the people,"
as Templer famously said, and winnow the population most sympathetic
to the Communist Party cause away from the insurgency.

The Cold War political goal of containment to prevent the spread of
communism—articulated in the United States by the Truman administration in the late 1940s—found parallel expression in the containment of
populations in discrete enclosed communities. Containment was a founding principle of postwar new town dogma, and evidenced the discourse
and logic shared by urban planning and military policy. Consequently, it
was only a short step for Templer to become a new town planner. From
1952, Petaling Jaya was expanded into a satellite town to include the thousands of squatter families living on the fringes of the capital; the land office was besieged with applications. Built on the model of the British new
towns, specifically Stevenage, Petaling Jaya contained all the requisite features of orderliness and improved appearance of the neighborhood unit
ideal. Neat rows of one- and two-story bungalows equipped with running
water and electricity were centered on a school and a *padang*, or open field.
Following the convention of self-help, the timber houses with galvanized
zinc roofs were built by the squatters themselves.[5] For British authorities,
the neighborhoods had coalesced into a town of happy residents far from
the tentacles of communist terrorists.

Templer's policies were lauded as a sterling example of modernization strategy as a force for political inoculation and stabilization. Petaling Jaya became a blueprint for the future, and a key location for Malaysian industry and land development. With the help of government financing and tax holidays, thirty-three factories were providing employment for the community by the mid-1950s. The town center had shops and a movie theater. By the time Malaya achieved independence in 1957, government offices, a hospital, and a campus of the University of Malaya were being built. Kuala Lumpur's middle class arrived to buy up brick cottages with picture windows and flagstone terraces along with "the family cars and television aerials, which give the impression to the visitor of a well-to-do western-style suburb."[6]

The first modern shopping complex, restaurants, and banks opened their doors in a New Town center north of the original area, which was then designated as the Old Town. These labels evinced the accelerated temporality of the modernization ethos. By the mid-1960s, Petaling Jaya was a veritable boomtown with a population of well over thirty-five thousand, mainly ethnic Chinese. Yet the new town reiterated customary social divisions: Malaysian Muslims lived in their own residential area. Over 250 factories, some owned by Western superstars Pepsi-Cola, the Singer Company, and Colgate-Palmolive, provided ten thousand jobs.[7] Petaling Jaya was the image of modern Western living and a dramatic symbol of Western developmental modernism as a road to independence. And the entire endeavor was carried out under the auspices of British military occupation and the shifting geopolitics of the Cold War.

And Petaling Jaya was hardly alone. Although their governments were only just coming into existence, newly independent nations nonetheless launched extravagant master planning schemes almost immediately. New towns proliferated everywhere as the signature of sovereignty. They were a deliberate and fully self-conscious utopian program, and an ideological practice that received enthusiastic support and aid from the Western powers. The Middle East in particular was flooded with architects and planners hired by the regimes installed there by the United States and Britain at the war's end. A throng of international aid programs set up offices and handed out money for projects. A new generation of home-grown young technocrats aided by foreign experts began to carve out modern national territories. Each new settlement, no matter how small or insignificant, was acclaimed as a symbol of reform, especially in poor, neglected areas that would now be touched by state beneficence. The settlements would root out and extirpate the past.

Arguably, all this passionate reformism precipitated the centralized planning that brutally implemented a Western-mandated political geography, regardless of the needs of local populations. However, the discourse followed a more complex course that emphasized civil society and self-help as a strategy for development. It was a compact between newly formed independent governments and people, who would become proactive, participatory citizens in a modern nation-state.[8] New towns created a spatiality of settlement myths and civic formation for these new nations. International organizations such as the UN and the Ford Foundation followed this line of reasoning, as did social reformers and town planners, who were eagerly sought after for their expertise. They were heroic agents of change, armed with social scientific profiles and officially sponsored research on the fast track to economic growth and modern life. Through education and self-help, the poor would participate fully in democratic governance and modernization. In a reiteration of Western planning philosophy, new towns were the physical space for this educational process.

Settlement Planning in Israel

While new towns were a response to refugee crises and overcrowded conditions in old cities, they were also state instruments for social cleansing and ethnic engineering, especially in the new nations of the Middle East. Israel portrayed itself as a melting pot of refugees from war-torn Europe, North Africa, and the Middle East. But in the case of its Arab settlements, new town status meant a systematic elimination of their Palestinian features and a makeover into identifiably modern Jewish urban places. After the failed UN partition plan for Palestine and the establishment of the Jewish state in May 1948, the unrestricted transfer and relocation of people caused immediate calamity. Jewish immigrants languished in refugee camps and hastily set-up transit towns, while Arabs were driven into exile in the Palestinian refugee camps. It was a devastating exchange of population from which the Middle East has yet to recover.

Among Israeli political elites, there was general apprehension that the masses of arrivals to the new Jewish state would congregate along the overcrowded coastal strip and in the three cities of Jerusalem, Tel Aviv, and Haifa, leaving the remainder of the country empty of Jewish settlement. To counteract this trend, the government instituted a Physical Plan for Israel for redistributing a projected population of 2,650,000 (the population was 870,000 in 1948) across the territory in a rationalized geography of planned communities. It was an operation in territorial sovereignty. According to Arieh Sharon, the plan's author and head of the Planning Department, comprehensive planning for settlements, industries, and services was considered both "imperative from the national and defence standpoints"[9] and a way to assert Jewish presence across the land. Israel was divided into twenty-four planning regions. With the full backing of Western governments, construction began on some thirty *ayarot pituach*, or development towns, in a flush of excitement over Zionist utopian possibilities. Some of the projects were located at what had been large transit camps, or *ma'abaras*,[10] for new immigrants; others were developed around the extraction of natural resources. A string of new settlements acted as military outposts to solidify control in border areas. They acted in the tradition of the *bastide*: little ideal cities that secured the frontiers created by the postwar political settlements. New towns would create a militarized, defensible territory.

Israel's new town strategy was lifted almost directly from the planning manuals in Great Britain and then transported into the geopolitical turmoil of the Middle East. The appropriation of British regionalist and new town ideology began with Patrick Geddes and his work on the master plan for Tel Aviv in the 1920s. In the early 1940s, Israeli statesman David Ben-Gurion organized a group of experts to work out a plan for the postwar migration of hundreds of thousands of people into Israel that provided an interchange with British planners. The Association of Engineers and Architects in Palestine also served as a meeting place for Jewish and British professionals to debate postwar settlement. Throughout these discussions, the 1940 *Barlow Report* and Patrick Abercrombie's 1944 *Greater London Plan* were repeatedly referenced.[11] Abercrombie also made frequent trips to Israel and was invited by Sharon to act as adviser on land-use planning and the new town scheme. Sharon himself was a graduate of the Bauhaus, and helped introduce the modern functional architecture and simplicity of design to Israel that was inculcated in the construction of Tel Aviv.

In 1950, the Physical Plan for Israel was made public in a series of press conferences and newspaper articles, and with a town planning exhibition at the Tel Aviv Museum. As part of the well-planned media campaign, the museum exhibit featured exquisitely rendered watercolors of the imagined promised land of settlements. They are atmospheric and emotional portraits of the future. Simple whitewashed bungalows and apartment buildings are interspersed with gardens and open spaces. That green garden cities and Bauhaus functionalism could be created and maintained in the middle of an arid desert was symptomatic of the abstracted vision and fantasies nurtured by state propaganda. As was the case with so many new town plans, the images aestheticized and romanticized urban design as modern utopia. There was a mandatory optimism and enthusiasm for all things new. The museum depictions of new towns were a way of blotting out the territory's Arabness and spotlighting the future of Israel.

The existing network of Arab settlements in Palestine, places such as Acre, Tiberias, and Zefat, was occupied by the Israeli army as the residents fled or were forced out during the 1948 Arab-Israeli War. These indigenous places were then labeled anodyne new towns and prepared for development. The United States signaled that the vast desert region of the Negev in the south was ripe for Jewish settlement under a plan by American engineer James B. Hayes, who had worked at the Tennessee Valley Authority. The Arab population of Beersheba in the Negev fled as the Israeli army moved in. The town was then designated as the Jewish capital for the area, and the buildings and water supply reported in good working order. Although the UN pleaded for Beersheba and the Negev to be a neutral zone between Israel and Egypt, several hundred Israeli pioneers quickly set up households. By 1952, ten thousand young Jewish Israelis were rebuilding Beersheba as a new town.

The new towns that once had been no more than camps in the Negev were now dreams coming true. In the United Palestine Appeal propaganda film *Song of the Negev* (1950), all previous civilizations in the region have vanished; there is nothing left of the past but ruins. The land has been exculpated. The desert is empty, ready to be developed. As in all new town visual imagery, young people are building their future. Regardless of the hardship, there is the fantasy of community and solidarity. In the Negev, they quickly clear away the debris of the past with their shovels and construct their settlement with hammers and nails. They stand fast, refusing to leave in the face of attacks by Arab forces.[12] A *New York Times* reporter visiting Beersheba in 1952 captured the pioneering atmosphere wrapped in bitter sectarian politics on a Saturday night:

> You will see the young people in the central café dancing the polka and the hora with plenty of Wild West abandon. They are a lean, sun-tanned, cheery-looking crowd, certainly cheerier than any crowd I have seen in Tel Aviv. By day you will see them stripped to the waist laying pipes, drilling wells and digging mines, building dams, roads and whole new towns. You will see mighty few of them toting briefcases and a great many carrying rifles. They don't go out of town without arms. There are still too many Arabs, including 50,000 Bedouins, who are not entirely reconciled to Israel's reconquest of the Negev.[13]

Mastering the land and building new towns would create an imagined Zion. The settlements were thought of as the urban equivalent of the kibbutz. They were a landscape of assimilation for new immigrants to become full members of the emerging nation. Moreover, the new town program was a way of managing and organizing the flood of immigrants and fixing their points of asylum and residence. By 1954, 56 percent had been diverted to newly developed areas of the country and to new towns. As forecasts for population growth reached 4 million, Israel updated its national plan for population distribution six times during the 1950s.[14] Planners were obsessed with creating an ideal distribution of population across the new territory. The settlement pattern was formalized with statistical abstraction into a strict hierarchy by population size: Order A for villages and kibbutzim of 500 inhabitants, Order B for rural centers of 2,000 inhabitants, Order C for rural urban centers of 6,000 to 12,000 inhabitants, Order D for medium-sized towns of 40,000 to 60,000, and finally Order E for large cities exceeding 100,000 people.[15] Ranked ordering of this kind was promoted by town planner Eliezer Brutzkus, who had worked in the 1940s with the Jewish Agency and eventually became director of Israel's Town and Country Planning Department. He argued that urban colonization was a strategy characterizing other frontier societies such as Australia, Canada, and the United States.[16]

Israel's comprehensive planning and its new town program enjoyed wide acclaim in the West in these early years. They were considered a powerful case study of developmental modernism, worthy of imitation worldwide. The settlements were fictionalized as communal life-worlds. In the 1961 propaganda film *Ashdod* about the new port town of that name (which architect and planner Albert Mayer had helped to design) south of Tel Aviv, the camera pans a forlorn settlement of government-built housing and an electric generator plant lost in a sea of sand, but the narrator states that Ashdod is "a city which has a future, and that is what really counts."[17] The film shifts temporality into a state of perpetual becoming. It treats the immigrant residents, mostly from North Africa, as pioneers, conquering the wilderness and creating a new world day by day.

But it was difficult to keep up the utopian pretense. The overly spread-out settlements created a jumble of disembodied units divided by dead zones. Investments were sparse, and they quickly dried up; overcrowding and unemployment were rife. Early settlers left what were reputed as "bad towns" in droves, while other towns were considered "good."[18] The rivalry between places grew heated as each vied for scarce resources. Yet regardless of the problems, these frontier worlds dutifully followed Western planning doctrine and were pictured as the bucolic paradise of Israel.

New Towns and Geopolitics in Iraq and Pakistan

Comprehensive land-use and settlement plans proliferated across the Middle East as the geopolitical map began to take shape and the Cold War superpowers jockeyed for power. The oil beneath Arab states became the focus of foreign policy and ideological positioning between the Soviet Union and the United States and its allies. Stability and security in the Middle East were an absolute priority as the desert refineries became the lifeblood of Western economies. The populations considered most threatening to this stability (and those most susceptible to the siren call of communism) were those caught in the teeming slums of the big cities.

For instance, the flux of people and ensuing Sunni-Shia sectarian tensions in places such as Baghdad, Iraq, were unprecedented. From the 1920s through the 1940s, the city's population tripled, reaching eight hundred thousand in 1947. By 1957, the population had doubled again, to 1.3 million.[19] Centuries-old city centers such as Baghdad's were surrounded by a sea of derelict neighborhoods and squatter settlements. The ominous political turmoil and the immediacy of the humanitarian emergency fed into the long-held conviction by Western urban reformers that old cities offered nothing but desperation and squalor. The solution lay in population dispersion.

The Hashemite monarchy reinstated by British occupation in the newly formed country of Iraq embarked on a Five-Year Plan that included dispersing people across the country in a web of New Settlements. The rationale was to achieve some measure of stability and control by moving people out of the nefarious slums of Baghdad, Mosul, and Basra. Many of the New Settlements were self-help rural communities involved in a vast land reclamation project to bring thousands of acres under cultivation for the new nation. The first application of the New Settlements program took place in the Greater Musayyib area along the Euphrates River, where some five thousand families occupied newly drained and irrigated agricultural land. The regional plan included "fundamental villages," "large villages," and "market centers" in an urban hierarchy that imitated Walter Christaller's central place theory, all connected by roads and canals. Rural extension and community development programs such as these were financed directly by the Ford Foundation, the UN, and USAID (United States Agency for International Development).

The *myth* of settlement was arresting. It would colonize the land, solve social problems, and launch the nation into modern time and spatiality. The development programs "will give a new form and a new life to the country," according to a Ministry of Development brochure. "As he witnesses such a project, the citizen of Iraq is filled naturally with satisfaction and optimism. He is able to imagine the magnificent picture which this country will present to the observer in the near future thanks to the efforts of the Government."[20]

Among the worst postwar conditions were those resulting from the August 1947 Partition of India. Millions of Muslims streamed across the border from India into the new Dominion of Pakistan under harrowing circumstances, while some 10 million Hindus and Sikhs fled the chaos and massacres by crossing into India. Families were torn apart; households were shattered. Cities in both newly sovereign nations were devastated by the displacement, the riots and ruthless ethnic violence, the ghettoization. Millions were made homeless. Muslim refugees arriving in Pakistan moved into the grossly overcrowded cities of Karachi, Lahore, and Hyderabad, where they eked out an existence in vast slums with little hope of obtaining a livelihood or decent housing. Caught between calamity and high hopes for the future, the Western-installed Pakistani government immediately embarked on a plan for the country's modernization. It prepared its first Five-Year Plans with the help of the Ford Foundation and consultants from Harvard University's Development Advisory Service.[21]

Pakistan's upheaval would be calmed by anchoring its nation building at the village level. As part of the country's Five-Year Plan, the Village Agricultural and Industrial Development Program (V-AID) embarked on initiatives such as the Thal land reclamation project in the Pakistani Punjab. The Ford Foundation and the United Nations Economic Commission poured money and technical assistance into the scheme, which was an attempt to put 2 million acres of desert into cultivation and settle refugees in twelve new towns and a thousand new villages. Projects such as this shaped the new national territory and solidified volatile border regions. They produced a landscape of modernization and utopian ambition.

Spare but decent homes were planned as part of new communities that included schools, health clinics, and local services. Unemployed refugees were put to work planting trees and constructing village roads and clubhouses, dispensaries and school buildings. The villages were grouped into regional development areas under the guidance of Western-trained extension workers, who taught everything from laying drains to making household improvements. Although these communities were to preach the rhetoric of self-help, each village's layout and housing as well as its social life were designed from above by planning experts who saw them as educational tools.

Despite government officials' efforts to root national identity in a village system, hundreds of thousands of people still flooded into Pakistan's towns and cities in search of jobs and a new life, especially in West Pakistan.[22] The main dilemma was the newly designated capital of Karachi, which was overrun by refugees and partially destroyed by vicious fighting between Hindus and Muslims. By the late 1940s, its population had quadrupled to nearly 1.5 million, and by the late 1950s it had exceeded 2 million. An early plan for a linear extension of the city was prepared by Swedish architects S. Lindström and B. Ostnas. But given the political instability and the porous territorial boundaries of the new nation, nothing could be accomplished. Only with the arrival of the military regime of Ayub Khan in 1958 was any progress made in resettling tens of thousands of homeless and slum dwellers.

Ayub Khan's Plan for the Greater Karachi Region split the overrun metropolis into five areas, one of which was Korangi. In the desert southeast of the capital, Korangi was constructed as a showcase community for the resettlement of five hundred thousand desperate people (fig. 3.1). It was

3.1. Muslim refugees from India unloading their belongings in Korangi, Pakistan, 1959. Photograph by Keystone-France / Gamma-Keystone. © Getty Images.

KORANGI MASTER PLAN

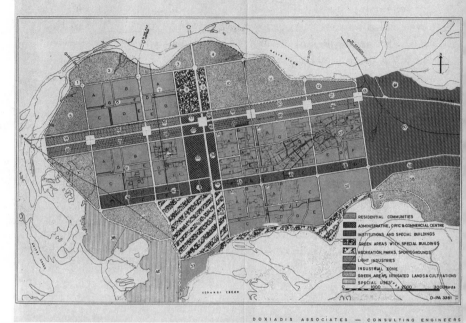

DOXIADIS ASSOCIATES — CONSULTING ENGINEERS

3.2. Master plan for Korangi, Pakistan (DOX-PA 147), November 18, 1961. Courtesy of Constantinos A. Doxiadis Archives. © Constantinos and Emma Doxiadis Foundation.

meant to eventually become a self-contained urban community attractive to the new Pakistani middle class. As a signal of US commitment to the new nation, the Korangi project received funding in 1958 from USAID and the Ford Foundation.

The planning firm of Constantinos Doxiadis was hired as project consultant. Doxiadis Associates was one of the largest planning firms of the early postwar era. It will be referred to repeatedly in this book, because it played such an outsized role in the new town movement. Doxiadis had the support of both the Ford Foundation and the World Bank for his many projects throughout the Middle East. Indeed, Ford spent more money on Doxiadis Associates than on any other private consultancy, and it was the foundation's most influential and visible partner. The company was enmeshed in a myriad of housing, resettlement, and "satellite town" projects in Pakistan.

Korangi was the Ayub Regime's signature modernization venture. As the largest urban rehabilitation project in Asia at the time, it was enthusiastically touted to the international press. Doxiadis planned a model town

3.3. Housing at Korangi, Pakistan *DA Review* (Doxiadis Associates) 8,
no. 80 [April 1972]). Courtesy of Constantinos A. Doxiadis Archives.
© Constantinos and Emma Doxiadis Foundation.

with schools, markets, and mosques in each of its neighborhoods (fig. 3.2).
It would be the "fulfillment of community needs," and offer a "harmonious
and balanced way of life."[23] The first simple concrete houses with verandas
and gardens were slapped together in less than half a year and offered on a
rent-to-own basis (fig. 3.3). Ayub Khan himself handed the keys to the first
refugee family to arrive in the settlement. And the photo opportunities were
endless. US president Dwight Eisenhower visited Korangi in 1959. It was a
stop on the Pakistan itinerary for both Queen Elizabeth and Prince Philip,
and US vice president Lyndon Johnson in 1961. The celebrity visitors po-
litely ignored the absence of basic infrastructure during their walking tours,
for they were there at the moment of the town's creation. As was the case
with all nascent new town projects, Korangi was experienced simultaneously
in the present and in the future. It was visualized as what it would become.

A Nation of New Towns: India

In newly independent India, the government under prime minister Jawaharlal Nehru likewise set off on an elaborate regional development and
modernization campaign to promote progress and solidify the nation. Historian Gyan Prakash argues that it did so with a divided ideological logic[24]
that in many ways mirrored Nehru's foreign policy of nonalignment with
any alliance or bloc and also the willingness of postcolonial elites to combine elements of both US and Soviet experiences. On the one hand, the Soviet Union served as an instructive example of what could be accomplished
in a socialist society. Five-Year Plans became national crusades to develop

resources, raise living standards, bring about law and order, and transform culture and society. On the other hand, a new Indian elite, metropolitan in orientation and Western in education, advanced an agenda of Western-style modernization on a multitude of fronts.

The Nehruvian vision insisted on newness as the defining characteristic of postcolonial India. Modernism meant an escape from the tradition-bound, conservative mores of the precolonial as well as the colonial past. This temporal rupture left India free to chart its own modern course. At the same time, these aspirations needed to be imagined as uniquely Indian. Nehru believed that science and technical expertise were the instruments for discovering India as an authentic modern nation. He laid out his vision: "We want . . . to build up community life on a higher scale without breaking up the old foundations. We want to utilise modern technique and fit it into Indian resources and Indian conditions."[25] The realm of technical experts and planned development shifted authority to the state as the means for fulfilling the nation's future and offering a distinctly Indian experiment in modernization. New towns were a mechanism for both jump-starting and disciplining this radical transformation.

The Delhi metropolitan area was in chaos when the new government of independent India took office. The capital city was overwhelmed by hordes of refugees following Partition as well as by a flood of poor rural migrants. Its population shot up from 700,000 in 1941 to 1.4 million in 1950, and to 2.3 million by 1951—an impossible situation foisted onto an already teeming city. Temporary refugee camps were constructed where thousands existed on the edge of despair. Housing colonies were erected in the suburbs.[26] But these gestures were immediately derisory. Cramming into ancient, ethnically segmented neighborhoods and squatter settlements of mud and straw *jhuggie* huts were the only other alternatives. Delhi's Islamic monuments were taken over by Hindu refugee squatters in an act of political defiance. A horrific jaundice epidemic played into the image of that city as filthy, overcrowded, and chaotic, and a hotbed of disease. In response, Nehru's government addressed the crisis by using the capital as a laboratory for metropolitan regional planning. In line with the state's scientific outlook, a civic survey was conducted and a master plan developed that was among the most ambitious in Asia.

The plan was carried out under the aegis of the Delhi Development Authority and the national Town Planning Organization, both new institutions that would chart the course for urban India. Delhi's master plan was funded by the Ford Foundation, which sent a team of British and American social science and planning consultants that had been pulled together by Albert Mayer.[27] An active leader in the American garden city movement, Mayer had worked in India building airfields for the US Army Corps of Engineers during the war. Afterward, while working on projects such as Kitimat in British Columbia with Clarence Stein, he returned to India to become a friend and close adviser on urban planning to Nehru.

Except for Mayer, the Ford Foundation's team had little familiarity with India. Nonetheless, its members were eager to help the country modernize, and their combined expertise was imposing. The team actually held its first meetings at the University of California, Berkeley, where it ruminated over India's intractable urban problems from the safety of a Western academic institution. Mayer then traveled to India to arrange a viewing of the celebrated 1939 documentary film *The City* to an audience of Indian ministers and the Town Planning Organization. The moment was prodigious, and the instruction unashamed. Moreover, Delhi was the first major exercise in their newly independent country for Indian planners, most of whom had been educated in the United States and Great Britain. They were euphoric about the possibilities of independence, and ardent in the belief that physical design could shape social goals.

The master plan was code-named "the Delhi Imperative." Although Delhi's informal neighborhoods were already highly efficient, self-functioning communities, they were denounced as overcrowded slums, and the "undesirable mix of land uses almost everywhere in the city" was condemned for good measure. All these traditional forms of urban living would be weeded out and simplified. After a public outcry, large-scale slum clearance was tempered. In a compromise, the population in the old walled city would be only gradually reduced to "urbanizable limits" and contained by a greenbelt. On the city's periphery, the haphazard proliferation of squatter settlements, housing colonies, and townships was to be brought under control. People and industries would gather in six major satellite towns of seventy-five thousand to one hundred thousand each that were connected by a system of express roadways: Ghaziabad, Gurgaon, Bahadurgarh, Loni, Narela, and Faridabad.[28]

Plans for Faridabad, for example, provided resettlement and jobs for some 40,000 destitute refugees. The town of Narela, with a population of 11,000 and a sprawling refugee camp, was revamped with schools, recreation facilities, and a new industrial zone linked to the proposed highway system. Eventually, the town would absorb 60,000 people. In all, the satellite towns around Delhi were dream texts of national integration and the harmonious coexistence of ethnic and religious groups as modern citizens. These brand-new places were unconstrained by history or by locality, and would carry Bharat Mata, or Mother India, into the modern world.

Beyond the satellite towns around Delhi, some thirty other new towns were initially constructed in India. Four satellite new towns were built around Hyderabad, three in the outskirts of Madras (the city now known as Chennai), and two at Calcutta (now known as Kolkata). Partition meant creating the new state capital of Gandhinagar in Gujarat, Bhubaneswar for Orissa, and Chandigarh in the East Punjab.[29] The Punjab had lost its principal city of Lahore to Pakistan. Its replacement, Chandigarh, was Nehru's fantasy city, a shimmering symbol of India's modern future. It catapulted India to international stardom. Albert Mayer was initially asked to create a plan for the new capital. He worked with Clarence Stein to fashion a garden city on the Radburn principles with its residential neighborhood units. But Nehru cast aside what was already becoming an outdated concept too closely attached to India's colonial past. Instead, he endorsed the transformative potential of Le Corbusier's bold modernist vision. Chandigarh introduced to India a new urbanity, a breathtaking handling of urban form and space. It was an escape from everything traditional and conservative, everything Nehru tried to shed in his quest for India's future.[30]

Some of the largest new towns adhered to a conventional function: they were steel towns meant to jump-start India's industrial economy. Steel was invested with symbolic as well as strategic importance in Indian nationalist discourse, and it was also evidence of continued Soviet influence. The steel towns were developed as part of the second Five-Year Plan under the auspices of the publicly held Hindustan Steel Corporation. Among the most important projects were Bhadravati in Mysore and Jamshedpur in Bihar. Rourkela in Orissa province was selected for the construction of a massive steel plant. Bhilai in Madhya Pradesh, Durgapur in West Bengal, and Bokaro in Bihar followed immediately after.[31]

The Soviet Union collaborated on the development of the steel mills at Bhilai and Bokaro, sending hundreds of engineers and technicians as well as funding. Of course, this turn of events dramatically stepped up aid to India by the United States and its Western allies. Britain collaborated on the steel town of Durgapur, and the Federal Republic of Germany on Rourkela. The venerable German steel combine Krupp and Demag Company provided financing and technical expertise for the Rourkela steel plant. Thousands of technical personnel were sent from West Germany, while workers from all over Asia came to Rourkela for jobs in construction. By the late 1950s, the population of both Rourkela and Jamshedpur had already shot up into the hundreds of thousands. According to historian Srirupa Roy, the utopian steel town's identity as an exemplary national space was linked to its distinctiveness; the times, spaces, subjectivities, and practices associated with it would be manifestly different.[32] Along with steel towns came a web of single-industry and resource new towns for aluminum, cement, mining and oil refining, and engineering in an entirely new industrial geography.

Despite this crusade for industrialization, a vast majority of Indians still lived in rural areas and were desperately poor. Modernizing the rural economy and transforming India's illiterate peasants into model citizens were key development goals. To carry them out, the Indian government proposed building at least three hundred new settlements by the end of the twentieth century. Model urban villages were set up. They were imagined as mechanisms for law and order and for a new egalitarianism. Ordering urban development according to defined space and population size was seen as the solution to the impoverishment and dire conditions characterizing the countless informal settlements strewn across the country. Town building would anchor civil society and mobilize everyday life. Ordinary people would become self-reliant citizens, partaking in modernization and public affairs.

To that end, Albert Mayer plunged into the modernization of settlements in the northern Indian state of Uttar Pradesh, along the border with Nepal; there he launched a pilot project covering 100 villages and over 50,000 people in the district of Etawah. Later, the project expanded to include some 112 additional villages with 79,000 people in the neighboring district of Gorakhpur. Mayer viewed these communities as social test sites for cooperative self-help and democratization as well as a springboard for development—projects that would unlock India's potential. He was also influenced by Gandhi's conceptions of the village as a self-sufficient, intimate collective that was creative as well as economically productive. Mayer also filtered these ideas into his new town projects in North America. In a speech at Swarthmore College in 1952, he explained, "Statistical information is important, but actually harmful when it tends to become a substitute for the vividness of a tramp over the ground, for personal talk with the people where they live, for personally seeing what they do, how they do it and why, for direct sparking to a solution."[33]

Mayer's self-actualizing "help the villagers help themselves" approach was enthusiastically reported in the Western press.[34] With the help of Ford Foundation and US Point Four funds, the pilot project was expanded into a major Indian government initiative. The Ford Foundation sent social workers into the newly formed village communities to teach farming techniques, artisanal crafts, roadwork, and sanitation, and for women, modern methods of cooking, sewing, and hygiene. Douglas Ensminger, who directed the Ford Foundation's work in India for nineteen years and was himself a rural sociologist, supported Mayer's pioneering ideas.

Mayer's concept of the urban village was actually a composite of Gandhi's village concept and the neighborhood unit ideal. It was a laboratory for trying out social and community organization techniques embedded in physical planning.[35] "The aim," he argued, "is to reorganize the physical frame and rebuild it, and hence the social life of the village as well. . . . The present village layout is typically uneconomical, with excessive winding roads and lanes, useless and unsanitary open areas cluttered with ruined walls and malaria-breeding depressions."[36] Mayer's initial projects were usually the construction of a community center and a latrine. Villagers would be taught cleanliness and cooperation. Mayer imagined the *panchayat ghar* (meeting place for the village's representatives), the school, and the seed store built around a little civic square where trees would be planted in some dignified pattern. Settling people around a core community area was understood as a way to create modern civic values and modern people attuned to national imperatives.

Nehru's government set up pilot projects for the new town of Nilokheri in the East Punjab, Kalyani in West Bengal, and Ulhasnagar in West India, as well as plans for the new port town of Gandhidam on India's west coast. Created by American-trained engineer and social visionary Surendra Kumar Dey (who eventually served as Indian Minister of Community Development), Nilokheri was a model self-help cooperative community for Hindu and Sikh refugees from West Pakistan (fig. 3.4). It arose from jungle marshland in the East Punjab north of Delhi, where Dey worked in refugee camps training men and women in artisanal crafts. With Nehru's support, he began organizing a new town to rehabilitate seven thousand displaced refugees.

The settlers rolled up their sleeves, cleared out the tropical wilds, and constructed Nilokheri in a symbolic act of collective responsibility. "Humming with the music of muscles," Nehru called it. Dey picked up the mantra and made it the tenet of faith in a new trinity: "Muscles can do it, muscles can be trained to do it, conditions can be created to do it." Rows of mud-brick houses with running water and electricity were rented at low cost to families, who would assume ownership in thirty years. Roads radiated outward from a compact town center "that would enable people to know one another."[37] There were health clinics, schools, markets and shops, work centers and practical training in crafts. Sports and an open

3.4. Members of a United Nations Technical Assistance Administration
study tour interviewing villagers in Nilokheri, India. The textiles produced by
village women are displayed, while the women remain on the rooftops with
their children. November 6, 1955–February 12, 1956. © UN Photo.

air-theater were organized. Nehru called Nilokheri the town of his dreams.
"You have presented a model to India," he declared. "It will change the
very face of the country."[38]

Members of India's parliament toured the refugee-built town to see
its progress. It became a mecca for social idealists. In 1952, US ambassa-
dor Chester Bowles toured Nilokheri and Albert Mayer's urban villages in
Etawah, and offered $50 million in aid to expand community develop-
ment in India. By the early 1950s, Nilokheri was a bustling community
whose industries provided jobs for seven hundred families, while more
made their living in construction and farming. Places such as these were
a form of state experiment in ideal urban typology and social utopianism.
The pitiful refugee would transmute into a self-confident, heroic citizen.
Through discipline, shared labor, and cooperation, a modern community
would wipe out the terrible scourges of caste, class, and illiteracy. This
ideology of self-help, or self-reliance, was crucial to the postwar model of
development in the newly formed nations of the Third World. Utopian
showplaces such as Nilokheri were part of a bold vision for nation building. Modern India would emerge as a countrywide mosaic of thriving, self-

sufficient townships. Initially, they were certainly no more than tent camps, but the rhetoric and plans for them were couched in superlatives.

Not surprisingly, these projects were initially dominated by the British town planning tradition and the tried-and-true morphology of the garden city, with single-story houses surrounded by gardens, parks, and open space. The root for this pedigree was once again Patrick Geddes, who worked in India from 1919 to 1925 and wrote an exhaustive series of planning reports about Indian cities. He collaborated with Indian poet Rabindranath Tagore to develop educational opportunities and circulate his ideas. In 1947, Geddes' son, along with architect Jaqueline Tyrwhitt published an edited volume of Geddes' reports as *Patrick Geddes in India* to promote the relevance of his concepts to India's modernization and town building schemes, particularly Gandhi's notions of democratic self-governance and the village as a fusion of work and life.

Tyrwhitt secured a spot as the UN's Technical Assistance Advisor to the Government of India. In that capacity, in 1954 she organized a UN Regional Seminar on Housing and Community Planning in New Delhi that welcomed a delegation of American planning experts from the University of Pennsylvania and MIT's Center for International Studies. Also in attendance was Constantinos Doxiadis, who was establishing his consulting firm. At the same time, Tyrwhitt was piecing together a working model of an Indian Village Center that was an amalgam of Geddes' principles and CIAM's vision of an urban core set amid experimental housing. The simple buildings made of sun-dried bricks enclosed an open space for village festivals with a nearby school, health clinic, and craft center.

The main Geddes disciple in India, however, was German architect Otto Koenigsberger, Tyrwhitt's friend and fellow traveler. In Berlin, Koenigsberger had studied with Hans Poelzig and Bruno Taut and briefly worked with Ernst May at Frankfurt. He fled Germany in 1933 and arrived in India, where he quickly moved through the ranks to become chief architect of the progressive princely state of Mysore, and then director of housing for the Indian Ministry of Health. This last position made Koenigsberger responsible for housing the hundreds of thousands of refugees fleeing Pakistan. He went on to chair the nascent Indian Board of Town Planners set up in 1949, which by 1951 became the Indian Town Planning Institute.

It was in the cosmopolitan atmosphere of late-1940s Bombay (now known as Mumbai) and the hopes for an independent India that Koenigsberger and the Indian intelligentsia, led by Mulk Raj Anand, founded the Modern Architectural Research Group (MARG) along with its professional journal of the same name. *MARG* magazine became the country's premier conduit for pathbreaking architecture and urban design and attempted to bridge the gap between Indian and Western architectural debates. In the premier issue, Anand penned that "planning is like dreaming—dreaming of a new world."[39]

Koenigsberger also was involved with the planning of numerous new towns in India, including the Tata and Sons steel town of Jamshedpur, the capital cities of Bhubaneswar and Chandigarh, and the new towns of Faridabad, Rajpura, Gandhidham, Sindri, and Nilokheri. He attempted to fuse Geddes' concepts and the garden city ideal with those of the modern movement to create a distinctly Indian variant of modernism.[40] For the architects gathered around MARG, India's traditional architectural heritage was tainted by its use in British colonialism and consequently unusable by an independent nation. This legacy was made more acceptable by mixing accustomed building designs with continental European and American scientific approaches in fabricating an urban future for India.

Koenigsberger routinely worked closely with the UN and US development agencies on his new town projects. He outlined his planning strategy in a long article in the British *Town Planning Review* in 1952. In his emerging language of tropicalism, Koenigsberger saw local climate and environment, landscape and topography as the defining national constituents of a modern style. He and Tyrwhitt, along with Albert Mayer, promoted a place-specific modernism that was rational and progressive while at the same time an expression of indigenous culture. They regarded India's refugee towns and villages as the merging of self-reliant "folk planning" and "homesteading" in the tradition of Patrick Geddes with Gandhi's strategy of creative self-governance and empowering the masses at the village level. Such a vision, they thought, would alleviate India's religious and social strife, and improve economic conditions. Local environment and native traditions would be the basis of development.[41] In point of fact, construction in developing countries was dictated mainly by the scarcity of materials and the need to build cheaply. Under these conditions, vernacular practices were the most practical, and were indeed imagined as a strategy for creating local identity as well as training in the building trades.

In Koenigsberger's worldview, garden cities and neighborhood units could be made Indian by inserting them into the local setting. Hence he did not conceive of them as foreign. There was little compulsion to review Western norms and standards in relation to the stark poverty of India's hinterland. Western ideals could simply be transplanted to India and synthesized with the indigenous cultural tradition and environment. An admirer of Clarence Perry and of Patrick Abercrombie's *County of London*

Plan, Koenigsberger introduced the neighborhood unit concept to India in his extension plan for an industrial facility in the state of Mysore. Low-density neighborhood units of three thousand to four thousand people were settled around primary schools, which then combined to form a high school district. The correct number of inhabitants was determined by the mathematical calculation of density of persons per acre. Rational computation replaced the religious and social divisions that had kept India's cities segregated.

The Indian steel town of Bokara was another experiment in neighborhood unit transference. The primary school was the center of activity for each of its neighborhoods, or housing clusters, of 750 families apiece. Neighborhood units would promote neighborliness and family contacts, and diminish class frictions. Their size roughly corresponded to the *mohalla*, or traditional Indian residential area, and was an attempt to translate the Western ideal into indigenous cultural patterns.[42] As in Mysore, each housing cluster would then join adjacent neighborhoods in a secondary school district. These districts would in turn combine to form an urban area of 18,000 to 20,000 inhabitants, with shopping and services, health and community centers, and parks. As in Clarence Stein's Radburn model, the boundaries for each neighborhood unit were drawn by the outer ring road, while its access roadways were for local use. Generously laid out greenbelts and parkland upheld the notion of natural landscape as the necessary theater for the realization of community life, and in fact took up excessive surface area in many of these early model communities. Although this quickly resulted in urban sprawl and extremely high infrastructure costs, nonetheless the utopian vision of the garden city was followed precisely. It was a means to apply Western planning techniques and make India modern. Based on this conviction, the garden city became a template applied across the developing world, merged into imaginaries of nativism.

These idealistic aspirations and model communities belonged to the heroic years immediately after independence, and were constructed amid intense social, ethnic, and political turmoil. Ultimately, the neighborhood unit was depicted as a quasi-magical space that would create a casteless, secular India.[43] Otto Koenigsberger's trajectories evidence the eclectic influences—German modernism, Geddes and the British garden city tradition, self-help and the developmental creed of Western international agencies, tropical architecture and planning, the American-style neighborhood unit—comprising the globalized landscape of midcentury planning. Historian Rhodri Windsor-Liscombe also points to Koenigsberger as an illustration of the contradictory stance of midcentury modernists who worked for the very regimes they were attempting to transform. He had the patronage of the Maharaja of Mysore; the last British viceroy, Lord Mountbatten, and the Nehru government; the Tata steel family; and the universalizing authority of the UN: it was "an uneasy mediation of imaginary future solution with imperative past condition."[44] The new town and neighborhood unit idioms were translated into local Gandhian practice, mixed with indigenous architectural styles, and reinvented as omnipresent reformism, fully accepted by state officialdom as the path to modern development. While they were construed as homegrown, they were also a homogenizing influence that offered stability and restraint, containment and control, a vision of democratic empowerment and happiness in the collective of the village.

Eventually, Nehru turned aside this Gandhian vision as too traditional and instead came under the spell of high modernism at Chandigarh. Koenigsberger, Jaqueline Tyrwhitt, and Albert Mayer all lost influence. But the town building ideal remained a strategy for social management and modernization. The entire geography of new towns throughout the developing world was defined according to a carefully calibrated population and territorial logic. The discourse was framed around discipline over people and space through classification and hierarchy, optimal size and dimension. Ordering and management of population movement would ensure a balanced dispersal of cities and people across newly acquired national lands.

Although this fixation on spatial typology was by definition a vital part of new town rhetoric, it was particularly zealous in newly formed nations. New towns rarely tested traditional social and ethnic boundaries. The need for stability was too urgent. The unending sectarian and social turmoil, the vast movement of people fleeing chaos and violence, cities overwhelmed by squatter camps and slums filled with despairing people open to the appeal of communism—these were the immediate postwar realities that underlay the movement to regulate and control population and territory. It revealed deep-seated apprehension about refugees, immigrants, and unruly population flows wandering unrestrained across what were hotly contested political territories.

Moreover, the promotion of futuristic dreamscapes did not necessarily mean an end to abysmal conditions. Provisional settlements and makeshift colonies somehow became permanent fixtures. A good number of these places remained nothing more than sterile compounds. Among the worst were the Palestinian refugee camps established after the 1948 Naqbah— the United Nations Relief Agency initially attempted to set up large-scale public works projects to promote economic development, but to no avail. Across the Middle East, thousands of people raised families and lived out their lives in sordid settlement camps glorified as utopian experiments. Model communities were forgotten and neglected once the fervor of the reconstruction years died down and state priorities shifted. The quest for perfection could slide toward collapse with amazing ease. Just below the mesmerizing vision of the future lay the dark shadow of dystopia and the searing reality of living in ignominy.

Oil and the Garden City as Dystopia

Just how quickly the rose-colored vision of garden cities could fracture into an ugly reality is evidenced by the oil towns of the 1950s and 1960s. They were the urban loci of global petroleum production. The oil business was an indelible part of colonial empire building. The scramble for oil and gas deposits throughout the Middle East and Africa was intensified by decolonization as private companies and newly independent governments fought for control over precious natural resources. The oil production network had its own particular geography built up over the twentieth century as exploration intensified, oil fields were discovered, and petroleum was extracted. Oil-rich regions had long been blotched with rudimentary boomtowns and work camps established around oil wells and export terminals.

But once their hegemony was disrupted by independence movements and political crises, oil companies brandished the garden city ideal as a way to prove their good intentions to emerging nations. This was the case with the Anglo-Iranian Oil Company,[45] which operated a network of some ten oil sites in the Khuzestan region of present-day Iran. The company had ignored amenities at its massive oil refinery in Abadan that had been producing black gold since 1908 and exporting it by pipeline. It was an overcrowded nonurban place on a mudflat in the Tigris-Euphrates estuary on the Persian Gulf, and the largest oil facility in the world. But when the Allied powers occupied Iran and deposed Reza Shah Pahlavi in 1941, the company began a "new propaganda of architecture and urbanism" at Abadan as a gesture of good faith and to quell any discontent. It was essential to secure the oil fields and Allied supply lines for the war effort.

By the mid-1940s, some sixty-five thousand people lived and worked in the sprawling refinery site (figs. 3.5, 3.6). Construction materials were shipped in by barge and rail. British architect James Wilson (who had worked as Edwin Lutyens's assistant at New Delhi) laid out the showcase garden suburb of Bawarda with an axial avenue and town center replete with *rond-point* and gardens, a cinema, restaurants and social club.

3.5. Abadan area, Iran, 1946–51. Image from the BP Archive, © BP plc.

Neighborhoods were vigilantly segregated according to rank and status, as were company services and recreational facilities. Company managers lived upriver from the refinery in neighborhood compounds with tight security, spacious homes, manicured lawns and date groves, and swimming pools. A patchwork of mud-brick row houses and dormitory estates were constructed downriver and separated European from non-European oilmen. Wide stretches of open terrain "disciplined social pathologies" and prevented intermingling. Non-European refinery workers were left to their own devices in squatter settlements and shantytowns along the town's outskirts, such as Segoush-i-Braim and Bahar.[46] In the Western mindset, these zones were havens for prostitution and criminal drug rings. As informal

3.6. Housing for married staff in Abadan, Iran, 1950. Image from the BP Archive, © BP plc.

places, they stood in stark contrast to the affluent and ordered garden suburbs constructed for oil company officers and trained personnel along with carefully selected native senior staff. The exclusionary tactics and the conditions in these ghettos were among the justifications given by the Iranian government for seizing the oil industry in 1951, when the British and the Anglo-Iranian Oil Company were summarily kicked out of the country.

For all their utopian aspirations, garden cities and new towns could embody some of the most unsavory, corrosive social practices. This was especially the case in Africa. If the adaptation of the ideal could produce little more than cardboard versions of utopia in developing nations such as Iran, the genuinely dystopic qualities lurking beneath the surface were painfully visible as the West scrambled for Africa's natural resources.

The extraction of mineral wealth from the African Continent was of course a long-standing colonial imperative. Shocking conditions in mining camps were an established feature of the colonial landscape, as were forced relocation schemes that moved people from traditional villages into bare-bones settlement camps. Garden cities were attached to this sorrowful legacy, which was further complicated by postwar decolonization and the establishment of independent African nations. If anything, it made clear that the garden city and neighborhood unit were not inherently altruistic idioms, and that no planning form is either innately liberating or innately oppressive.[47] As design approaches, they became immutable emblems of Western modernization, incriminated in the shifting geopolitics of the postcolonial world and the increasing scale of resource extraction. This was particularly the case once large reserves of crude oil, natural gas, and uranium were discovered in Africa. Oil companies scrambled for concessions to develop the continent's extraordinary raw materials. In 1963, there were already eleven refineries, with drilling opened in another seventeen fields that would double output.[48] The oil economy created a new network of boomtowns and heavily fortified oil compounds where strict ethnic-racial segregation and the exploitation of native populations were everyday policy.

The first oil "gusher" in the Sahara was at Hassi-Messaoud in the eastern part of the French colony of Algeria, about six hundred miles into the desert from Algiers. Discovered in 1956 in the midst of the Algerian War, it was the site of one of the Sahara's largest potential oil reserves. Charles de Gaulle welcomed it with the announcement that France "may have found a new destiny."[49] Every effort was made to protect it along with the four hundred miles of pipeline that brought the extracted petroleum to the Algerian port of Philippeville for export. The French army carried out military sweeps around the oil fields to drive out the revolutionary Algerian National Liberation Front and any locals cooperating with them.

Working together with the Algerian government on a plan for the future city, the two French companies developing the oil fields imagined Hassoud-Messaoud as a place "of habitability, hospitality, and humanity" where "man would be liberated in the desert."[50] The French SN REPAL oil company sent in tons of soil, fertilizer, palm trees, and plants, and constructed a concrete water tower as the town monument as well as massive windshields, a eucalyptus garden and a citrus grove—all to create the illusion of a garden city oasis for the young *pétroliers* who took up residence. Flowerbeds and lawns were meticulously watered every day. The Compagnie Française des Pétroles Algérie constructed a company complex dubbed the Maison Verte that was an entirely air-conditioned model French village (fig. 3.7). The customary post office, police headquarters, and mayor's office stood alongside the bakery and *tabac*. Both camps at Hassi-Messaoud (about ten miles apart) were outfitted with running water and electricity, a public garden, swimming pool and sports center, movie theater and restaurants.

That Western companies could create an entirely artificial movie-set Eden in the emptiness of the Sahara epitomized their power as modernizers. In the scramble for oil, the garden city ideal had become little more

3.7. Aerial view of the Maison Verte base camp at Hassi-
Messaoud, Algeria, date unknown. © Bernard VENIS.

than green gadgetry. Both settlements at Hassi-Messaoud, which comprised
about six thousand people in 1960, were heavily militarized and encircled
by sand-dune embankments and barbed wire for security. The oil crews
were organized as military units and armed with rifles; a French military
camp was permanently stationed nearby. A separate tent camp for Arab
workers outside the security perimeter was left unplanned and neglected.

The new town of Arlit, while not an oil town, followed the same logic.
It was established in the 1960s by Niger's military regime when one of the
world's largest uranium deposits was discovered in the forsaken northern
reaches of the country between the Sahara and the Aïr Mountains. The de-
posit was considered key to the country's economic future, and its first uranium mining company, SOMAIR (Société des Mines de l'Aïr), was financed
by French investors.[51] The French interest was dictated entirely by the need
for enriched uranium for France's atomic arsenal. The nation's Atomic Energy Commission and mining companies held the majority stake in the
company.

Arlit was one node in the global cartography of atomic weaponry—a
system of mining camps, uranium enrichment facilities, research laboratories, and atomic plants set out as showcase settlements (including Deep
River in Canada). By the 1970s, it was a boomtown with a population of
some twenty-five thousand mine workers, traders, fortune seekers, and

expatriates enjoying the momentary good life of uranium production. Known as "Petit Paris," Arlit boasted a cosmopolitan atmosphere of cafés and night life, shops and supermarkets, a modern hospital, a swimming pool, and recreation facilities carved out of the sands of the Sahara. Trees and bushes were flown in as cosmetic touches for the uranium pioneers. The residential compound for the European managers and engineering staff was on the opposite side of the town center from that for locally trained staff. Thousands of black miners, many of whom suffered radiation poisoning, cobbled together shacks in townships along the periphery.

Examples such as these stood in full relief across the African resource landscape, but they were particularly devastating along the Niger delta, where dirt-poor African villages such as Oloibiri stood alongside massive oil refineries. Commercial quantities of petroleum were first discovered in the Niger delta at Oloibiri in 1956. Spearheaded by the Shell BP Company, neighboring Port Harcourt became an international boomtown overnight, with thousands of oilmen arriving to exploit the Niger's black gold. By 1963, Port Harcourt's population jumped by 370 percent, with 250,000 people living in the city and the sprawling hinterland.

The town had been a colonial coaling station and terminal for British military operations during the First World War. In the 1940s, an urban plan was laid out, with tree-lined avenues and a sequence of neighborhood units for the colonial administrative staff and the European coal brokers who controlled the town. One of their first municipal acts was to construct a golf course and country club. This was enough for the British to christen Port Harcourt a garden city, which was separated from the African area of town by parkland. Just as it was emerging as the center of the oil economy for the independent nation of Nigeria, the British staged one last ceremonial Regatta at Port Harcourt in 1956 to welcome Queen Elizabeth and the Duke of Edinburgh during their tour of West Africa. They were met by a royal salute from a flotilla of ceremonial war canoes. Although the Regatta reveled in imperial spectacle, the war canoes were a magical display, according to historian Andrew Apter, of the coastal region's emerging power in the new Nigerian nation.[52] Three years later, in 1959, a master plan drawn up by Israeli planner Y. Ellon included a vast new industrial and residential estate at Trans-Amadi that welcomed the giant Gulf, Elf, and Mobil oil facilities. Shell BP built an entirely new neighborhood for its administrative staff in the Rumuokoroshe area; a veritable "city within a city." The Senior Staff Club, with its golf course and swimming pool, was the center of its social life. The company's offices were the largest and most conspicuous building in the town.

In the meantime, African (mainly Ibo) laborers and the area's poor lived in wretched slums along the gigantic waterfront complex or in squalid villages outside the fortified, heavily securitized refinery sitting on one of the world's largest wetlands. Pollution, toxic waste, and garbage-heaped slums were the reality of life in Nigeria's celebrated garden oasis. Half the households in Port Harcourt lived in one room in abject poverty, dependent on bucket latrines and shared kitchens without piped water.[53] Clean water, waste disposal, and basic sanitation were nonexistent in the dystopia situated next door to the pristine garden city for foreign oil companies in league with the succession of corrupt political juntas governing Nigeria. In Africa, the garden city imaginary did little but prop up the racial distortions and the lethal combination of repression and naked corruption of the postcolonial oil landscape. The result was escalating violence, rage, and disillusionment that made oil towns a breeding ground of political radicalism and armed struggle. The brutal impoverishment and environmental despoliation in the wretched backwater of Oloibiri provoked the violent insurrection led by Isaac Adaka Boro in the 1960s.

Ekistics and the World City

The astonishing success of Constantinos Doxiadis and his ekistics movement is made comprehensible by this midcentury exhaustion of the garden city ideal in the developing world. The garden city played out as inefficient and unusable, and in the case of Africa as a spatial strategy for racial domination and a radical new imperialism. Doxiadis appeared to offer development institutions such as the Ford Foundation an innovative, more powerful alternative.

Established in 1952, Doxiadis Associates became one of the largest engineering, architecture, and planning consultancies in the world, with projects ranging from housing programs to new towns in over forty countries. Doxiadis himself was the jet-setting impresario of the modernizing regime. He produced dozens of books and hundreds of articles and planning reports about his urban concepts. He is most closely associated with his design theory known as ekistics, or the science of human settlements. Ekistics was also a promotional tool: the World Society for Ekistics, along with a host of international sponsors, pumped out a steady stream of publicity for Doxiadis's visionary ideas.

Extensive research has been carried out on Doxiadis and his city-building theories.[54] Two aspects of his prolific work deserve attention in the context of the new town movement. First are his new town projects themselves. Second, Doxiadis and his World Society for Ekistics became the magnet for an assembly of charismatic futurists probing alternative visions of global human settlements; many of them attended his famous Delos Symposia and wrote regularly for his *Ekistics* journal, begun in partnership with Jaqueline Tyrwhitt in 1955. The Athens Technological Institute (founded in 1958) and its Center of Ekistics became the command center on the future of cities. Its seminars and conferences were a catalyst for high-spirited discussion with, for example, Lewis Mumford and Clarence Stein, who attended in 1960. Celebrated British architect Richard Llewelyn-Davies and American engineer and visionary Buckminster Fuller were among the seminar leaders in 1966. Government ministers from the new nations of the Middle East and Africa trekked to Athens as the mecca of ekistics planning techniques. In a real sense, intellectual debate about new towns in the late 1950s and 1960s was driven by Doxiadis as prime mover, along with his global network of planning luminaries.

For Doxiadis, a settlement was a continually evolving organism at once biological and technological, rationally organized into a hierarchy according to function and scale. He identified five elements of human settlement: man, nature, society, networks, and shells (meaning buildings). These concepts were initially derived from Walter Christaller's central place theory. Doxiadis received his academic training in both Athens and prewar Berlin, and was initially influenced by both Christaller and settlement concepts outlined in Gottfried Feder's *Die neue Stadt* (see chapter 1). He subsequently served as head planner in Athens and as US Marshall Plan coordinator for the reconstruction of Greece, where he began working out his ekistics ideas.

Doxiadis viewed the science of ekistics as entirely interdisciplinary. Modeling his methodology on that of Patrick Geddes, his site surveys were based on a deep reading of statistics, texts, photographs, ecology and environment, and spatial and visual cues. These surveys took into account not only an area's landscape, geography, and current condition but also its history and traditions. Doxiadis trained his camera on the past as much as the present, looking for clues that would reveal the secrets of space and place. The results of his surveys were produced in sophisticated schematics accompanied by an inventive technical jargon. He then derived his urban designs from this data. Because Doxiadis imagined the spatial patterns of a site as elaborate matrices and information flows, he produced his urban designs as calibrated grids, tree diagrams, and abstract charts rather than traditional artistic renderings.

At the same time, his ekistic concepts offered a convincing strategy for

Western-style modernization. They formed a prepackaged development scheme, coproduced in conjunction with institutional sponsors and the modernization regime.[55] Doxiadis elaborated his ideas as a global brand distinct from that of his competitors. He offered an alternative not only to the garden city idea (he considered satellite cities and small towns a dangerous backslide) but also to the uncompromising modernism of Le Corbusier. In addition, his theories advanced solutions to the "population problem" and the plight of overgrown, hydra-like cities spreading across entire regions.

Ekistics was a middle ground that seemed efficient and practical. It imagined a blend of indigenous urban patterns and Western models of progress that matched nation-building agendas and the developmental thinking of Doxiadis's sponsors, especially the Ford Foundation. They were eager to stabilize populations susceptible to the lure of communist insurgencies in the newly formed nations in Africa and the Middle East. These regions were Doxiadis's proving grounds. His work was shaped by the Cold War and decolonization and exuded the rhetoric of social control and harmony. As he argued in a letter to Paul Ylvisaker, Ford Foundation president, "In seeking . . . to create for the less developed nations a way of life and a pattern of thought that would lead them to 'identify themselves with the West,' we should conceive 'Development' to include economic as well as socio-cultural development designed to raise living standards and create conditions for a better life."[56]

Doxiadis argued that modern cities should be dense, low rise, and mixed use, and built around public transportation systems. They should be designed on a grid plan, with superblocks and clusters of small-scale community sectors (an adaptation of the neighborhood unit ideal). This pattern would produce livable neighborhoods featuring public spaces and squares, schools and community centers, and daily opportunities for social interaction. His designs created "community spirit." They would, according to Doxiadis, be "real communities where people can live happily." In line with development thinking in the 1950s, he stressed the importance of self-help and local activism, a combination of modernity and democracy that held enormous appeal for governments in the Third World as well as the Western funding agencies that bankrolled mammoth infrastructure schemes. Doxiadis's Greek background and his settlement theories distinguished him from the swarm of Western consultants and advisers who were streaming into developing countries. The proof of his planning approach lay in his years of successfully coordinating the Marshall Plan in

Greece, and then in his active engagement with policy makers in Washington, DC, and the UN.

For Doxiadis, the real dimension of cities was not space but time. The main thrust of his argument concerned evolutionary processes, the patterns of urban growth and spatiality. He predicted that the "population problem" would stabilize only after two centuries, by which time chaotic urban growth would have spread to the ends of the earth. This frenzied expansion had to be scientifically managed and stabilized before it destroyed the earth's ecosystem. It should spread in one direction, forming a gradually widening linear city known as *dynapolis*. Settlement cells and community sectors could be endlessly generated along a dynamic, interconnected grid.

Doxiadis's terminology was steeped in the regimented, indeed militaristic, practices for controlling hostile populations. Grouping people in isolated, self-contained cells along road grids for monitoring and easy troop maneuvering was a classic strategy of urban warfare. *Dynapolis* could easily be construed as a design for defense and security coordination. But this subtext of an urban vision implemented entirely in the politically volatile territories of the Third World was hidden from view. Instead, *dynapolis* was portrayed as producing a harmonious, balanced city without boundaries. It was a world-encompassing urbanity described by Doxiadis as *ecumenopolis*.[57] It would cover the entire earth as an urban garden—a universal system of life—extending itself in a continuous interwoven network of nature, settlement cells, and transportation and communication flows.

Both ekistics as an overarching concept and the imagery of *ecumenopolis* corresponded to postwar visionary thinking along the worldwide holistic frontier. Although Doxiadis's urban theories are usually compared to the modernist ideology of CIAM and Le Corbusier, they are better placed among the futurists assembled on his Delos Symposia cruises. It was there that Buckminster Fuller, Canadian communications guru Marshall McLuhan, and Doxiadis met and tussled with visions of global networks and complex systems.

Fuller, the most venerated visionary of the age, fully embraced Doxiadis's ideas, was a frequent visitor to the Athens Technological Institute, and was president of the World Society for Ekistics from 1975 to 1977. Among his many trailblazing ideas, he proposed the end of urbanism as it was understood, and believed "the notion of self-contained permanent settlements obsolete." Instead, he outlined an urban future of "unsettlement" that consisted of a network of hypermobile nomadic forms operating across the world and connected by invisible radio links.[58] It approached the imagery of *ecumenopolis*. Covers of the *Ekistics* journal featured Fuller's Dymaxion Map of the entire surface of the earth as a shared habitat.

Both Fuller's and Doxiadis's concepts were in keeping with the postwar modernization regime's ambition to shape the totality of the human environment. Both men offered an optimistic, all-encompassing vision of global development through sophisticated systems design and technological ideals. For them, new towns were the seedbeds for this future. New settlements were a generative map of possibilities, an experimental terrain for realizing their exhilarating prophecies.

Constantinos Doxiadis's initial foray into actualizing this global ideal began in the late 1950s with his City of the Future project, funded by the Ford Foundation. It was a concept design for Islamabad, a new capital city in northeast Pakistan. Despite the foundation's suspicion that it was funding "a promotional piece for Doxiadis Associates,"[59] Islamabad morphed into a celebrity demonstration project. The Athens Technical Institute became a hotbed of international experts thrashing out ideas for the capital as "an entirely new man-made city to be created on virgin ground" that could be a model for the world.[60] The Ford Foundation sent promising young Pakistani planners directly to Athens for training in ekistics planning techniques. They became bona fide "Ekisticians,"[61] a new Pakistani technical elite ready to modernize their nation. Nonetheless, they were expected to be compliant, acting as junior collaborators that would carry out Doxiadis's wishes.

The decision to leave Karachi behind and create an entirely new capital city for Pakistan was made by General Ayub Khan after his bloodless takeover of government in 1958. The inevitable political squabbles began almost immediately. East Pakistan was less than satisfied with the selection of another site in West Pakistan for the federal capital. But the nation's army wanted the capital to be close to its headquarters in Rawalpindi. Others argued for maintaining the capital at Karachi. The rebuttal was that civil servants in Karachi were easy prey for corruption. Islamabad would be a morally healthier climate, away from the disorder and internal strife that deadlocked politics. Doxiadis argued that Karachi was never meant to be the capital. Rather than spend millions attempting to remake an already beleaguered place, he held out the vision of pristine space that would be invested with symbolic meaning, to say nothing of the desperately needed jobs that could be generated in construction. Pakistanis would learn new skills and modern building techniques.[62]

It was a difficult picture to resist, and a measure of Doxiadis's developmental and promotional genius. His relationship with local Pakistani officials was fractious, but his far-reaching schemes clearly resonated with the grandiose imagery of heroic nation building, progress, and modernization characterizing so many of the new countries in the Middle East and Africa where he experimented with his ideas. Utopian projects emerged at moments of historic liminality, especially as part of the cognitive mapping of new nations.[63] Thus, Doxiadis received the blessing of the Ford Foundation for his projects, precisely because they matched international development strategies so well. His new towns represented the distinctive modern transformation of what were perceived as backward, unstable societies while still respecting both environment and culture. Ayub Khan also worked with Doxiadis in order to legitimize his rule with the West.

In sum, Doxiadis's plans for Islamabad provided an outward appearance of national identity, economic prosperity, and political stability.[64] President Eisenhower's administration quickly embraced Ayub Khan as a key ally against the Soviet Union's and China's communist influence in the region as well as a foil to increasingly socialist India. A bounty of funding from the Ford Foundation and the World Bank flowed into the Islamabad project. British architects and engineers arrived under the Colombo Plan[65] to provide technical assistance and training.

The new Islamabad was paired with the adjacent town of Rawalpindi and a national park in what would become a *dynapolis*.[66] In the original concept, the two cities were understood as the same urban space along the Grand Trunk Road, which ran through the valley below the Margalla Hills. Rawalpindi would be the commercial and military hub, and Islamabad the administrative center. Doxiadis laid out the city on a triangular grid plan (fig. 3.8). It was intended to grow dynamically westward along a linear axis paralleling the hills, eventually becoming a huge metropolis of 3 million people. As with all new town projects, planning began with a detailed survey of the site. But the natural setting took second place to staging the capital in an "ideal landscape" that would heighten its architecture and symbolic grandeur.[67] The administrative center was placed atop three knolls for added monumentality. Broad lawns and some seven hundred thousand trees were newly planted to provide a garden atmosphere.

Every square foot of the future city was allocated for a specific purpose. The city's grid was divided into "community cells" one mile square. Fifteen distinct cell types were classified along a logarithmic scale. Each higher order of community cell encompassed all its lower-order communities. Each Community Class V of Islamabad was self-sustaining and self-supporting. At its heart were a community center, mosque, market, and health clinic. It was then subdivided into smaller community cells, or

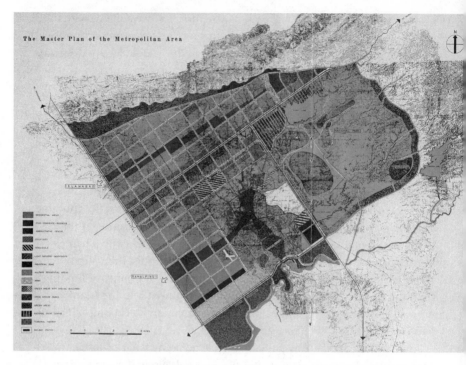

3.8. Islamabad, Pakistan, master plan of the metropolitan area (*DA Monthly Bulletin* [Doxiadis Associates], no. 64 [March 1964]). Courtesy of Constantinos A. Doxiadis Archives. © Constantinos and Emma Doxiadis Foundation.

Community Class IV, according to the income of its twelve thousand residents. The plan was based on strict social segregation worked out through data collection and then the spatial mapping of wage distribution and income categories. People were assigned a residence and lot size according to their income. High government officials lived in one community cell, lower-order bureaucrats in another, and workers in their own community sector. The assignment of dwellings took place according to the same strict social categories. Doxiadis planned every detail, down to the tiny bungalows built with local materials and the traditional courtyards for office sweepers (fig. 3.9). Each Class IV community was defined by a secondary school. It was then further divided into Community Class III for around three thousand people, each with a primary school, and then a further division into Class II, with children's playground. The same scaling took place for social services, recreation, and sport. This systematized pecking order of spaces would guarantee the correct functioning of urban society. Islamabad would then grow by adding on more community cells and extending the grid outward in one direction.

Inevitably, revisions were made to the City of the Future by the Pakistani commissions and subcommittees tangled up in the planning process. Rather than becoming an integrated *dynapolis*, Islamabad and Rawalpindi were functionally separated in one of many compromises. The Pakistani embrace of Western planning models was contradictory: at times overexaggerated, at times defiant. Ekistic schemes were manipulated within an embryonic national consciousness, and with ideas about cultural authenticity.

Doxiadis completed extensive surveys of vernacular built form during his projects in Iraq and Pakistan that provided the basis for his understanding of Islamic culture. His most important collaborator on Islamic style was noted Egyptian architect Hassan Fathy,[68] who pioneered the merging of new technologies with vernacular building traditions. Fathy was an ever-present figure at the Athens Technical Institute and at the Delos Symposia. While the two operated as coproducers of a modern design appropriate to the Islamic city, what constituted proper Islamic style was elusive. In a 1967 *Architectural Design* critique, the private buildings in Islamabad were described as exhibiting "the worst clichés of the postwar middle-eastern pop art."[69] The Pakistanis wanted romantic curved roads following the contours of the topography rather than a grid. Doxiadis nearly quit over the authorities' refusal to have the ceremonial National Avenue constructed as a two-tiered structure separating traffic and pedestrians.[70]

3.9. Islamabad, Pakistan, pedestrian street in a low- and middle-income neighborhood, date unknown. Courtesy of Constantinos A. Doxiadis Archives. © Constantinos and Emma Doxiadis Foundation.

To position the new capital as a hotbed of Islamic modernism, a host of celebrated Western architects were invited to imagine the buildings at its governmental heart (fig. 3.10). Modern functionalism was mediated with an orientalist motif. The Secretariat Buildings for some six thousand civil servants were designed by Italian architect Gio Ponti. Proposals by Danish architect Arne Jacobson and by American architect Louis Kahn for the Presidential Palace and Parliament House were summarily rejected by Ayub Khan. Instead, the stark white Presidential Palace was the work of American architect Edward Durell Stone. Gerard Brigden, head of the British Colombo Plan team, judged the Mogul pastiche of the government buildings "as remote from Muslim culture as a sauna bath." He bemoaned the difficulties of teaching Pakistani construction workers precast concrete techniques, and carped that Doxiadis's layout for the administrative center "showed just how irrational 'rationalist' planning can be." Then Brigden summarily changed it.[71] Modernization and urban planning were hybrid affairs yielding an assortment of compromises, mediated results, and good and bad judgment. Islamabad was the result of these fallouts and concessions.

The ekistic concept was next applied to the new town of Tema in Ghana, one of Doxiadis's most noteworthy demonstration projects. New nations such as Ghana in West Africa embarked on behemoth capital-intensive megaprojects financed by Western powers and the World Bank to develop natural resources and industrial production. The scale of these modernization schemes was enormous and required an entirely new urban infrastructure. On the face of it, the Tema endeavor was no different from territorial acquisition and the construction of resource towns such as Kitimat in western Canada for Alcan or Kwinana in the state of Western Australia at the giant oil port of Perth. The postwar years were driven by the search and extraction of vital resources for modernizing processes *everywhere*. Yet clearly the case of Africa was pierced with blatant racism and the gross exploitation of native populations. The conditions in the uranium town of Arlit and on the oil-rich Niger delta at Port Harcourt made these injustices obvious. The new African regimes were also caught in the global rivalries of the Cold War, manipulated by both sides in their jockeying for alliances and vital resources.

The Kainji Dam project on the Niger River exemplified these colossal development schemes accompanying independence in Africa. It was a cornerstone in the new nation of Nigeria's Six Year Development Plan to provide hydroelectric power for its step into modernization. Called "Nigeria's future version of the Tennessee Valley Authority,"[72] the project was financed

MINISTRIES—ADMINISTRATION BUILDINGS

PRESIDENTIAL PALACE
MOSQUE

PARLIAMENT
SUPREME COURT
ARMY–AIRFORCE HEADQUARTERS
NAVY REPRESENTATION

BUILDINGS OF NATIONAL IMPORTANCE

NATIONAL INSTITUTIONS

PERSPECTIVE VIEW OF THE ADMINISTRATIVE SECTOR OF ISLAMABAD

3.10. Perspective view of the administrative sector of Islamabad, Pakistan (*DA Monthly Bulletin*, no. 28 [May 1961]). Courtesy of Constantinos A. Doxiadis Archives. © Constantinos and Emma Doxiadis Foundation.

by the World Bank, Great Britain, the United States, Italy, and the Netherlands. Its blueprints included bridges and highways, a railway head, a marine port, an airport, and new town. Town building was used as a highly efficient strategy for targeted economic development. It required submerging
the old market town of Bussa in the lake formed by the dam's construction:
a literal drowning of the past and its replacement by a utopian future. The
local population of sixty thousand was forcibly resettled in "New Bussa",
where they were joined by construction workers, expatriate engineers, and
skilled technicians employed by the Niger Dams Authority. Although segregated construction camps were well established at the site, the government set about creating a model town as a symbol of the new Nigerian
society and the city's future as a tourist center.[73]

New Bussa was designed by the British husband and wife team of Jane
Drew and Maxwell Fry (fig. 3.11), who were among the most influential

3.11. Jane Drew and Maxwell Fry with a model of one of their many
buildings for the Gold Coast (now Cape Coast). Education Department,
Ghana, ca. 1950. © RIBA Library Photographs Collection.

and productive architects working in British territories both before and after the Second World War. In Iran, Jane Drew designed residential and office blocks for the Anglo-Iranian Oil Company at the Masjid-i-Suleiman refinery, and planned the new oil town of Gachsaran for the company. Both she and Fry had been involved with the plans for a garden city at Port Harcourt. During their long careers, they completed some seventeen projects in West Africa in an officially sanctioned modernist design, and a number of educational commissions there on modern architecture. They were members of the midcentury generation of architects and planners immersed in both avant-garde modernism and the garden city tradition, adapted to colonial territories through the lens of tropicalism.

For Fry and Drew (as well as for Otto Koenigsberger), the flawed policies of colonialism could be corrected by demonstrating sensitivity toward native cultures and society. Functional modernism could be sympathetic to indigenous architectural styles, native customs, and family life while still emphasizing hygiene, rational planning, and a transformative future. Interpreting modernist design within local climate as well as native topography and society made the colonies into a magical setting.[74] The result was often a curious mix of exoticist forms that became a normative midcentury model of improvement. Tropicalism could easily slip into British colonialist discourse of social betterment in its overseas territories based on empathy with native subjects, and the belief that Britain was uniquely capable of enhancing ethnocultural diversity. This attitude became even more pronounced during the postwar shift of former British territories into independent status.

New Bussa was clustered around the emir's palace, the mosque, and the site's administration building. Some eight thousand trees were planted in the arid climate to provide garden city landscaping. The construction camps were redesigned on a rational grid, with buildings alternating direction to avoid monotonous rows of concrete-block housing. The residential compounds were distributed according to traditional social status. Although they were imagined as multiracial, they quickly converted into ethnically segregated areas. But in a groundbreaking sign of progress, all the compounds were connected to the town's electrical, water, and waste systems. The government could claim that it had successfully spread material abundance and happiness. New Bussa pointed the way to the future.

Constantinos Doxiadis's plans for the new town of Tema can best be seen as a response to this colonial tradition of tropicalism on display at New Bussa. His approach offered an exhilarating alternative to what had become by the 1960s the rather moribund "garden city in the tropics" phantasm. Built under the regime of Kwame Nkrumah, Tema was situated on Ghana's Atlantic coast seventeen miles east of Accra. It was a megaproject that trumpeted the dreams of an independent nation. Modernization blended into the formation of national identity as a manifestation of empowerment and patriotism. The shift from British colonial dominion to independence also internationalized the project and set it within the geopolitics of the Cold War.

When African ventures of this scale began flirting with Soviet models of development, the World Bank, along with the Eisenhower and eventually the Kennedy administrations, stepped in and dangled financing carrots. Nkrumah traveled to the United States to court American lenders, eagerly seeking US assistance for his scheme.[75] The Ghanian Tema Development Corporation coordinated work contracts with British construction firms, foreign technical advisers, and a host of companies that included Russian and Italian interests, the British I.C.I. Chemical Company, and the American Kaiser Aluminum and Chemical Corporation. Doxiadis Associates was part of this constellation of Western interests. Doxiadis's scientific planning offered a depoliticized approach that fit hand and glove with Western companies investing heavily in Africa. It did not insult local leaders by association with the colonial past, nor did it stoke radical African politics that swung toward the communist Eastern Bloc.[76] Doxiadis positioned himself away from all these threats to profitability, and instead offered ekistics as a bridge between the developed and developing worlds.

The new town project had even further significance in that it was seen as a test case of democracy in the first independent country of sub-Saharan Africa. The story of Tema, according to the Ghana Development Ministry, was "the story of planners and of contractors, of simple folk and of social workers, of industrialists and politicians" working together for "peaceful and progressive purposes."[77] Tema was to be the nerve center of Ghana's economy. It was part of the mammoth Volta River Project to construct the Akosombo Dam and the "largest manmade lake in the world." Around it was planned an industrial complex and port for the production and exportation of bauxite and aluminum fueled by the dam's hydroelectric power. The largest manmade harbor in Africa was designed by the British to handle 2 million tons of cargo traffic each year. This massive infrastructure required the removal of the local population as well as the settlement of African work crews brought in for construction. They would find new lives in fifty-two planned new towns.

Initial plans for Tema in the early 1950s were drafted by Maxwell Fry and Otto Koenigsberger, and were grounded in the British garden city tradition with the tropical twist of native design. Doxiadis threw aside this colonial typology and instead offered his vision of Tema as a modern *dynapolis* for one hundred thousand people, with all the amenities of contemporary life from superhighways to libraries, movie theaters, and luxury hotels.[78] Eventually, the town was to include Ghana's racecourse and sports stadium. Tema would form one point in a triangle of dynamic growth and modernization with the cities of Accra and Akosombo. These extravagant dreamscapes and the investments that made them a reality were entirely in line with the vast scale of the Western modernization enterprise as it was applied to postcolonial Africa. The garden city concept was left behind in a radical shift to grandiose development schemes. Doxiadis's ekistics framework was perfect as their urban component.

Tema was publicized as a "daring, perhaps unique, experiment in urbanisation, a bold, but carefully planned attempt to unite people of differing backgrounds and beliefs in an integrated social, commercial and industrial community."[79] It was a phantasm applied to all of Doxiadis's new town projects. In keeping with his ekistics model, the town was broken into seven communities, each made up of compact neighborhood cells classified according to income level (fig. 3.12). Wealthier residents would live in high-rise buildings close to the town's civic and commercial heart with its splendid esplanade. Workers would dwell in tight low-rise blocks of wooden huts near the port, oil refinery, and aluminum smelter. Although social segregation was deeply embedded in the design, Doxiadis made the argument that it would nonetheless foster a limited but appropriate intermingling between social classes.[80] It was tailored to Western apprehensions about political radicalization in that it would allow an immediate leap into modernization without threatening social or sectarian turmoil. Social reform would take place gradually, and within an interconnected system of stable, carefully controlled communities.

Each community cell was to contain its own health clinic, schools, community center, and social services. Thousands of trees and shrubs were planted to create a garden city ambiance. Thirty-two different community design typologies created a mosaic of leafy footpaths, squares, parks, and playgrounds. Dozens of social clubs and recreational facilities took shape, from the YMCA and Tema Boys' Club to soccer fields and golf course. The ten thousand local villagers displaced by a modern hotel were rehoused in a "Tema New Village" across the lagoon from their ancestral home. At this village scale, the tropicalism of Maxwell Fry and Jane Drew was permitted to play a part. Their design predictably replicated traditional Ghanaian tribal society, with gender-separated compounds organized into four sec-

3.12. Low-income housing and a pedestrian path in Tema, Ghana, date unknown. Courtesy of Constantinos A. Doxiadis Archives. © Constantinos and Emma Doxiadis Foundation.

tors with the added benefit of maternity clinics, schools, and community centers. A new canoe beach and fishing harbor promised the continuation of the village's customary livelihood (fig. 3.13).

Tema's elaborate thoroughfares were designed to connect the town center, the port, and the industrial areas. Transportation engineers set out a web of multilevel highway interchanges as a sign of the town's modern economic vitality. But the result of this kind of grand-scale spatiality was isolated, congested neighborhoods where rents far exceeded the means of the work crews who had built the city. The immense shantytown of Ashiamang, the "problem child of the Tema Development Corporation,"[81] took root along the Accra-Tema highway. By the mid-1960s, some twelve thousand people lived there in abject poverty, surviving with only six public water taps.

But these slums were ignored, as were the people who lived in them. Instead, Tema made the landscape and urban spaces of independent Africa legible in Western epistemological terms as a standard of developmental modernism. The town was imbued with the invented imagery of an indigenous people made modern. Promotional material for Tema was filled with photos of smiling young Ghanaian workers building a heroic future. New towns didn't just construct model cities—they constructed model people for a democratic nation.

It was precisely this arresting imagery that captured the imaginations of the visionaries assembled around Constantinos Doxiadis and his ekistic movement. The vision of modern Tema could be reiterated across the global landscape in *ecumenopolis*: a universal city of man, the ideal urban settlement of the future. In a series of lectures entitled "Dystopia and Utopia," Doxiadis explained that the "present city—without reason, without dream—leads to dystopia and disaster." The only road forward was toward

3.13. Ghanaians plying their fishing business in Tema New Village across the lagoon from the Volta Aluminum Company storage dome, 1968. Photograph by Pamela Johnson. © The World Bank Photo Collection.

entopia, an ideal that would give practical form to the coming *ecumenopolis,* the universal city of man. Constructing it "relies on scientific, systematic knowledge of the situation as it is now in 1966." It is "firm and hard ground from which to take off": a "place that satisfies the dreamer and is accepted by the scientist," and can be realized through programs and plans.[82] It would be built from a rational hierarchy of community cells, each of which would correspond in miniature to urban utopia. Within such cells, "we can save man from the city that will crush him: it is within them that the community can have complete freedom for its expressions, and man for his life."[83] This new town concept was far different from the garden city with its ideal neighborhood unit. Doxiadis's theories marked a transition from earlier social reformist notions of new towns. His city of the future was conceived on an entirely different scale, tied to developmental modernism and a metanarrative of globalism, and driven almost entirely by social science techniques.

These were concepts also on the mind of French geographer Jean Gottmann as he grappled with the "megalopolis" he termed along the Eastern Seaboard of the United States, the Great Lakes Megalopolis around Detroit, and the Tokaido Megalopolis at Tokyo—all of which both he and Doxiadis had studied. Gottmann mused that perhaps a future *ecumenopolis* was not impossible. A schema of this kind "could be proposed of a transeuropean megalopolitan belt crossing the continent from the Mediterranean to the North Sea and the Irish Sea. . . . We could visualize an 'urbanized isthmus' from Rome (or Naples?) and Venice in the South to Amsterdam and Hamburg, jumping even over the Straits of Dover to include most of England. Such a formation may call for a new term, such as *megistopolis.*"[84] Planning for such an urban monster was a topic of unending discussion by urban futurists. And new towns were the testing grounds for this future.

FOUR

Cybernetic Cities

Tema, Islamabad, Korangi—Constantinos Doxiadis was seemingly everywhere, pitching his global ekistics brand. In the 1960s, he was a vocal participant at UN meetings on human settlement, where he admonished the gatherings that they had to protect humanity's future, before rampant urbanization and the exhaustion of critical resources left no other path but doom. It was in part his frustration with the UN's stumbling response to his arguments that prompted him to organize gatherings of his own.

The Delos Symposia began in the summer of 1963, when Doxiadis welcomed a host of luminaries aboard a cruise ship in the Aegean Sea to discuss urgent problems posed by rapid population growth and urbanization. Underwritten by the Ford Foundation, the symposia were held twelve times (the last in 1975) and assumed both celebrity status and the mantle of futuristic thinking. They also were the zenith of Doxiadis's remarkable career. Among the participants: architects Siegfried Giedion and Jaqueline Tyrwhitt of CIAM, as well as Edmund Bacon (director of Philadelphia's Planning Commission), historian Arnold Toynbee, futurists Buckminster Fuller and Marshall McLuhan, and anthropologist Margaret Mead. Jean Gottmann of megalopolis fame and distinguished town planner Colin Buchanan of *Traffic in Towns* joined the A-list cruises, as did designer of British new towns Richard Llewelyn-Davies. Robert Matthew (in charge of numerous British new towns) and British new town enthusiast, economist, and journalist Barbara Ward were also on the invitation list. Together the Delos participants spent mornings in vigorous debate about topics such as the "crisis of human settlements," the "city of the future," and the "practice of regional planning" (fig. 4.1), and then the afternoons relaxing and socializing at dinners in Nafplion, on Mykonos and Hydra.[1]

Doxiadis's guests were a galaxy of visionary superstars known for hyp-

4.1. Session aboard a cruise ship during the Delos Symposium of 1964.
From left: Constantinos A. Doxiadis, Lord Richard Llewelyn-Davies, Sir Robert
Matthew, and Margaret Mead (*with back to camera*). Courtesy of Constantinos
A. Doxiadis Archives. © Constantinos and Emma Doxiadis Foundation.

notic visions of the future that echoed the exhilaration and anxieties of
the 1960s. Their ideas aligned with those of Walter Rostow in his book
The Stages of Economic Growth (1962) and the Western model of progress.
Nations, especially poor nations, could achieve prosperity through modernization. Doxiadis and his circle of high-profile intellectuals ruminating
over the evolution of human settlements shared a faith in the power of science and technology, in democracy and modernization, and in global cooperation to create the world of tomorrow they imagined. They also shared
a cosmopolitan "radical idealism," as Buckminster Fuller called it. Atomic
power, computers, and telecommunications augured a future of untold
possibilities. Cities would be in balance with nature. Environmentalism
and the zest for technology were fused together in their thinking. Social
transformation would flow automatically from land-use planning and allow future generations to live in peace and harmony.

Fuller and media theorist Marshall McLuhan met for the first time at
the 1963 Delos Symposium. McLuhan had just published *The Gutenberg
Galaxy* (1962), in which he postulated that communications technologies

and electronic media constituted a new consciousness. The electronic age would collapse spatial and temporal boundaries and unify the globe into a single, harmonious system. Both men originated the idea of the global village, and understood the world as a single place wired by electronics and the instantaneous flow of information. We would all be connected—One World, in the mantra of the age. Operating within this refined prophetic atmosphere, Doxiadis himself imagined a City of the Future as a biomorphic organism spreading out across the earth until it finally became a single planetary *ecumenopolis* linked in a network of communications.

Fuller, along with the other idealists and nonconformists assembled at the Delos Symposia, moved 1960s thinking onto a planetary plane. As part of this global awareness, they preached unflappable enthusiasm for the promises of the Space Age. Space—whether on Earth or in the heavens—was the canvas for the imagination, and the universe was the ultimate environmental home. Satellite imagery of the planet hurtling through the galaxy, the technological wizardry to launch humankind into the cosmos became cultural fixations. It was believed that technological know-how could surmount any challenge.

Early space missions were featured on the cover of Stewart Brand's great countercultural manifesto, *The Whole Earth Catalog.* Fuller coined the term *Spaceship Earth.* The Earth itself was a space settlement: "We are in space and have never been anywhere else," he declared. "We are already a space colony."[2] Fuller's single-mindedness about freeing terrestrial dependence took the form of mammoth floating cities. "Cloud Nine" was the name for his proposed airborne habitats created from giant geodesic spheres, which levitated by heating their air above the ambient temperature.

It was an upbeat, audacious confidence dimmed by the dark clouds of the Cold War and the space race. The shadow of dystopia lay just behind the optimistic prophecies. People were simultaneously optimistic and terrified, and the vision of new towns lay somewhere between the extremes of hope and dread. The world lived under the fearsome threat of nuclear Armageddon. The language of "survival" and "saving the planet" coursed through 1960s thinking. There was growing alarm about the population explosion, food shortages, resource depletion, radioactive fallout, air and water pollution, environmental degradation.

The earth was at the mercy of the evil consequences of science and technology. Colonizing space was a tactic for survival once Earth was devastated by this litany of impending disasters. The wasteful overconsumption of resources in the developed world could not be repeated as the Third World embraced modernization. Without planning, the global village could degenerate into an abyss. The planner had to guarantee human existence. To that end, at the conclusion of the 1963 voyage, the Delos group solemnly signed the Declaration of Delos—written by Jaqueline Tyrwhitt and Barbara Ward and modeled after CIAM's Athens Charter—in a candlelit ceremony. The charter declared, "We are citizens of a worldwide city, threatened by its own torrential expansion and that at this level our concern and commitment is for man himself." The urban crisis would outstrip all other problems other than nuclear war. But "man can act to meet this new crisis . . . modern technology permits the mobilisation of material means on a wholly new scale."[3] Without rational and dynamic planning, there would be unforgivable waste, undermining of civic order, and destruction of civilization. The Declaration of Delos, then, was a call to save the earth.

Cybernetics and systems analysis were the answer to that call. They were international blockbusters that crossed national cultures and political boundaries. Cybernetics has been defined in a variety of ways, but most simply, it is the discovery and study of systems, whether these are mechanical, physical, or social systems. During the 1960s, it found its intellectual expression in communication and feedback mechanisms, theories of computation and applied mathematics, and computer modeling. These tools allow systems to be analyzed, controlled, and automated. This chapter examines the impact of cybernetics on urban planning and the utopian ideal of new towns, and how it was incorporated into these projects on both sides of the Cold War divide: in the United States as well as the Soviet Union and the Eastern Bloc countries of Poland and the German Democratic Republic.

The rise of cybernetics was also fundamental to the Space Age. The computerized systems theorized in cybernetic science were behind the complex machinery required to launch humans into the cosmos.[4] Spacecraft were synonymous with the onboard computers that monitored data and managed their complex guidance, command, and control systems. The result was that cybernetics exuded a hallucinating fantasy of computers and robotics, astronauts and cosmonauts, and futuristic cities that functioned like spaceships.

In sixties scientific thinking, it was imagined that the technological breakthroughs fulfilling the promises of space travel could also be used to save Spaceship Earth. On both sides of the Cold War divide, cybernetics and systems analysis were seized as a universal formula for activating social and economic transformation. Urban utopias could be fashioned with the aid of information theory and the number-crunching computer. New towns became the testing ground for land-use and transportation models, for an ideal city conjured from systemwide data analysis and control. The tradition of planning surveys begun by Patrick Geddes already advocated the use of quantifiable data that could be measured, classified, and mapped. These gave a snapshot of settlement patterns. With the cybernetic revolution, planning entered a new stage. The discourse on urban life was rewritten in the language of mathematics and computers. This kind of diagnostic capacity opened the prospect of rationally arranging not only cities but large-scale metropolitan areas and whole regions. It marked a dramatic swing away from new towns as old-style garden cities. Even the ideal of neighborhood units as the bread and butter of community life took a backseat to an aggressive engagement with new towns set out in a cybernetic matrix.

All these techniques were deeply entwined with the Cold War and the military-industrial complex, and heightened the prestige of American think tanks and schools of planning. They were exported in a wave of enthusiasm to western Europe, the Soviet Union, and the communist East, and from there across the world. In short, cybernetics became a new channel for progress and idealism in the 1960s. The result was a striking enlargement in the scale and complexity of new towns, and an intense debate about social improvement and its expression in urban form.

Atomic Warfare and the American Research Regime

The threat of atomic destruction gripped the American psyche in the early postwar years. The nightmarish image of smoldering ruins of cities targeted for annihilation is readily traceable in the nation's films and television shows, novels, newspapers, and pseudoscientific reports of the 1950s and 1960s.[5] Fears of nuclear apocalypse reinforced the argument for population dispersal that had been part of the intrinsic connection between wartime and new towns since the early twentieth century. They also triggered deeper collaboration between military strategists, urban reformers, and government officials. The result was an intimate link between the emergent Cold War security state and a new class of urban experts. Innovations originally designed for military defense became, according to historian Jennifer Light, "the weapons of choice in battles to solve urban problems and maintain security in the nation's cities."[6]

For example, a popular pocket book entitled *The Atomic Age Opens*, published shortly after the August 1945 bombings of Hiroshima and Nagasaki, predicted "the end of urban civilization as we know it . . . cities of the future may have to burrow downward instead of upward; dispersion, rather concentration." But if the highly enriched uranium of the atomic bomb was a terrible destroyer, it also "can be harnessed to produce the Utopia that men have dreamed of." It "will cause our cities to spread out all over the countryside, even to territories barren and now uninhabitable. . . . Humanity might well become a single, uniform community, sharing as neighbors the whole face of the world," according to a physicist at the California Institute of Technology.[7] It was imagery similar to Constantinos Doxiadis's planetary *ecumenopolis* and Jean Gottmann's megalopolis. Nuclear physicist Edward Teller, known colloquially as the father of the hydrogen bomb, wrote in 1946 that the only solution to the threat of the atomic bomb was to disperse populations from the deathtraps of congested cities and settle them in small towns arranged either in clusters or along a grid.[8]

American urban planner Tracy Augur, one of the most vocal and well-known evangelists of dispersal, repeated that space was the best military defense against the bomb, and invoked the vision of the British garden city as the most efficient strategy for decentralization. His prewar career, in fact, had been devoted to advancing regional planning and the garden city concept. He worked with Clarence Stein and the Regional Planning Association of America as well as with the Tennessee Valley Authority, where he was planning director for the TVA new town of Norris and consulted on the plans for Oak Ridge. For Augur, dispersing the endangered multitudes into satellite towns of fifty thousand residents each, separated by open country, would not only protect the nation but "secure a much finer environment for home and work than the average citizen now dares dream of."[9]

Both Clarence Stein and Lewis Mumford remained active campaigners for the regional city cause, testifying regularly before the US Congress that national security offered the opportunity for a federally supported new town program. They viewed Washington's anxieties over the atomic bomb and the Korean War as a chance to secure federal support for their long-standing regionalist project. Stein maintained that the dispersal of industry to low-density communities surrounded by open country was the only realistic protection against atomic attack.

University of Chicago sociologist William Ogburn joined the chorus in his article "Sociology and the Atom," in which he argued that urban civilization would be better off with well-planned smaller cities and towns. "We could have better health, fewer accidents, wider streets for automobiles, more parking places for automobiles, landing places for helicopters, more sunlight, space for gardens, more parks, less smoke, more comfortable homes, efficient places of work, and, in general, more beauty," he maintained. It was the fantasy of new town utopia, and in fact nearly matched the quixotic drawings of Milton Keynes, England, being produced at the same moment (see chapter 5). But the exact physical designs of these places did not matter for Ogburn as much as transportation: "any redistribution of cities means a redistribution of transportation." In a signal of what would fast become the priority for regional scientists, he insisted that "much research on transportation and the location of the new cities is needed."[10]

Not unlike the case of the British new town movement, war and military defense gave the ideal of planned decentralization official legitimacy and urgency. In both the United States and Great Britain, it enabled state interests and powerful national organizations to take up the cause of urban and regional reform. In the United States, however, unnerving anxiety about atomic destruction and scattering a panicked population out to the countryside yielded few immediate results. Rather than embarking on a new town movement as had been done in Great Britain, Americans got in their cars by the hundreds of thousands and drove out to suburbia. In the 1960s, it was a land of identical tract homes ("little boxes," in a protest song of 1962) spread out into the distance without rhyme or reason other than the profits made by real estate development. For urban reformers, the horrifying vision of sprawl only added to the litany of anxieties, and to the urban crisis that was weakening the nation.

Cybernetics and the Systems Revolution

There was, however, another solution. In 1948, mathematician Norbert Wiener (who worked for the Office of Strategic Services, forerunner of the CIA) published his classic book *Cybernetics or Control and Communication in the Animal and the Machine*. The field of cybernetics had taken off during the Second World War in connection with automated anti-aircraft radar and artillery systems capable of calculating input and producing feedback to accurately predict combat targets. Wiener defined cybernetics as the study of systems and how they communicate through messages and networks and interact with other systems. Everything meaningful could be gleaned from the pattern and flow of information. His worldview was comprehensive: machines, human brains, cities, and societies were all connected *systems*.

Cybernetics and information theory were instant sensations, and helped shape the conversion of the US wartime economy to civilian uses. Cybernetics was heralded as the great leap forward, a metascience that could be applied to a multiplicity of fields. One of the participants in the legendary Macy Conferences on the cybernetics revolution had the sense that it "was destined to open new vistas on everything human and to help solve many of the disturbing open problems concerning man and humanity."[11]

In *The Human Use of Human Beings* (1950), Wiener argued that the cybernetic revolution would trigger a revolution in thinking. It now was possible to create new cities and urban systems that could "think" and operate intelligently through information streams. To demonstrate this, Wiener drafted an impromptu civil defense plan for American cities for alleviating the dense concentration of people and industry that he believed made them both vulnerable to atomic attack. A brief version of the plan appeared in *Life* magazine as part of an interview in December 1950. Wiener proposed a visionary scheme of new towns on the metropolitan periphery, with each town surrounded by a "life belt" of feeder roads, highways, and rail lines that would act as a communications grid and escape route. The new towns would arise at the greatest stress points along the grid. In Wiener's estimation, cities were nodes "where railroads, telephone and telegraph centers come together, where ideas, information and goods can be exchanged."[12] The purpose of civil defense was to protect these information and communications networks.

In his 1964 book *God and Golem*, Wiener went further in suggesting that these systems could be replicated. The complex components that make up urban life could be made to operate as a machine and then analyzed, manipulated, and reproduced—the dream of unifying knowledge and grasping the totality of the urban experience from one optic. With the computer's capacity to integrate and process data about the behavior and transactional patterns of the urban system, planners could solve problems, forecast outcomes, and control the future of cities.

Ebullient from the power of the cybernetic dream, scientists, engineers, and planners leaped into the urban vortex. Input-output analysis and mathematical models were developed that simulated the behavior of urban systems. Once the variables and data were loaded into a computer, planners could predict the patterns and flows, the social and economic consequences of policy decisions. The computer was not just a tool but an environmental model. Mathematical equations and graphic depictions of interactive urban components structured planners' thinking about cities and created the conceptual framework for urban policy. The city as machine could be designed and built in a predictable fashion and made to perform in whatever way it was programmed. Cybernetics defined a new planning grammar and offered the possibility that planners could create an ideal urban type—an intelligent city or an early version of a "smart city" that could then be replicated across the landscape. The production and circulation of this system's discourse were vital to the legitimacy and ascendance of the planning profession in the 1960s. In sum, cybernetics provided a mechanism to concretize utopian aspiration through unprecedented scientific tools and technologies.

Urban planners newly instilled with systems thinking found kindred spirits in the military strategists at the RAND Corporation, the new think tank emerging from the Douglas Aircraft Company and the United States Army Air Forces in the early years of the Cold War era. Along with engineers, mathematicians, and economists, RAND employed a range of social science professionals, and systems analysis was at the core of their research. The aim was to develop a science of war and an entire system of national defense based on cutting-edge research in both the physical and the social sciences. This inquiry was taking place when the United States discovered that the Soviet Union had detonated its first atomic bomb. This and the 1957 launch of the Soviet Sputnik I, the first manmade satellite, which circled Earth every ninety-five minutes, were a titanic shock. Suddenly, the US military realized that computers were essential to national defense.

In response, RAND's researchers aggressively pursued the development of computer tools such as linear and dynamic programming, systems simulation, game theory, and artificial intelligence.[13] The rise of expert technocracies such as those at RAND (in the United States as well as in the Soviet Union) was a response to the space race and the harnessing of technological innovation for state purposes. RAND became synonymous with the systems approach and the blazingly fast analysis enabled by computer modeling, all meant to endow the United States with decisive military advantage.

Although RAND continued as a defense contractor, with its break from Douglas Aircraft and the Air Force in 1948, warfare ceased to be the organization's guiding mission. RAND scientists began switching their expertise to domestic urban issues and offering systems analysis "as a scientific method of planning and decision-making" to governments besieged by problems.[14] It was an untapped market that could ensure RAND's future. After all, to its way of thinking, the escalating urban crisis was a national security emergency. The readiness to defend the country against attack appeared to be undermined by the deplorable state of American cities. The initial forays into urban planning focused on transportation and land use. Systems analysis seemed ready-made for simulating patterns of urban and regional development around the circulation and flow of vehicles.

As a result, RAND became a hotbed of enthusiasm for the quantitative revolution in geography and the emergence of American spatial science. It enjoyed its heyday precisely in the Cold War years of the late 1950s and

1960s, and reinforced formal mathematical models, hard data, and the use of cutting-edge technology in these disciplines. Economic geographer Ira S. Lowry was one of the earliest proponents of computerized simulation models for enabling planners to evaluate and forecast land-use patterns. Lowry joined the staff of RAND in 1963 as part of its research program in urban transportation, and produced a wide-ranging series of reports about land-use planning, housing, and population issues. The federally funded Community Renewal Program in Pittsburgh offered him the first real-world chance to experiment with his techniques. Completed in 1962–64, the landmark Lowry Model was the first large-scale regional simulation model to become operational. It was a breakthrough in location theory and quantitative growth models.

In explaining "A Model of Metropolis," as his report was called, Lowry sided with "social physicists," who studied empirical laws of social interaction and mass behavior. His model built the metropolis from scratch and attempted to simulate "complex physical, biological, and social systems"[15] in order to understand the impact of public spending decisions on population dispersion and the retail trade in the Pittsburgh metropolitan area. It was a "set of simultaneous equations whose solution represented 'equilibrium' in the pattern of land use and in the distribution of employment and population."[16] In other words, the model invented the city and the urban system in a steady-state mode. It was a sensational tool for policy makers and public officials grappling with governance and the quagmire of urban politics. As a turning point in computer-based location analysis, Lowry's model instantly attracted copycats that steadily upgraded and enhanced simulation performance with submodels and sophisticated matrix formulations.

As it had with the Community Renewal Program in Pittsburgh, the RAND Corporation worked hand in hand with the federal government to develop systems analysis as a tool in urban planning. At the first Conference on Information Systems and Programs for Urban Planning held in Los Angeles in 1963, RAND analyst Edward Hearle noted the abrupt change in semantics brought on by cybernetics. Where the vocabulary of planning officials had only recently been ruled by words such as "long-range, development, growth, renewal," now the most oft-repeated words were "system, model, integration, simulation, interface."[17] Town planning had shifted to computable sets of equations, abstract logic diagrams, and flow charts. The equations were themselves treated as aesthetic objects equivalent to the spaces they were imagined as representing. The entire interwoven fabric of urban life was reduced to a set of functional components rendered in diagrams as cells, and interrelated by flows, feedback loops, and circular causality to become coherent entities.

The logic diagrams were a visual semiotics for the new science. They decoded an elaborate mathematical rhetoric and were endowed with almost magical qualities. This systems way of knowing interpreted the real in order to transcend it. It functioned as the utopian leap outside, an alternative and even subversive perspective on urban life.[18] An exaggerated formalism took root. It was a brutal rupture from the tradition of physical planning as *art* that the mid-twentieth-century generation had produced, or even from the survey and mapping techniques that had been the state of the art during the reconstruction years. The physical character of the city and its layout had less authority. By the early 1960s, this older discourse was belittled as dated and inadequate to the urban crisis at hand. The planner's customary toolbox—deeply embedded in spatial illustration—lost legitimacy or was itself transformed.

Perhaps the most cogent illustration of the shift in visual perception was the work of mathematician and urban theorist Christopher Alexander. His hallmark 1965 essay "A City Is Not a Tree" condemned outright "artificial cities" and pointed to some of the most celebrated garden cities and new towns as the worst specimens: Greenbelt, Maryland, and the new town of Columbia, Maryland; Abercrombie and Forshaw's *Greater London Plan*; Le Corbusier's Chandigarh in India along with Lúcio Costa's Brasilia in Brazil; and Paolo Soleri's Mesa City project in Arizona. One by one, they came under Alexander's disparaging scrutiny. He described their design plans as "trees" that tragically disconnected the cities' various functions and led to their failure as urban places. The structural simplicity of trees "is like the compulsive desire for neatness and order." Instead, he offered a vision of the city as a semi-lattice with overlapping subsets in a complex and comprehensive urban system.[19]

Alexander used mathematical criteria to organize data into sets, and then he combined the sets into an overall structure. Urban design became mathematical and discoverable in a patterned language with the help of complex information processing and the computer. Ultimately, Alexander had more influence over the advancement of computer science than architecture. His reasoning was a powerful stimulus to program-language design and software engineering in the 1960s. Most important, it was symptomatic of the systems revolution that a creative thinker such as Alexander could see the parallels between architecture, the city, and systems theory and visualize them in computerized schematics. He believed that this approach would bring urban design closest to the spontaneity and rich complexity of "natural cities" that were the ideal.

Authority shifted away from architects and toward Young Turk experts in the new spatial sciences and, ultimately, in mathematics and computer programming—or in the case of Alexander, experts in all three areas. They performed cybernetics and systems analysis as a 1960s avant-garde techno-futurism shaped in alliance with state technocratic authority and American research institutions. Planning peppered with analogies from cybernetics and systems theory reflected the preoccupation with integrated teamwork during these years. Together, this regime of expertise shared a deep confidence in science and technology and in the coherent design of the urban realm.

The creed of modernization produced a rationalizing imagination and an idealism that found its praxis in urban systems. In his remarks at the Conference on Information Systems and Programs for Urban Planning in 1963 in Los Angeles, Charles Zwick, who worked for RAND and served as director of the Office of Management and Budget in the Johnson administration, tried to smooth over the unfair criticism by young cybernetic wizards that earlier planning was "Pre-Darwinist." Nonetheless, he carried the flag for systems analysis and the new systems culture. In formal models, he noted, "we are concerned with abstracting the essential interrelationships of highly complex phenomena in the hope that if we confront the model with a specific environment or context, the model will behave in a manner that approximates the process of urban change in the real world."[20] The best policy scenarios could then be offered to public officials. This formalism and modeling immediately deflected attention from the urban environment as messy lived experience. Planning became a matter of technical expertise, computer simulation, the design of metropolitan areas for optimization and profit maximization. It produced an all-encompassing vision of standardized intelligent cities built from scratch and capable of being automatically plugged into any metropolitan landscape.

The Rational Imagination: Urban Systems and Superhighways

Suitably, Zwick used transportation planning as his example. As Americans purchased cars in record numbers and fled to suburbia, the entire nature of land-use planning changed. Millions were gleefully taking to the road in the course of their daily lives. There was a general outcry against snarled traffic on antiquated roads, the congestion and pollution, the chaotic battle between pedestrians and motorists. But the automobile was the future.

It was a masterstroke of mobility, and a new spatial order was required to accommodate it. After the Federal Highway Act of 1956, a 42,500-mile interstate network of highways was put in place across the United States. American progress was dramatized by this modern highway system. It was a gigantic public works program that radically altered the country's geography. It made the national territory coherent and efficient, and wrapped the country's metropolitan areas in a web of highways. The idea of a single new town, or even a group of new towns, as a cloistered, picture-perfect world now made no sense. Instead, the ideal was fixed on highway systems in an open and fluid regional nexus.

Already in 1957, the impact of highway construction on cities and metropolitan regions was being examined at a conference held at the new Connecticut General Life Insurance headquarters in Bloomfield. Entitled "The New Highways: Challenge to the Metropolitan Region," it assembled the *beau monde* of American urban thinkers. Lewis Mumford attended, as did Victor Gruen, Carl Feiss, Albert Mayer, and James Rouse. The conference proceedings were published as the 1959 book *Cities in the Motor Age*. The subtitle for the conference was "How Can We Increase the Efficiency and Livability of Our Cities through the National Highway Program?" No doubt it was the highway program that led to the suburban sprawl enveloping every city in America. But for the conference participants, this was hardly a sign of either efficiency or livability. They were "in full agreement that expressways and other public works should be constructed in accordance with a master plan for the metropolitan region."[21] The highway program would be an opportunity to redesign the urban environment—and demonstrating a comprehensive transportation and land-use strategy for an entire metropolitan region was a prerequisite for the funds dangled by the 1962 Federal Highway Act. Such a strategy would build new towns from a regional systems approach.

The US highway program was a remarkable example of the bond between state-driven modernization and scientific research, and its development coincided with the revolution in geography and regional planning as professional disciplines. Regional science was coproduced in parallel with state projects and incorporated into the practices of territorial rationalization.[22] Economist Walter Isard was the key champion of spatial science in the United States in the 1950s and 1960s. He merged quantitative geography with German location theory and hardheaded neoclassical economics to create the academic discipline of regional science. Publication of his *Location and Space Economy: A General Theory Relating to Industrial Location, Market Areas, Land Use, Trade and Urban Structure* (1956) was a defining moment in regional analysis. The research was funded by the Social Science Research Council (one of the premier promoters of modernization theory) and the Resources for the Future think tank (initially supported by the Ford Foundation and with strong ties to the Truman administration). Equally important as Isard's pioneering research was his founding of the Regional Science Association in 1954. His research at the University of Pennsylvania and then his launch of the *Journal of Regional Science* in 1958 established the university as a center of scientific inquiry and a petri dish for regional planning aspirants. Researchers at Penn were behind the famous Penn-Jersey Land Use Model that was a masterstroke in urban systems analysis derived from computer simulation. It laid out the entire Philadelphia metropolitan economic system as a network of activities, nodes and flows, and then forecast future land use. Tellingly, it was funded directly by the State of Pennsylvania and its Bureau of Public Roads.

All sorts of metropolitan area transportation studies and models materialized in university laboratories. They were computer simulations of the way cities functioned,[23] and made metropolitan regions into open matrices for the flow of people, goods, and services. The Detroit Metropolitan Area Study (1955) and the Chicago Area Transportation Study (1956) were groundbreaking achievements in analyzing urban transportation as a system and coding data into a computer to forecast logical land-use patterns. Studies such as Lowdon Wingo's *Transportation and Urban Land* (1961, funded by Resources for the Future) and William Alonso's *Location and Land Use* (1965, a product of Penn research) built an entire corpus of quantitative research, systems analysis, and modeling in regional science. Technocratic expertise, modeling practices, and the use of the computer, the very meaning and discourse of regional science, were constructed in conjunction with government policy, which essentially acted as a political agent for rationalizing and modernizing national geography.

These glory days of quantitative geography, regional science, and transportation modeling produced an explosion of utopian fantasies. The American love affair with car culture and the highway was, after all, unprecedented in the 1960s.[24] Cars brought liberation and progress. Futuristic superhighways and jet age automobiles mesmerized the American public. The Ford Motor Company and General Motors pavilions at the 1964 World's Fair in New York were spellbinding glimpses into a motorized future. At Ford's Magic Skyway exhibit, fairgoers stepped into a sleek Mustang convertible for a drive on the "turnpike of tomorrow." General Motors brought back its spectacularly successful Futurama exhibit from the 1939 World's Fair in an updated version. A model city of glimmering skyscrapers was crisscrossed by computer-controlled superhighways. Giant roadbuilders hacked through the most inhospitable environments and churned out highway systems spanning the continents. The exhibit's message was clear: Highways bring progress, prosperity, and a future of boundless promise. The world they create is a world of movement, flow, connectivity.

The imagery of the Futurama exhibit was reiterated in popular magazines and comics, such as Arthur Radebaugh's syndicated Sunday comic strip *Closer Than We Think*. Radebaugh worked for the Detroit automobile giants and auto industry magazines, devising dream cars right out of *The Jetsons*, the phenomenally successful Space-Age television cartoon from the 1960s. His comic strip reached an audience of 19 million that was captivated by his depiction of the future. In 1965, *National Geographic* produced a two-page, full-color spread of a radiant "City of Tomorrow."[25] It was a visual banquet of urban glamour. Gleaming city lights, sleek highways, rocket-cars—a romanticized tomorrow derived from car culture and the Space Age. Cities and highways would conquer the world and colonize the galaxy.

This utopian virtuosity was on official display in the US Department of Housing and Urban Development (HUD) report entitled *Tomorrow's Transportation: New Systems for the Urban Future* (1968).[26] A massive scientific study undertaken by university researchers and transportation engineers, it reviewed some three hundred feasible projects for the future. The object was to use the "transportation system to enhance and improve the total city system."[27] However, the findings focused less on transportation technology breakthroughs than on implementing systems analysis and computer simulation to better plan cities and metropolitan regions. These tools would open the door to the dreamlands displayed at the World's Fair and in the pages of *National Geographic*. The report featured sketches of "personal rapid transit systems" connecting cities and suburbs. Computer-controlled traffic systems monitor the highways. Automated monorails, pod cars, and people movers glide between glittering high-rise towers. Stylish transit link stations welcome passengers. All this required a regional perspective that imagined highways and transportation networks spreading like silken webs out from central cities to suburban growth centers and satellite towns. The viewpoint of highway design was aerial and wide-angled.

The dreamscapes of American avant-garde architects such as Alan Boutwell, Michael Graves, and Peter Eisenman reproduced this officially sanctioned imagery. Their drawings featured continuous linear cities stretched out along highway corridors and fantasy megastructural road towns. Architecture itself became mobile and fluid, capable of imitating highway linear-

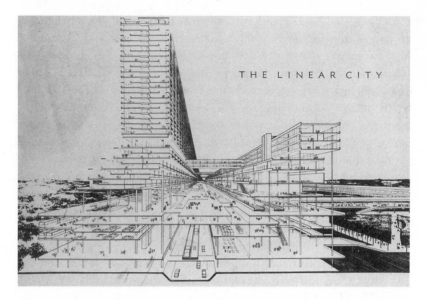

4.2. Peter Eisenman and Michael Graves's Jersey Corridor Project, 1964–65. Published in *Life* magazine, December 24, 1965. © Eisenman Architects.

ity in thrilling science fiction depictions of the future. Alan Boutwell's *Continuous City—USA* (1969) spanned the North American Continent from New York to San Francisco along the exact route of Interstate 80, which was constructed during the 1960s. Architects Michael Graves and Peter Eisenman designed a linear city along a proposed I-95 highway extension to connect New Brunswick and Trenton in New Jersey (fig. 4.2). Called the Jersey Corridor Project (1964–66), it was published in a special issue of *Life*, "The U.S. City—Its Greatness Is at Stake."[28] Graves and Eisenman imagined an endless stream of megastrutures; these would eventually connect with similar linear extensions along the Eastern Seaboard from Maine to Miami to form a continuous *system*. Industry would be on one side along the highway strip, while an "endless downtown" of homes, shops, and services would be on the other. Alongside it in the pages of *Life* was a marvelscape devised by architects at the University of Illinois for colossal platform towns situated along highway spines radiating from Chicago. In a futuristic vision by transportation experts at the Cornell Aeronautical Laboratory, automated "urbmobiles" and "century cruisers" streak along expressways designed for traffic moving at one hundred miles per hour. The visions of avant-garde architectural circles and university scientists were merging in a dazzling performance of utopian highway theater.

Nowhere did this highway reverie come more alive than in the state of California, and nowhere were more official new town projects constructed than in Southern California. Of twenty-one official new communities in California listed in 1969 by the US Department of Housing and Urban Development, fifteen were located in the southern part of the state, mainly in Los Angeles and Orange Counties.[29] At the same time and as part of the Federal Highway Acts, California set in motion the construction of 12,500 miles of freeways (fig. 4.3). Every new town was bisected by, bordered by, or adjacent to a major freeway. Freeway design had become an art form. Cloverleaf and spaghetti interchanges, five-level freeway matrices with fly-overs, sweeping bypasses, curved entrance and exit ramps, reiterated the fantasies of the future and with them the lyrical aerodynamic language of architecture and design.

They were the World's Fair's Magic Skyway come alive. The visual language of systems analysis was both mathematical and aesthetically expressionist. The streamlined, curvilinear appeal of 1960s modernism was a semiotics of systems modeling. Until the freeway revolts of the 1970s, these highways were considered refined engineering marvels—a design performance that was racy and seductive. Drivers could relish the thrill of aerodynamic order and geometry, and of speed. The highway engineer had become an urban stylist.

The colossal 130-square-mile new community of Irvine Ranch in Orange County south of Los Angeles was the most compelling case of urban development within a matrix of California freeways (fig. 4.4). Started in the early 1960s and with a projected population of nearly five hundred thousand, it was the largest of the US new town projects and was subsidized by the federal government. The Irvine Company that owned and developed the property had close ties to the Eisenhower administration (in 1960, the company's president had been Eisenhower's secretary of the Navy). The master plan was created by legendary Los Angeles architect William Pereira, who was known for futuristic megastructural buildings inspired by his love of science fiction. He was steeped in Southern California's variation on Space-Age culture. Pereira's signature style was on display in the Theme Building for the Los Angeles Airport, the Disneyland Hotel in Anaheim, and San Francisco's Trans-America Building. He also became one of the most prolific of the new town builders.

Irvine was the handiwork of a new breed in California: the master planner. Pereira was the epitome of the rugged individualist, the visionary bent on propelling the American city into the future—a persona that landed him on the cover of *Time* in 1963. Irvine was an unprecedented opportunity to

4.3. Santa Monica Freeway and Harbor Freeway Interchange, downtown Los Angeles, 1962. Photograph by Dave Packwood. © Automobile Club of Southern California Archives.

design utopia from scratch: it was "a godlike game."[30] Pereira's master plan spread the town's functions out as "activity corridors" and "residential villages" along the Santa Ana, San Diego, and Newport Freeways and their feeder roads. Studies of the regional transportation network and computer modeling of trip loads based on land-use scenarios determined urban development.[31] The highway access and regional airport immediately attracted a spate of industries, as did the new University of California campus at Irvine. Eventually, the Irvine Ranch contained eleven different town districts laced into the Orange County highway system. It was a developer's dream. Yet it seemed to carry out the ideas of urban planner Kevin Lynch's famed *The Image of the City* (1960) on a large scale, as it radically counteracted the Los Angeles Basin's suburban sprawl. Each village was designed with its own landscape and architectural features to give it a distinctive environment and "imageability."[32] Sophisticated marketing strategies reached out to singles, families, young marrieds, divorcees, empty nesters, active retirees—each category provided with its own housing choices in one of "tomorrow's cities."

Cities and towns were imagined as nodes or cells along transportation

4.4. William Pereira (*center*) showing the plan for Irvine Ranch, California, to Daniel Aldrich, chancellor of the University of California at Irvine (*left*), and Charles Thomas, president of the Irvine Company (*right*), ca. 1964. Courtesy of the Irvine Company.

and information webs. Entire metropolitan regions were envisioned as coherent *systems* linked by circulation flows. Melvin Webber, director of the Institute of Urban and Regional Development at the University of California, Berkeley, was one of the most influential urban theorists spreading this systems thinking. Like Pereira and his new town of Irvine, Webber championed what was labeled California car culture and the new networked society, and his concepts were rapidly taken up in planning circles. He pointed to the "quasi-science of futurism" that was supported by a flood of new research institutes such as Resources for the Future, the Ford Foundation, and RAND Corporation. Predicting the future, he admitted, was an age-old quest. What was different about the 1960s version was "the emergence of a legitimate and organized activity, explicitly devoted to systematic and normative interpretation of potential future histories."[33]

Webber's major contribution to this arena of prophecy was "nonplace realms." With examples such as the Los Angeles and San Francisco metropolitan areas, he argued that urban planners must throw away the old notion of place and focus instead on the material and informational connections that bound urban regions together like a giant switchboard. Future cities would be dispersed settlements relying on communication, transport, and information flows. In his *The Post-City Age* (1968), Webber predicted that a "new kind of large-scale urban society is emerging that is increasingly independent of the city." The glue that once held settlements together was dissolving. Instead, he foresaw a future "community without propinquity," a phrase he originally coined at a Resources for the Future symposium. Improvements in communications technologies, from television and long-distance telephone service to computers and real-time information, were reducing distance and making densely packed cities unnecessary.[34] The city would instead become a virtual space of information streams and electronic traffic. Webber's imagery of the networked future city captured the attention of new town planners such as James Rouse and William Finley at Columbia in Maryland, George Mitchell at The Woodlands in Texas, and ultimately Richard Llewelyn-Davies at Milton Keynes in England (see chapter 5).

Universities and research foundations were workshops where experts tinkered with the latest land-use models and computer programs. The models became increasingly elaborate in the quest to capture the totality of processes in each city and metropolitan region. Inventories were taken of the existing state of the system: population, employment and household income, industry and services, land rents, traffic flows, and so on. The data was input for analysis: computer programs calculated parametric outputs and simulated the behavior of the system. The models were economics- and engineering-driven, and shifted planning toward a corporate model of urban development. Social class was defined as a household with income and consumption constraints that could affect movement and transportation flow.

In these early models, the urban system was imagined as self-equilibrating. Urban activities were reduced to their elemental forms, aggregated, and then represented spatially. Amid the political volatility and social agitation of the 1960s, systems analysis seemed to offer the fantasy of predictability and control. There was an unflinching belief in statistical measurement and mathematical formulas as representing reality. The utopian focus shifted from ideal city to ideal planning process, which would result in an optimal city as output. It was a generic formula that could be produced universally. This organizational logic was programmed into the very substance of the city itself. At the University of Chicago, Brian Berry, one of the leading American geographers working on the new techniques, believed that symbolic models "provide idealized representations of properly formulated and verified scientific theories relating to cities and sets of cities perceived as spatial systems." Systems theory and the computer revolution meant the "virtual elimination of the once lengthy gap between problem formulation and evaluation of results, sharpening of the questions asked, initiation and completion of experiments of a size unthinkable under earlier technical conditions."[35]

Problems of data credibility and distortion were largely pushed aside or dealt with by making models even more complex. Unwanted externalities were disregarded. The city and its citizens were understood as purely passive, and planning as active. There were critics of this kind of resplendent scientific performance, chief among them Lewis Mumford. In 1965, Mumford warned that "abstract intelligence, operating with its own conceptual apparatus, in its own self-restricted field, is actually a coercive instrument: an arrogant fragment of the full human personality, determined to make the world over in its own over-simplified terms."[36] A sterile desert would result. Writing in 1968, Britton Harris, who became chair of the University of Pennsylvania's Department of City and Regional Planning, warned about the obvious: that the model was only as good as the data coded into it. While ranking the invention of the computer with such technological innovations as fire and the wheel, Harris broached that complex urban systems were in fact nonlinear, and that rather than enhancing creative solutions, scientists were shaping problems and generating ideas in conformity with the machine.

Harris's corrective was published by Constantinos Doxiadis's periodical, *Ekistics*, in a special issue on "computers in the service of ekistics."[37] Also contributing articles were such powerhouses as Margaret Mead and Walter Isard. Each essay demonstrated the feats of analytical wizardry made possible by computer technology. Isard's case study modeled the impact of a proposed new town of two hundred thousand people on recreation and ecology at Plymouth Bay in Massachusetts. He depicted both the social and the ecological systems as commodities in an input-output analysis.

In a basic sense, even Doxiadis's *dynapolis* was a complex system based on ordered hierarchies, and his *ecumenopolis* an interwoven network of settlement cells and transportation and communication flows. He began using computers in 1962 to develop mathematical models of his settlement ideals. Computer programming and modeling became the main undertakings of his Athens Institute, where all his survey data was typed onto punch cards and fed into a computer to discover patterns and flows.

By the end of the 1960s, American computer guru Jay Forrester developed a more dynamic simulation model for understanding the impact of policy making on urban systems. At the end of the Second World War, Forrester had worked at MIT's Lincoln Laboratory on the development of an advanced aircraft flight simulator that evolved into the Whirlwind digital computer. Whirlwind became the core of NORAD (the North American Aerospace Defense Command), which was the early warning system against atomic missile attack. By 1956, Forrester had moved to the MIT Sloan School of Management, where the application of his research to the field of organizational decision making broke new ground with early computer modeling languages such as SIMPLE and DYNAMO.

Then, in collaboration with John Collins, the former mayor of Boston, Forrester applied his computer modeling skills to the plight of American cities. Their study was undertaken in 1968, amid the riots and protests that besieged inner cities, and were a glaring signal that better urban policies were urgently needed. Published as *Urban Dynamics*, the results of Collins and Forrester's work were a milestone in systems analysis. It presented an entire science of cities. Forrester refuted the conventional wisdom that urban decay was caused by such factors as the population boom, dwindling fiscal resources, and suburban sprawl—all of which were beyond the control of municipal government. Basing his argument entirely on his computer-based simulation model, he instead made the case that most urban problems arise from complex, interrelated processes that occur within cities themselves.

Forrester's model was *dynamic* in that it simulated how urban policies played out over time. It could reveal hidden patterns in the system that obstructed policy or had pernicious side effects. When job training programs were added to Forrester's model, for example, they caused unemployment to rise. In contrast, policy choices that initially appeared incorrect or counterintuitive often yielded the best results. Low-income housing, Forrester argued, created pockets of poverty, while tearing down slums and public housing improved neighborhoods and the standard of living. The book provoked an immediate political backlash and outrage from the black community in Boston. To calm tempers, MIT and the US Department of Housing and Urban Development sponsored a series of case studies in the Boston area to validate the model.[38] When these yielded inconstant results, HUD refused to endorse the model.

The storm over Forrester's urban dynamics indicated the sweeping influence that systems analysis had in planning and policy making. The science augured a new age in which the entirety of human experience could be understood. Systems models were models of the world. This universalizing epistemology was symptomatic of Sixties futurism. Visionaries from Buckminster Fuller and Marshall McLuhan to Constantinos Doxiadis and Jean Gottmann worked out their concepts at a planetary scale. For example, Fuller had plans for a cybernetic World Game inside his famed Geodesic Dome at Montreal's Expo67, and had established a World Game Institute. World Game was a strategy contest that simulated world governance and challenged players to deal with complex scenarios, and then find solutions at a planetary level. During their deliberations, they could consult a computer-driven Dymaxion world map, with information continuously streaming from massive data banks.[39]

Jay Forrester shared this public stage with Fuller and other visionaries as a celebrity knowledge prophet. He created a series of "world systems" computer models to understand the connections between the multifarious predicaments facing humankind. The models, WORLD1, WORLD2, and WORLD3, were progressively enhanced and simulated the interaction between world population, industrial production, pollution, resources, and food. They also generated two books: *World Dynamics* (1971) and then the popularized version, published as *The Limits of Growth* (1972). This last was an instant blockbuster, sold several million copies, and was translated into thirty languages. If Forrester's earlier *Urban Dynamics* was a talisman of progress, these later publications evinced the pessimism of the age. They painted a stark picture of the unintended consequences of exponential population growth and runaway development that his models predicted. The outcome was catastrophic: the collapse of the natural environment and the planet's ability to support human life. The only way out was to recover the equilibrium between humankind and the earth.

It is precisely in this atmosphere of unbounded hopes for the future shadowed by a sense of impending doom that the search for the ideal city should be situated. The optimism of the 1960s was haunted by anxieties over global decline and nuclear eradication, urban decay and environmental degradation. Cybernetic optimizing and computers were prerequisites to the exciting promises of space travel, but they were also behind military weapon systems. The atom bomb itself was the outcome of computing technologies. Computers, modern telecommunications, futuristic highways, and people movers were the signatures of progress, but the urban crisis threatened to derail this dream. A chorus of alarm rose about overpopulation, food shortages, pollution, resource depletion, and of course atomic warfare. Systems analysis and the work of pacesetters such as Jay Forrester seemed to be the way out of the storm. They would use cybernetics for the cause of good. Society could be orchestrated and controlled. Nothing would be left to chance. Planners stressed the significance of their expertise by associating it with high moral principles, public service, and saving the earth from chaos. They spoke movingly of their moral obligation to construct a new urban world.

Ultimately, however, this utopian vision of urban systems assumed consensus and centralized decision making. Melvin Webber admitted that the extraordinary capacity of systems planning was mainly being exported internationally, because "it seemed to require far more centralization of authority and control than is tolerable or possible in this nation."[40] For all the emphasis on formal models and self-regulating cities that could be duplicated unendingly, the application of cybernetics depended on the politics of planning culture and on the national resources dedicated to investing in and learning the newest computer technologies. As a result, cybernetics and systems analysis spread across national borders with profound modifications. Each country promoted its home-grown version and tackled the intricacies of the new planning science from the viewpoint of its own aspirations. The upshot was a fertile intellectual encounter with utopia and a range of new town experiences.

Soviet Cities of the Future

For all the excitement it caused in the West, cybernetics as a metascience was actually most instrumental in the Soviet Union, where it became the ultimate means for achieving socialist society. From the mid-1950s, Soviet premier Nikita Khrushchev signaled a radical change of course for the USSR, and promised that it would catch up with and overtake the West. This was a contest not just of ideologies but of efficiency, productivity, and technological innovation. Systems analysis, new techniques in urban planning, and the mass production of housing were at the forefront of this campaign. In the process, the Soviet Union became the undisputed leader in the construction of new towns, and the systems approach became the framework for a new ideal of *sotsgorod*, or socialist city.

It is worth remembering how horrific the situation was in the Soviet Union's cities after the Second World War, and that it remained so well into the 1950s. The wartime devastation had occurred on an unimaginable scale. According to historian Tony Judt, 70,000 villages and 1,700 towns were destroyed, along with 32,000 factories and 40,000 miles of railroad track.[41] Dreadful living conditions were a fact of life across the USSR. People eked out an existence in ancient, dilapidated lodgings without water or waste hookups, and with little heat. In the vast expanses of the Urals and Siberia, families crowded into universally dismal dormitories and barracks.[42] The wretched circumstances in which so many lived made the construction of genuine towns with housing and infrastructure an absolute emergency.

In a dramatic shift in policy, Khrushchev promised to solve the housing crisis in twenty years. Socialist realism in architecture was denounced as excessive and costly, the Soviet Union would launch an all-out drive for standardized mass-produced housing. Building functional prefab places to live in and building larger and faster would solve the acute housing shortage.[43] The Communist Party Central Committee issued a resolution "on eliminating superfluities in design and construction." The heroic ideological discourse of "happy socialist cities" faded into the background, and with them went the idealized *microrayon* neighborhoods for young worker families. Like their Western counterparts, Soviet planners were faced with residents who made their everyday lives with little use for official ideas on how to conduct themselves in urban space. Given the energies that went into fabricating perfect neighborhoods, there was a subversive nonconformity to these commonplace practices that is hard to deny. It was a good reason for shedding the *microrayon* as a planning priority.

Two theorists in particular set the tone for this new version of *sotsgorod*: Stanislav Strumilin and Boris Svetlichny. Strumilin (one of the deans of Soviet economics) had earlier introduced the *microrayon* concept to Soviet urban planning. He survived the Stalinist purges to become a ranking member of Gosplan (State Committee for Planning) and the Academy of

Sciences. In his influential article "Family and Community in the Society of the Future" (1961), Strumilin envisaged collective "palace communes" that catered to every service for young families and fulfilled the need for "joyful comradely communion." Towns would be graced with schools, hospitals, stadiums, and theaters; with pools and skating rinks. They would function as an "economic and social organism."[44] Their citizens would live in modern, spacious dwellings. Strumilin's narrative aestheticized Khrushchev's building program as everyday paradise. It was part of the media blitz that accentuated the distinct characteristics of the *sotsgorod* as different from downtrodden cities in the West. Regardless of the political posturing, however, the fantasy of community and solidarity was a universal motif of the new town movement, as was the hope for a better life.

For his part, Boris Svetlichny was a deputy chief at Gosplan and the leading mouthpiece for Khrushchev's reforms. He captured official enthusiasm for the new course in his 1959 article "Cities of the Future." Beginning his utopian reflections with Marx and Engel's original condemnation of the capitalist city, Svetlichny next laid out the dream of the communist city as a new stage in human settlement. The hypertrophic metropolis beaten down by capitalist greed was finished. In its place, a *system* of ideal cities and towns, large and small, was the optimal form of urbanization. This meant satellite towns, or *sputnik*, as they were called, in the metropolitan regions of Moscow, Leningrad, and the largest cities of the Soviet republics. New towns would stimulate development even in the remotest parts of the USSR's vast territory. Each of these spanking-new places would feature an array of housing set in light, air, and greenery. They would be open to everyone through prefabrication and standardized construction techniques: 15 million new apartments in towns and worker communities, 7 million new homes in rural areas. Svetlichny applauded Russian experts for inventing state-of-the-art prefabricated housing—this despite the fact that Soviet construction relied heavily on French large-panel building systems such as Coignet or Camus. But custom architectural designs, cheerful colors, and assorted textures would make each building unique and aesthetically beautiful. Heated glass passageways would protect residents as they daily ambled to communal eateries and services. The housing blocks would be shielded from industrial pollution and from the automobile by greenbelts and gardens.[45]

In pursuit of these utopian fantasies, some twenty-five to thirty new towns were churned out every year by Gosplan during the late 1950s and 1960s, with more than half constructed in Siberia. This effort was entirely consistent with development strategies across the postwar resource landscape. New towns were the leading edge for industrial production, extraction of raw materials, and massive hydroelectric and nuclear power projects vital to Soviet economic output. There were generic plans for petrochemical towns, for metallurgical towns, for machine-tool towns. Many of these places were highly specialized military and scientific research facilities. Most were built up from existing small settlements and labor camps and became official new towns under state auspices. About one third were constructed from scratch on completely vacant sites.

Altogether by the mid-1960s, the Soviet Union had famously constructed some 900 to 1,000 *novy gorod* of one kind or another, although the definition was ambiguous. Nonetheless, they signified that a massive resettlement eastward had taken place. New towns represented over 60 percent of the Soviet Union's urban places. Some 30 million people lived in these sprouted-up urban worlds, most of them in the long, squat, five-story apartment buildings known as *khrushchyovka*.[46] The new towns were imagined as a bright new communist future in which social progress was permanent and the most far-flung utopian ideals could be achieved. Some had populations that ballooned to 250,000, while others remained smaller, with populations of 100,000 down to 25,000, while still others were no more than dusty crossroads. As both planning successes and planning failures, they constituted an entire system of urban places stretching across Siberia and the Volga-Urals region to Kazakhstan and south central Asia. As the frontrunner in new town construction, the Soviet Union hosted the United Nations Seminar on Physical Planning Techniques for the Construction of New Towns in 1964, and again in 1968.

It was an audacious new course. It also meant applying the cybernetic lessons of the West. Work on cybernetics and systems analysis in the Soviet Union was monumental: it was declared a key instrument in building the communist dream. Computers—or EVMs, as they were known in Soviet parlance—were a playground of experimentation. "This was a period," wrote Andrei Ershov and Mikhail Shura-Bura, the patriarchs of Soviet computing, "of unlimited optimism, a kind of computer euphoria—a childhood disease, like a pandemic, that enveloped all countries of the world."[47] In the pages of technical journals such as *Problemy Kibernetiki* (Problems of Cybernetics) as well as in I. A. Poletaev's runaway best seller *Signal* (the first popular book on computers), the Soviet Union was pictured as a dynamic organism managed by intelligent computers and robots. The Soviet press began heralding computers as "the machines of communism."[48] The entire economy was seen as a complex cybernetics system. Input-output analysis, automation, and artificial intelligence consumed the imaginations of experts ready to use the new tools for planning on a national scale. Reflecting in 1962 on machines that outstrip human beings in their ability to think, mathematician Sergei Sobolev (who was behind the effort to build the science city of Akademgorodok) prognosticated, "In my view the cybernetic machines are the people of the future."[49]

This imagery was best captured in the figure of the cosmonaut. The mania for spaceflights and the cult of the cosmonaut exceeded even those for astronauts like Neil Armstrong and John Glenn in the West. The launch of the Sputnik satellite in October 1957, Yuri Gagarin's first manned spaceflight in April 1961, the first space walk, and the Mars probe were all Soviet triumphs. These spectacular advances in aerospace relied on cybernetics. Computers and the automation of complex engineering systems made the space capsule possible. They merged with the cosmonaut onboard into a cybernetic man-machine, a human automaton capable of both interacting with the complex automated computer systems and actually flying the spacecraft. The superhuman cosmonaut became the prototype for the New Soviet Man. The self-regulating, flawlessly engineered spaceship was the model for the new town in which he would live. Cybernetics was the path to communist utopia and the road to a world where everyone would live in dignity. It was an aesthetics of technologism, a cyborg bionic fantasy shared across the avant-garde spectrum in the 1960s.

Like their Western counterparts, Soviet planners of the 1960s shared the belief that cybernetics and systems thinking could revolutionize urban life. These were the tactics for controlling resources and optimizing urbanization and productivity, and carrying out Khrushchev's reforms. The possibility of creating not just individual towns but systems of towns linked by communications and transportation grids seemed within reach. In *The Ideal Communist City*, Alexei Gutnov, Ilia Lezhava, and the New Elements of Settlement (NER), a group of young architects at the University of Moscow, took up the challenge. They attempted a scientific projection of the model communist life by integrating the Marxist conception of social relations with cybernetics, information theory, and human engineering. The chaotic growth of cities under capitalism would be upended by a dynamic *system* of urban settlement. The emergence of rationally planned areas (industrial, scientific, residential) would indicate "that we have moved into a new stage of conscious urban development ultimately aimed at uniting the planet into a single system corresponding to a new kind of social organization and to the growing potential of modern technology."[50] The narrative paralleled that of Western futurists such as Buckminster Fuller and Constantinos Doxiadis.

This portrait of the ideal communist city reached France as a special issue of the journal *Architecture d'aujourd'hui.*[51] It imagined each city as a complete unit with an exact size and population. These urban settlements were corpuscular colonies, or cells, situated along transportation networks. This vocabulary was taken up by Soviet planners in their descriptions of towns as "multidimensional cells" and "ganglions" in socioeconomic space. They would function as living, self-regulating "cybernetic human-machines."[52] The new town became science fiction, a metaphor for the space station in the wildly successful Soviet space program.

This kind of techno-futurism symbolized the triumph of communism. The ganglions were put in place in a series of settlement systems across the Soviet Union, all of them based on new towns. They stretched from the Naberezhno-Chelninskaya system on the Volga River to the Abakano-Minusinskaya system in the Far East, and the Achinsko-Itatskaya system in Siberia. Siberia was the prototypical settlement system, and galvanized urban planning in the USSR. All this land-use planning culminated in the *General Scheme of Settlement Structure on the Territory of the USSR* (1975), which aimed to rationalize and equilibrate the vast geography of the Soviet Union through the scientific distribution of towns and cities. It detailed the spatial structure of Soviet ideology, sanctified and controlled through Gosplan. At the same time, a plan to link research centers and professional organizations, factories and construction sites into a "single republic network of computer centers" would systematize the entire Soviet economy.[53] Transportation would be entirely automated. The urban network would become a hyperfuturistic symbiotic flow of information and communication. For Soviet planners, systems thinking solved the incongruities between city and country, between natural and built environments. Architect Alexei Gutnov, lead author of *The Ideal Communist City*, claimed that the visualization of these vast schemas was found in the experimental mega-structures of the late 1950s and 1960s (see chapter 6), which the Soviet Union fully shared.[54]

To a great extent, authority over urban planning in the USSR shifted from the architect as creative virtuoso to state engineers and scientific experts. The urban systems were produced at the Central Research Institute for Town and Regional Planning using input-output analysis and state-of-the-art computer modeling. Soviet planners dove into systems thinking. They fixated on Jay Forrester's urban dynamics.[55] Rather than the rigid ideological version of the ideal *sotsgorod* of the Stalinist years, new towns were seen as laboratories of experimentation for fleshing out the principles of cybernetics and systems analysis. At the UN seminar on new towns held in

Moscow in 1968, Soviet experts boasted that "mathematical methods and electronic computers have been applied to problems of population distribution, land use, economic priorities, traffic and transport networks, systems of cultural and welfare services." What was needed in the future was computational study of social processes and social communication, groups and associations, and leisure time. "The application of modern mathematical techniques of statistical inference can help to formulate the mathematical models whereby the probable behaviour of people in towns of different planning patterns can be determined."[56]

These ideas were put to the test in an extraordinary network of secret cities, or ZATOs,[57] that were a direct response to the Cold War and fears of nuclear annihilation. They were aerospace or missile sites, military and space surveillance installations, or atomic research centers, each hooked into a dedicated computer network.[58] They did not appear on any official map and were deeply covert places in Soviet geography: a majority (though it is not known how many) were located beyond the Urals in the inhospitable reaches of Siberia and the Far East. The ZATOs were concealed from public view and more or less strictly regulated, with admission only possible with special permission. They were phantom cities under different regimes of control—places of extraterritoriality in the literal meaning of *utopia*. They were known in colloquial parlance as "blue cities," according to historian Barbara Engel. It was a term popularized by a song in the Soviet film *Two Sundays* that depicts the dream of a better life in the new cities of the East.[59]

ZATOs were mythical beacons in the wilderness where the field of possibilities was open. The inhabitants (some of the best scientists, engineers, and technicians in the Soviet Union) were restricted from contact with the outside world in return for high wages and a cornucopia of privileges. The sites were scrupulously standardized, their town plans and building typologies nuanced for specific climate zones. In the United States, Oak Ridge, Tennessee; Los Alamos, New Mexico; and Hanford, Washington, fell into this category of secret cities related to atomic research and the arms race, as did Deep River in Canada.

The ZATOs were also resource towns meant to unearth the USSR's extraordinary natural wealth and develop extractive industries. In one example among many, the petrochemical city of Angarsk was established in the wild Taiga region of eastern Siberia in 1948 to drill for oil deposits. Designed by a team from the Leningrad Planning Institute as a model linear city along the Angara River, Angarsk's residential neighborhoods were protected from the giant petrochemical complex by an elongated greenbelt. It was a rendition of Nikolai Miliutin's original *sotsgorod* that had been used in the plan for Magnitogorsk (see chapter 1) and evidenced the long heritage of linear design in the Soviet Union. By the 1960s, a city center with a Lenin Square and Karl-Marx-Allee along with sophisticated promenades, parks, and public spaces made Angarsk into a Soviet civic dreamscape. Initially built for thirty thousand people, the city quickly mushroomed to well over two hundred thousand residents.[60]

Also on the Angara River deep in eastern Siberia was the new town of Bratsk. Its origins lay in the construction of a three-mile-long dam across Padun Gorge and the colossal Bratsk hydroelectric complex (fig. 4.5). The town grew quickly from fifty thousand to over two hundred thousand young workers. At 3 million kilowatts, the hydroelectric station was the world's largest power producer. Prominent Western policy makers and journalists were invited on official visits to gawk at the outsized symbol of communist technological prowess. Its size and capacity, the human energies that flowed into its construction and that of the new town arising around it (fig. 4.6)—all signified the Soviet New World. Yevgeny Yevtushenko, Soviet poet and spokesman for the Sixties generation, penned the well-known ode to the legendary Bratsk complex in 1966. He heard the history of social revolution in its roaring:

> Take a look,—
> > on the vanes of my turbine,
> bubbling,
> > shimmering,
> > > bursting,
> reuniting,
> > pushing one another,
> disappearing and rising again,
> amidst the spray
> > in the blue hum
> flows vision
> > after vision . . .[61]

The secret city of Sillamäe[62] on the Gulf of Finland in Estonia was built in the hope that local shale oil might yield the prized uranium ore. The closed town of Yangiobod took shape near the uranium mines in the Tashkent region of Uzbekistan. Neither settlement appeared on any maps; there were no place-names on the town plans. Yet these were socialist dreamscapes of the atomic age. Secret cities such as these, in the words of historian Asif Siddiqi, "represented aspirational spaces, idealized urban formations

4.5. Building the Bratsk hydroelectric complex, USSR, 1962. Photograph by Nikolay I. Perk. Courtesy Bratsk Joint City Museum and Bratsk State University.

4.6. The new town of Bratsk, USSR, ca. early 1960s. Photograph by Nikolay I. Perk. Courtesy Bratsk Joint City Museum and Bratsk State University.

where the day-to-day amenities of life were seemingly abundant and plentiful."[63] Their secrecy fortified the cult of science, and the scientist as a new type of human being.

Actor Alexei Batalov recalled making a film about the life of nuclear physicists in 1960 during which "filming took place in a world no one had known about before. . . . No one knew . . . how they lived, how they worked, what they talked about."[64] Just behind this fantasy stood staging grounds where workers toiled in the construction of an off-limits paradise. These were a parallel system of wild mobile settlements and work camps where people lived in containers and tents for five to ten years in harsh environments, setting up industrial sites and constructing ideal *sotsgorod*. Here was the hidden dystopian world that always lay beneath the surface of utopian ambition—the shade in an enigmatic image of mystery. But the ZATOs' position as Shangri-La gave them mythological status in the Russian imagination.

This was especially the case for the partially closed science cities of Zelenograd northwest of Moscow and Akademgorodok in Siberia, which were especially relevant to the cultural significance of computers and cybernetics in the Soviet Union. Zelenograd was founded in 1958 along the Leningrad Highway for the development of microelectronics and computers. The location was already the site of the early model community of Kriukovo and the model Kruikovo Colony prison camp, where Western visitors were frequently taken for tours. Utopia revealed its dark shadow in a landscape of meadows and pine forests. The place was a demonstration site for the full extent of Soviet state power, both good and bad. Zelenograd was a *sputnik* of Moscow, and the resonance was not by happenstance. New towns such as Zelenograd were envisioned as part of the Soviet launch into the Space Age; they were machine towns in the image of Soviet spacecraft.

Electronics engineers and scientists began their work at spanking-new research facilities and the enormous Elion electronics plant in a utopian landscape of modernist prefab buildings. The facilities and plant were constructed by the Zoblin worker brigade, a youthful labor battalion that symbolized the building of communism. Prefabricated panels bolted onto reinforced concrete skeletons cut construction times in half while labor productivity rose by 40 percent in a new version of hyperproductive Stakhanovism. The scale of the city shifted upward to buildings of twelve and fourteen stories. Standing tall on an elevated square alongside the shopping center and highway interchange was the city's administration building. The surrounding prefab housing estates and modern research laboratories were the metaphors for catching up and surpassing the West, and the

Soviet version of Silicon Valley. By the late 1970s, Zelenograd was producing a computer chip rivaling that of Intel.

Similarly, Akademgorodok, set amid Siberian forests on the manmade
Ob Sea had been proposed by computer scientist Mikhail Lavrentev as a
showcase for cybernetics (fig. 4.7). By 1964, nearly fourteen thousand scientists and researchers arrived at the new town, supposedly an idyllic utopia freed from the stultifying bureaucratic atmosphere of traditional acad-
emies.[65] Scientists had access to Western publications and colleagues. They
could travel abroad.

In 1967, a *New York Times* reporter given official access to Akademgorodok described "young science sophisticates who love Western music and
British detective stories."[66] The town was the capital of a Swinging Sixties
generation of well-educated and talented professionals who were hip, cool,
and idolized as the New Soviet Man. With fourteen scientific institutes including the Siberian division of the Academy of Sciences and the Institute
of Mathematics with its Computer Center, it offered a unique environment of free thought, cultural diversion, and scientific research. At the local

4.7. Soviet premier Nikita Khrushchev visiting the Institute of Hydrodynamics
SB RAS with the plans for Akademgorodok, USSR, 1959. Photograph by
Rashid Ibrahimovic Ahmerov. Courtesy of the M. A. Lavrentiev Archive at the Institute
of Hydrodynamics SB RAS and the Photo Archive of SB RAS at www.soran1957.ru.

student club, the same reporter "confronts a bearded young man, guitar in his lap, strumming an American folk-rock song of protest. For a moment the visitor thinks he is back in Berkeley."[67]

Mathematics and computation were the methodological glue of Akademgorodok as a new town, and central to its vision as urban place. The city was constructed using the newest prefabricated mass production techniques—the largest concrete factory in the Soviet Union was built in Siberia to churn out the ready-made panels for slapping together into new housing. Laid out to take advantage of the surrounding natural landscape of birch and spruce trees, Akademgorodok was the ultimate in progressive new town design. Linear ribs of academic institutions and *microrayon* neighborhoods paraded up from the Ob Sea shoreline. Their spine was a broad, tree-lined boulevard. On one side was the proverbial House of Culture along with a hotel and a shopping complex featuring a wide-screen movie theater; on the other were the Party and governmental buildings. Pride of place in the town center was given to the Scientists' Club with its library and club rooms, and to the University of Novosibirsk buildings. The names of the streets (Sea Street, Pearl Street, Science Street, Marine Avenue [fig. 4.8]) and those of the bus stops (Hydrodynamics, Nuclear Physics, Economics) spoke of a scientific paradise.

Of course, things were never quite this straightforward, whether at the gold-standard towns of Zelenograd and Akademogorodok or the ordinary resource towns out in the frontier tundra. The actual practice of Soviet urbanism meant the partial failure of these great hopes and visions for new towns—a truth that was openly discussed in the daily press of the time. Writing on Soviet urban planning, American planners Robert Osborn and Thomas Reiner pointed to "the processes of bargaining and criticism—often creative processes—that go on within this structure [of planning], in spite of its outwardly monolithic character." Turf wars and political battles erupted at all governmental levels, from Moscow down to local municipalities, as did skirmishes between local government and the industrial companies carrying out construction. This meant that "stadiums are built while the city is left begging for more humble facilities, such as public baths and laundries."[68] Planners consistently underestimated the population they were building for. Some new towns were welcoming two to three times the projected number of residents before the cement was dry. There were the inevitable construction delays and cost overruns, the shortage of skilled builders and poor quality control, the lack of services. Maintenance was often nonexistent. The half-done quality of Soviet new towns, the bleak housing estates, and the miserable deficiencies of everyday life became a

4.8. Morskoy Prospekt (Marine Avenue) in Akademgorodok, USSR, 1970.
Photograph by Anatoly Polyakov. Courtesy of the Archive Exhibition Center
of SB RAS and the Photo Archive of SB RAS at www.soran1957.ru.

moral lesson on the failure of communism. The gap between ideal and
reality was a bitter pill to swallow.

And yet, as historian Elke Beyer argues, "millions of square feet of living
space and an enormous infrastructure were created, and the basic foundations of planning and infrastructure for hundreds of Soviet cities were laid
out. In a process of global knowledge transfer, the Soviet urban planners
worked through the fundamentals of urban Modernism and many visions
of the future that were popular during the 1960s."[69] The apartments were
certainly plain, and the built landscape drab. But in comparison with the
pitiful communal lodgings and barrack dormitories of the past, they were
miraculous. Millions of families enjoyed inexpensive dwellings with kitchens and bathrooms, with schools, playgrounds, and health clinics nearby.
For Alexei Gutnov of *Ideal Communist City* fame, the new towns were the
tangible realization "of socialist society and its principles of equality and
social justice."[70] As a dreamscape of what this could mean, new towns and

cities such as Zelenograd and Akademgorodok registered the most successful legacy of progressive Soviet planning.

Polish New Towns and Threshold Analysis

Cybernetics and systems analysis swept across the Eastern Bloc like a cultural fever. The great Polish science fiction writer Stanisław Lem celebrated
cybernetics in his blockbuster novels *The Astronauts* (1951) and *Solaris*
(1961). Lem used Norbert Wiener's *The Human Use of Human Beings* as
the basis for his stories, especially for the cyberneticians who are the main
characters in these fantasy portraits of the communist utopian future. *Solaris* presents humans as cybernetic systems that store data information.
It was one indication among many of the romance with cybernetics and
computers shared in both East and West. Cybergeeks, astronauts, visionary planners: they were the future of humankind. They could create new
worlds on Earth and in outer space. Utopian thinking on this scale had
authentic meaning in the 1960s. Techno-futuristic fantasies could be put
into practice and flawless cities created through the magic of computers.
Input-output analysis and simulation models forecast optimum distribution scenarios for towns across a geographic region.

Polish architect and urbanist Bolesław Malisz of the Institute for Town
Planning and Architecture in Warsaw was one of the East's most prolific and
vocal advocates of the new systems planning. He was a member of the midcentury generation that came of professional age in the 1950s and 1960s,
ready to cut ties with the past and launch into the future of urban form. He
began his career as the manager of the regional plan for Pomerania in the
late 1930s, and headed the Polish Office of the National Plan after the war
before becoming the country's chief urban planner. Then in 1961, Malisz
participated in the influential United Nations Conference in Stockholm
on the planning of metropolitan areas and new towns, where he met with
Constantinos Doxiadis and British new town enthusiast Barbara Ward. He
was also in regular contact with urban planners and community activists in
the new town of Reston, Virginia. Their letters signal the exchange of ideas
across Cold War boundaries,[71] especially Malisz's theories about controlled
growth and the use of data collection and computer analysis for new town
development. His own planning technique, known as threshold analysis,
was presented for the first time in 1964, at the United Nations Symposium
on New Towns in Moscow.

Malisz attended as the author of *Poland Builds New Cities*, which set out
the national agenda for urban development. The first order of business

in the 1950s was rebuilding the devastated and utterly ruined capital of Warsaw, which would be "our great contribution towards the building of a better tomorrow; a better, happier world," in the words of Polish Communist leader Bolesław Bierut.[72] But the government also imagined a vast modernization program for the Vistula River basin as well as the northern regions of the country. Poland claimed that forty-nine new towns were built from 1946 to 1960 at the same time that the urban population of the country doubled. Certainly not all these places were new, since some were informal settlements finally given official municipal status. However, there were a substantial number of from-scratch settlements. The best known and iconic of the future were Nowa Huta and Nowe Tychy. What excited Malisz about them was precisely "the changed quality of the new town . . . the new conception of the city, the possibilities and the means for realizing this conception."[73] His book provided a tour of Polish new towns in the 1950s with comparisons to English, French, Dutch, and West German planning concepts, and acknowledged Lewis Mumford as intellectual guide.

Malisz's threshold analysis was an ambitious theory of settlement systems and a means of assessing the state of cities and their operation. Planners could then use this information to simulate future planning scenarios. The planning technique required the integrated analysis of a jaw-dropping array of socioeconomic variables along with the requisite flows and feedback loops. The imagination of Polish planners was saturated with mathematical models and abstract planning techniques. Quantitative methods and systems theory lit the way forward. First and foremost, the calculations produced efficiency indices to evaluate when urban development could or could not be justified. They optimized public investment.[74]

By the late 1960s, the priorities in Polish planning were financing and operating costs; fantasies of social equality and balancing regional growth were shunted aside, though not without political controversy.[75] New town development was absorbed into comprehensive regional development schemes that took advantage of economies of scale and capital savings. Nowe Tychy and the other new towns of Silesia in the 1950s were lavish productions built around coal mines. By the 1960s, however, the coal industry was fading, and there was little to save the region's depressed economy and loss of jobs. Planners turned to a far more systematic approach to Silesia that involved modeling settlement hierarchies and a diversity of economic activities. Malisz's threshold theory provided a rational quantitative method for deciding on the best investment strategies: "the problem is not whether we should build a new town or not, but rather how to provide a functional system of settlements."[76]

By 1968, Malisz was working as a UN expert in Greece, where he applied threshold analysis to a regional plan for the southern Adriatic. At the same time, his theory was employed in two major projects in Scotland, the Central Borders and the Grangemouth-Falkirk Plans, by his student and collaborator Jerzy Kozłowski. (Grangemouth and Falkirk were situated along the River Forth east of Edinburgh.) The threshold analysis ideograms and model that Kozłowski produced for the Grangemouth-Falkirk study are aesthetic objects in and of themselves (fig. 4.9). They visually depict

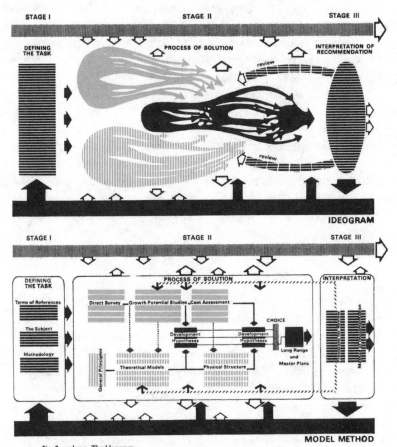

Fig. 2 — above. The Ideogram
Stage II in the diagram shows the 'process of solution' and its three main analytical streams: the direct physical analysis (orange), the process of synthesis (black) and the model structure (vertical hatched orange). The authorities (black and vertical hatched black) are shown to influence the whole of the planning process.

Fig. 3 — below. The Model Method
Diagram showing the more detailed application of the ideal theoretical framework in Fig. 2.

4.9. Jerzy Kozłowski, Threshold Analysis Ideograms and Model Method, Grangemouth-Falkirk Plan, Scotland. Originally published in "Threshold Theory and the Sub-Regional Plan," *Town Planning Review*, 39, no. 2 (July 1968). Image used with permission of Liverpool University Press.

the design methodology as an informational stream, from defining the task to the solution process and then interpreting the model outputs.[77] The drawings were meant to replicate computer-generated diagrams, and have an uncanny resemblance to Cedric Price's Fun Palace (1961) and Dennis Crompton's Computer City (1964), themselves improvisational portraits of cybernetics and systems thinking (see fig. 6.4).

The reiteration of this visual language across the urbanistic avant-garde was extraordinary. Logic diagrams and flow charts were a means of organizing and rationalizing a multidimensional problem, and then imagining the potentiality of computer technologies to solve it. As urban design patterns, they were nonfigurative and ignored local particularities and distinctiveness of place. The viewer could retreat from the complexities of the real and follow the symbolic language of form and color.

Threshold theory spread to Yugoslavia in 1967 and Ireland in 1969, where it was adopted to measure the potential for tourism development in various areas in both countries. By the 1970s, it was embraced in cities throughout Europe and had spread to Canada, Uganda, and the Cochin area of India. Kozłowski produced *Threshold Analysis Handbook* for the UN and transported the concept to Australia, where he taught at the University of Queensland. Planners in the Eastern Bloc had appropriated systems analysis as a creative field that gave them both professional authority and international influence.

Computer Cities in East Germany

In East Germany as well, *cost efficiency*, *rationalization*, and *cybernetics* became official buzzwords. Its failed Five-Year Plan of the 1950s led to the New Economic System (NÖS), which privileged scientific and technological innovation: "Take a stand for innovation and learn!" proclaimed president Walter Ulbricht.[78] The GDR became a pacesetter for the Eastern Bloc on scientific management, opening the way for a new generation of technocratic elites to be set in place with the NÖS. They adopted cybernetics and systems theory as fully achievable only in advanced socialist countries. Philosopher Georg Klaus asserted in 1961 that cybernetics was the most impressive confirmation of dialectic materialism. Technical experts were brought in to the Socialist Unity Party (SED) as advisers. Rather than stemming from ideologically driven directives, planning for the future would depend on scientific models. By the Seventh Congress of the SED in April 1967, Ulbricht began talking openly about systems thinking and the work of Norbert Wiener. An advanced socialist society could be thought of as a total system (*Gesamtsystem*) composed of self-regulating subsystems. Information and data processing enabled a scientific management of these complex societal structures "by the people for the people. In this context we will make full use of cybernetics,"[79] in Ulbricht's words.

Responding to the SED's call for an advanced socialist system, the Deutsche Bauakademie (German Building Academy) opened debates about a "new stage in urban planning" at conferences and study trips, and in numerous articles in the architectural press. The science of urbanism was based on four factors, of which the first was "atomic power, automation and cybernetics as the new productive forces." These, along with increased productivity, urbanization, and respect for history and the environment, would produce cities that were "larger, more differentiated, more complex, denser, higher, more mobile, lively, urbane, and humane. Of course, these progressive ideas can only flourish in the socialist city, where careful planning has replaced the profit motive." Among questions socialist planners should ask is how their work can be put into mathematical form, and how international concepts could be put to use.[80] The consequences were enormous: regions throughout East Germany were reimagined as unified systems of production, with cities and towns situated along planned transportation corridors.

In addition, each new town itself was planned as a unified *system* of carefully arranged buildings, spaces, and traffic arteries. Bauakademie planners began developing computerized "settlement structure models" known as MTTKOS, PET, and KOMS.[81] Rather than the earlier heroic archetype of Stalinstadt as a discrete, bounded place with its own inner logic, towns were now nodes in the circulation of people, goods, and information. Their form followed their functional role in the regional, and ultimately national, economic system.

Richard Paulick (fig. 4.10) was the premier new town builder in East Germany. As one of the leading architects in the GDR, his concepts were implemented in the new towns of Hoyerswerda, Schwedt, and Halle-Neustadt. Paulick arrived at systems thinking after a long career that spanned the most influential planning traditions of the mid-twentieth century. He entered his profession when modernism was cutting-edge architectural design, working as Walter Gropius's assistant at the Dessau Bauhaus. He was involved with the Metal Prototype House and the Dammerstock Housing Estate in Karlsruhe as experiments in the Neues Bauen (New Building) architectural style.

Paulick fled to Shanghai in 1933, when National Socialism rose to power in Germany. As the first to hold a university chair in urban planning in China, he prepared for the postwar world by studying Clarence

4.10. Richard Paulick (*center, pointing*) leading a tour of Hoyerswerda, East Germany, for members of the Polish parliament, June 30, 1959. Bild 183-65512-0001 / Photograph by Erich Zühlsdorf. © Bundesarchiv.

Perry's neighborhood unit concept, which by that time had entered into the worldwide professional canon. With the Shanghai Town Planning Office, Paulick initiated that city's first comprehensive metropolitan plan for the Nationalist Chinese government. It resembled Patrick Abercrombie's Greater London scheme and was structured around the garden city and neighborhood unit ideals.[82] Returning to Europe in 1949, Paulick traveled briefly in France and Italy, and then reached a shattered Berlin in 1950. He adapted himself to whipsawing GDR urban policy and quickly moved up the ranks to become vice president of the Deutsche Bauakademie.

Paulick's genius lay in his ability to incorporate the newest planning techniques into GDR urban policy. He was an internationalist, at ease with the latest Western trends as well as those from the Soviet Union, and was one of the GDR's earliest and staunchest advocates of mass-produced housing. The *Plattenbau* (prefabricated) housing estates that popped up across the GDR reiterated the landscape of construction that spread across Europe in the mid-1950s as well as in the new towns of the USSR. Outsized panels and plates were churned out in series directly in the factory as walls, floors, and roofs. They traveled to the building site on trucks and then were joined onto the building skeletons by huge cranes. Almost like magic, entire apartment blocks appeared. They were among the most joyous spectacles in postwar Europe: people would finally have decent homes.

The GDR's new towns were the testing ground for these system-built construction techniques, which solved the national housing shortage quickly and cheaply. Hoyerswerda, Schwedt, and Halle-Neustadt contained the largest stock of brand-new *Plattenbauten* in East Germany, and the demand for them was enormous. They were a major step toward planning entire cities as cohesive systems. Each of Paulick's new towns demonstrated this progressive incorporation of the systems ideal, from prefabricated housing to the use of computers in design.

The new town of Hoyerswerda was the original pilot project for system-built housing, and was one of the largest construction projects of the GDR's Second Five-Year Plan (1956–60). It was located near the new border with Poland, in the Lausitz region that had once been part of Silesia. The old town of Hoyerswerda had been heavily damaged during the war by the invading Red Army, and its population dwindled to only seven thousand. The value of the site, however, was the rich lignite coal deposits nearby. In 1955, the Schwarze Pumpe coal-fired power station was built there to supply gas for much of East Germany, and a new town was envisioned for some forty-eight thousand people northeast of the old town ruins.

Hoyerswerda was to showcase the planning and technological skills of the new socialist society. Population forecasts for the new town were worked out using the newest computer programs. "Under capitalism," wrote Paulick, "the town site would have been based on the needs of executives and white-collar workers. . . . In socialist society, such considerations play no role. We are not only able to regulate the development of productive forces, but distribute them among the regions of our republic in a properly planned way based on scientific principles, as well as layout the homes of our working population based on their needs."[83] The construction workers building the town and the skilled employees at the Schwarze Pumpe station would alter the social composition of the region. It would be made modern.

The groundbreaking ceremony took place in August 1955. The town's new power station included gasworks and electrical distribution facilities, administration buildings, restaurants, and club rooms for the employees, much of it designed by architect Hermann Eppler in the tradition of Neues Bauen formalism. The state-of-the-art technical standards for the *Plattenbau* housing complexes (fig. 4.11) at Hoyerwerda were described with relish in the pages of the journals *Deutsche Architektur* and *Architektur der DDR*.[84]

4.11. The residential district of Cottbus in Hoyerswerda, East Germany, 1962. In five years, the population of Hoyerswerda tripled. Bild 183-94925-0001 / Photograph by Horst Sturm. © Bundesarchiv.

Groß-Zeißig, the first fully mechanized, large-scale plate and panel manufacturer in East Germany, began production in Hoyersweda with a contract for seven thousand apartment units per year. Paulick took over leadership of the project with the goal of a completely industrialized city in mind: "Our objective in building Hoyerswerda is to unify technological production and assembly. The construction of Hoyerswerda is an experiment in economic, structural, technological and architectural planning. . . . The entire city will be prefabricated."[85]

Each of the seven self-contained residential complexes functioned as a complete urban system. Each would have its own shops and services, and its own commons and gardens at its center, with schools located along the edge. Although Paulick began developing a series of theoretical postulates on socialist design during his work on Hoyerswerda, he was also following in the footsteps of urban reformers from Clarence Stein to Swedish sociologist Alva Myrdal and architect Sven Markelius (see chapter 2). He saw the housing (or apartment) block as the smallest unit of social organization. Its significance lay not only in its physical layout but in the art, culture, and above all social life it sustained. The residential-complex ideal carried out the socialist emancipation of women by providing all the needs of daily life within its precinct. The apartment block would improve public health through strict zoning and protection from noise, traffic, and pollution. It would integrate the best of town and country with its fresh air and green areas for outdoor activities. And most important, interpreted as a system-built residential complex, it would solve the chronic housing shortage wreaked by capitalism.

The result was that Hoyerswerda's seven linear residential districts, each made up of their apartment blocks of varied height and length, were strung along Bautzener Allee in the Neustadt as discrete entities separated by open space. The segregation created by the flat, horizontal planes between them assured the distinctiveness of each district. The city center mirrored this sweeping perspective. The old socialist vision of the lavish ceremonial square so much in evidence at the GDR's celebrated postwar new town of Stalinstadt was cast aside. Richard Paulick instead looked toward Sweden's utopia at Vällingby for inspiration. As at Vällingby, Hoyerswerda's city center was planned as an open plaza at the intersection of the railroad station and the main avenue to the Schwarze Pumpe facility. This "flat center" was to be surrounded by great department stores, specialty shops, and services that "offered a complete product range in a fully developed socialism." Along with them would be cinemas, cafés and restaurants, and the House of Culture.[86] Although a Marx-Engels monument was planned as its focal point, the design of the socialist new town looked increasingly like its Western counterparts.

The production of full-scale models, diagrams, and illustrations for Hoyerswerda and Halle-Neustadt was extraordinary. Architects set out the *Plattenbauten* like dominoes across blank sheets of paper. Their spatial composition appeared to be decisive to the new town's success: housing blocks in parallel lines, at right angles, in diagonals, and in squares stretched across the fictional cityscapes. It was a matter of diagrammatic aesthetics and ethics. The nonhierarchical configurations represented a cooperative society in which all shared and participated equally within a socialist urban system. These were pure dreams, instruments of suspended reality, and a utopian apparatus.[87] East German architects fixated on designing a spatial realm that would create social cohesion— an obsession shared among architects and planners in the second half of the twentieth century. It yielded an overwhelming corpus of elaborate drawings and diagrams that defined planning as professional expertise.

Graphic illustration represented a new symbolic language. The design fantasies took on a life of their own—produced and reproduced as visual texts of the future. They also shaped a mutual futuristic fantasy of urban life. Illustrations of the British new town of Milton Keynes accompanied the epic descriptions of Hoyerswerda in the pages of *Architektur der DDR*.[88] Drawn by artist Helmut Jacoby, they are a glamorous mirage of an open city with futuristic roadways spreading out across the region. Hoyerswerda was the socialist version of that dream.

It is ironic that despite this passionate effort to build a socialist paradise, the *Plattenbau* complexes at Hoyerswerda came to symbolize the failure of modernism more than any other East German site. Despite Paulick's devotion to the system-built social environment, the repetitive prefab concrete buildings were a monotonous, stultifying landscape. Variety was sacrificed to "the socialist principle of order."[89] Each complex was home to some 4,500 and 5,500 people. The first playgrounds were not built until ten years after the town's founding.

East German author Brigitte Reimann moved to Hoyerswerda in 1960 to become a laboratory assistant at the Schwarze Pumpe power station. Her best-selling novel *Franziska Linkerhand*[90] is a searing indictment of the social isolation, indifference, and violence that became symptomatic of the housing estates. In the novel, a young draftswoman working in a local design office challenges housing policy and finds her views pitched against those of Schafheutlin, the chief architect of the town and a stand-in for Paulick. Franziska lives in the dystopian inverse of his system-built ideal town—a barren, deadening environment littered with garbage and trampled soccer pitches. The divergence between the planning fantasies and the disheartening realities of everyday life may have been Hoyerswerda's most salient feature, and certainly its most notorious one.

One cause for these conditions was the number of people streaming into the city for jobs at the Schwarz Pumpe facility. The computerized population forecasts for Hoyerswerda were far too low. By the mid-1960s, 35,000 people were already living there while construction workers were still slapping up panels on the building frames. That number jumped to over 53,000 by 1968 and 73,000 by 1980. The urban amenities, the cafés and terraces around the city's central plaza, were nothing but phantoms, dreamscapes on the drawing boards. Housing construction took precedence over everything. In the process, the miserable environment doomed the *Plattenbauten* even as they were being built.

Finally in 1968, the Centrum Department Store opened its doors to great celebration. Inside, consumers could find the household goods, especially the new refrigerators and washing machines, the radios and television sets that were finally becoming available in East Germany. The fashion-conscious could inspect the newest styles in GDR clothing. Even at its worst, Hoyerswerda was still the experimental site for the consumer meccas and commercial zones that were replacing the factory as a common socialist space.

However, it was the new town of Halle-Neustadt (or Halle-West, as it was initially called) that was most consciously designed and built from systems logic (fig. 4.12). It was a networked city par excellence in the transportation infrastructure sense of the term. Halle-Neustadt was a major transportation hub, with rail, road, and inland waterway connections spreading like veins throughout East Germany. It was the heart of a fluid circulatory system, a cybercreature imagined as an organism, yet created and run by machine. It was planned around the Leuna and Buna chemical plants on the outskirts of the manufacturing city of Halle, situated along the Saale River in Saxony-Anhalt. GDR president Walter Ulbricht and Socialist Unity Party officials were directly involved in the decisions about the new city's form. Richard Paulick brought in teams of urban planners and architects, economists and sociologists, engineers and computer experts to consult on

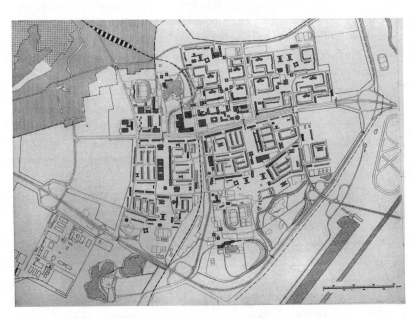

4.12. The plan for Halle-Neustadt, East Germany, date unknown. Photograph by Fototechnische Werkstätten. Courtesy of the Leibniz Institute for Regional Development and Structural Planning.

a master plan that integrated every aspect of urban life and infrastructure from a regional point of view. Mathematicians and computer engineers pored over technical data that was keyed into early computer programs to forecast transportation flows and create design models. Ideology was relegated to the sidelines. As in the West, systems analysis shifted decison making to cost-benefit analysis and the most efficient deployment of scarce resources.[91]

Halle-Neustadt was greeted with wild enthusiasm. It was one of the largest town planning schemes in the GDR, involving a hundred companies and some four thousand workers. Technical experts from Poland, Czechoslovakia, Bulgaria, and Yugoslavia were called in. The ever-present worker Youth Brigades hammered away. It was in building Halle-Neustadt that the young workers of the Artur Becker Youth Brigade were trained in the newest construction and engineering techniques. In exhibitions and film shorts, official publications and articles in the press, the city was flaunted as a futuristic apparition about to be made real. Fidel Castro visited in 1972, as did delegations from North Vietnam, the Palestinian Liberation Organization, West Nigeria, and France. The most spectacular iconography occurred during the visit in October 1965 by Ulbricht and Soviet cosmonaut Alexey Leonov. In an official photograph of their tour splashed in newspapers and on television, Leonov smiles broadly as he peers down at a scale model of the fantasy city on display.[92] The image production was a form of Pop culture and public education, an indication of how the Socialist Man was to live and work in the future.

The real embodiment of Halle-Neustadt was children and young people relishing their lived environment and the benefits it offered. "Halle-Neustadt was a symbol of the new," the official publication, *Halle-Neustadt: Plan und Bau der Chemiearbeiterstadt*, exclaimed. "It is a magnet for youth." In its photographs, the dreamscape of social utopia comes alive in the smiling faces of children in playgrounds, young couples promenading through the parks with their baby prams, and children planting trees, cavorting in fountains, walking to school. This imagery of a new socialist generation was codified in the symbolic laying of the foundation stone for the new town, which took place during the groundbreaking for the city's polytechnical high school. The city had developed, *Halle-Neustadt* continued, "into a complex aggregate of work, housing, culture, trade, administration, technical and organizational needs. . . . This requires a harmonious blend that forms the structure of the city." Creating an integrated system such as this had been impossible under capitalism. Under socialism, it could be achieved for the first time in history.[93]

As at Hoyerswerda, Paulick situated the residential districts along the linear paths of the main highway and rail lines. They were distinct, autonomous entities separated by open space. The scale of the *Wohnkomplex* at Halle-Neustadt was enormous in comparison with the earlier new towns, accommodating some fifteen thousand residents per district. Mammoth building sites and standardized apartments became the norm. As for building materials, Paulick and his team of architects experimented with everything from gravel and ceramic to advanced materials such as fiberglass and aluminum. They tested colored plastics, movable modular wall panels, and precast concrete shells in hyperbolic shapes for the roofs of public buildings. All these were physical manifestations of communism's access to and borrowing of Western advances. Paulick wrote avidly about American construction technologies in the GDR architectural press.[94]

The layout of each complex was codified in the configuration of interior spaces enclosed by the massive *Plattenbauten*. For Paulick, these spaces were essential to sustaining collective life. The neighborhood center for Wohnkomplex I (fig. 4.13), for example, was a sleek, modernist *Kompaktbau* shopping mall constructed in reinforced concrete masonry and surrounded by gardens and playgrounds. Praised for its economy and functionality as well as its sophisticated style, the *Kompaktbau* offered all the services required of daily life: supermarket and restaurant, post office, pharmacy and infirmary, hairdresser and cleaners, club rooms and auditorium. It was within walking distance of every apartment in the complex. More than just a shopping center, it was meant to be the collective heart of the community. Architect Erich Hauschild's nearby Buratino kindergarten, designed in a disk shape with curved sheet-glass walls, exuded the 1960s Space-Age aesthetic. The ensemble was a transcendent interpretation of Paulick's apartment block, or neighborhood social environment—the ideal that so captured new town planners across the twentieth century. The designs would "shape themselves into consciousness and contribute to an emotional identification with the environment." In quoting Kevin Lynch's *The Image of the City* (1960), the official description of Halle-Neustadt raised the hope that with the correct design, the harmonious, intimate feeling of neighborhood that Lynch described could potentially be extended to the city as a whole.[95]

However, it was the dreamscape for the town center of Halle-Neustadt that most clearly embodied systems thinking and its fusion with the communist new course. It exuded a hypermodernity of high-rise towers, shopping malls, and sophisticated urban life with cafés and restaurants.[96] It was a prestigious setting dedicated to the culture of consumption rather than to political ideology. From the mid-1960s, car production and highways

4.13. Apartment Complex I at Halle-Neustadt, East Germany, with the "Buratino" kindergarten by architect Erich Hauschild, date unknown. Bild 183-H0909-0009-001-1 / Photographer unknown. © Bundesarchiv.

became yardsticks of the GDR's economic achievement. Consequently, Halle-Neustadt's town center was sheathed in a cocoon of expressways and intricate road systems, overpasses, and roundabouts. They stretch out into the surrounding region. In architectural illustrations of the town center, the highways cut past sleek pedestrian plazas and multilevel shopping malls reached by escalators. These drawings comprise a sensational, glam modern pictorial gallery—the unveiling of a utopian world that combined urbanism and science fiction. The city's life streams between glimmer-glass towers in an entirely new public domain.

Halle-Neustadt was modernist eye candy identical to that of British and French new towns of the 1960s, and built from the cybernetics taken up by planners as scientific creed. On both sides of the Cold War divide, cybernetics became the future of cities. It conjured top-down utopias with the master builder at the helm, whether that was William Pereira in the United States or Richard Paulick in East Germany. At their side were legions of planning experts, computer jockeys, and construction engineers. Their new town fantasies were reliant on massive state investments in housing, transportation, and physical infrastructure.

More than exacting scientific method, the cybernetic city was also a broad intellectual production. It shaped a virtual knowledge that rendered metropolitan areas in mathematical and diagrammatic form. For the most avid partisans of cybernetics and systems analysis, designing cities was entirely a mechanical task, a linear correspondence between ends and means. The whole process of design, forecast, and evaluation could be done in one epistemic process called "end-reduction" that left no need to decide between strategies. The ideal plans would be provided by machine intelligence. Human settlements would enter a new age of optimization. The capacity to construct urban utopia inside the machine, and metaphorically as a cybernetic machine, was mesmerizing. It infused city building with new rigor and vastly expanded the authority of planners themselves. It created a mystique of planning as hard science and the planner as a kind of virtuoso akin to the mathematician. The precision and pure logic involved in the cybernetic process generated the sensation of power and control. From the new science were derived visions of the city, of social and economic modernization, of regional and urban studies, and ultimately of a future way of life.

FIVE

Towns of Tomorrow

Energized by the new systems dynamics, international planning enjoyed its heyday during the 1960s. A mini-industry of conferences and seminars, project collaborations, and professional organizations sprang up that gave planners the chance to soak up the latest in trailblazing techniques. This global production network operated as an expert workforce and key axis for the modernizing regime. Within it, new towns were invested with almost Olympian transformative powers.

Michel Foucault argued that modernization processes were both totalizing and homogenizing. State policies and the planning typologies and hierarchies that built modern cities were a disciplinary apparatus. They were a modern way of making the urban environment normative. Indeed, the new town seemed to be the very symbol of the unbounded authority that Foucault identified, reiterated as it was across the global landscape with a sameness that could be stupefying. Countries with the most imposing state apparatus carried out the most complete versions of the new town ideal, with the Soviet Union being the prime example. But even in those countries that eschewed top-down government planning, such as the United States, new town ideology was embraced with enthusiasm.

Despite this universal promotion of a cookie-cutter utopia, planning then as now takes place within the context of local circumstance. Once the surface sameness of the projects is pierced, even new towns are specific to place. The ethos of starting from scratch spread worldwide, with all sorts of idiosyncratic modifications. This chapter looks at "towns of tomorrow" experiences in four countries during the 1960s—Britain, France, India, and the United States—and the application of systems logic in land-use planning. The scope is comprehensive and expansive, across regions that were imagined as cybernetic webs of information and communication flows.

It places the new town movement squarely in this context of regionalist thinking.

Regionalism and metropolitan planning had been espoused and carried out since the days of Patrick Abercrombie's *Greater London* Plan of 1944. But the impact of the systems revolution was profound, particularly when it was articulated across regions from an aerial perspective. Vast areas became open to maximum visibility and complete mastery. In the case of Britain and France, new towns became nodes in regional land-use programs that increasingly resembled the circuits and synapses in a computer motherboard. In India, the new town of Navi Mumbai was set in place to expand and systematize urban territory. In the United States, the footprint of modernization logic and new town development was stamped by private real estate interests.

From the standpoint of an intellectual history, the four examples reveal the dynamic ways in which the currents of international knowledge that produced new towns were filtered through local circumstance, and served different political and social interests. Planners and policy makers tacked back and forth between a shared corpus of concepts and models, and their own particular urban contingencies. Each locale developed its own unique brand of utopia. At the same time, these locally tailored experiments in futurology were scrutinized and pulled into the vortex of planning knowledge worldwide. Like Tapiola in Finland and Vällingby in Sweden, the new towns discussed in this chapter were the megastars of the movement, enthusiastically studied in the 1960s as glimpses of the future. The architects and planners who created them achieved celebrity status. As well as being a concrete manifestation of the urban future, the new towns' very *ideal* became an apparatus of scientific exchange and an exercise of professional power. Thus, the new towns were both utopian dream and techno-scientific object. The outcome of all this was that the new town represented an achievement by a wide array of planning experts and institutions at composite scales: local, national, and international. This exchange and circulation of knowledge produced a kaleidoscope of experiences.

Most important, this knowledge production was undertaken by progressive-minded international organizations. The UN maintained a regular schedule of conferences and seminars on regional development and new towns. The first meeting of a Group of Experts on Metropolitan Planning and Development was held in Stockholm in 1961.[1] A second Symposium on the Planning and Development of New Towns was held in Moscow in 1964. The United Nations Summits on new towns and land-use policies continued in Nagoya, Japan, in 1966 and in London in

1973. They were great gatherings for thrashing out the latest policies and planning techniques from around the world, including on the communist side of the Iron Curtain. New town enthusiasts from the West, from the developing nations in Latin America, India, and Asia mingled in the rarified atmosphere of seminars and speeches, cocktail parties and general camaraderie. Local planners from poor countries struggling to enact urban reforms rubbed shoulders with the jet-setting elites of the new town movement. Constantinos Doxiadis, James Rouse, Richard Llewelyn-Davies, Derek Walker, and Barbara Ward were all regular participants at the UN conferences. They were among the brightest stars in the new town planning galaxy, waxing eloquent about their visions of the urban future to rapt audiences.

The accent in these meetings was on managing the explosive growth of the global population, especially in the developing world. The UN began crafting what became its stock message on urbanization: "Not only is the population of the world now doubling in size . . . but the larger share of this growth goes to the already urbanized areas, and its heaviest concentration is experienced in the great metropolitan regions." Hysteria over the "population bomb" was fueled by the 1968 publication of Paul Ehrlich and Anne Ehrlich's international blockbuster *The Population Bomb*, which predicted catastrophic upheaval and mass starvation in the 1970s and 1980s as a result of overpopulation. Cities would fall into complete chaos. Moreover, the glaring problems of overcrowded cities and overused infrastructure seemed to defy all efforts at improvement. Poverty and slums were normal conditions for millions of people. The remedy, according to the UN, was modern technology and science, which offered the prospect of population control and "a wider range of possibilities for the distribution of people and settlements, and for raising levels of living." Comprehensive national and regional planning provided the common ground for reform: "Our world society must learn to guide the process of urbanization."[2]

At the UN's 1964 Moscow symposium on new towns, Barbara Ward's speech underscored the coming upheaval and the urgent need for answers:

> Urban evils, unchecked, could mean the destruction of the very framework of stability on which the economic growth needed to end the evils must depend. Urban mobs, driven to despair by privation, can become so disorderly an element in national life that rational government itself is impeded. Modernization demands a certain stability, rationality and continuity. Leave the cities to degenerate into sinks of despair and nations may lose the means of ever climbing out of the pit.[3]

It was both a call to save humanity from the brink of disaster and the expression of hope for what Ward called a "world policy for urbanization."

British born and a baroness by marriage, Ward was an economist and journalist and a recognized authority on developing countries, especially India and West Africa. In her groundbreaking book *Spaceship Earth* (1966), she advocated a holistic approach to sharing world resources and argued that developed countries should donate a portion of their GNP to aid for the Third World. A member of the World Society for Ekistics and adviser to policy makers in the United States and Britain, Ward was at the pinnacle of her career and among the most vocal evangelists in the new town crusade. She and the experts assembled at the United Nations Summits in Stockholm and Moscow recommended that a central planning agency be set up in each country. They lobbied for building new towns as a method for the systematic distribution of populations at the local, regional, and national levels. Ward suggested that a number of developing countries be designated as official experimental sites for new town planning strategies. These projects would be based on the flow of international aid and the technical assistance of planning experts. The Moscow conference endorsed Ward's call and recommended that planning research centers be set up for "the systematic collection and analysis of data," and to "bring together specialists in all aspects of development, such as economics, sociology, public health, engineering, physical planning and architecture."[4]

The UN conferences were an indication of the dramatic power of the new town movement. By assuming the mantle of science, urban planners had been given tremendous authority. They were being called on to solve some of the world's most intractable problems, and new towns were their most potent tool. Systems analysis had crossed Cold War frontiers, passed into developing countries, and become a generalized instrument for modernization worldwide. Then in the mid-1970s, the International Association of New Towns (l'Association Internationale des Villes Nouvelles, or INTA) was formed, with headquarters in The Hague. It was the main organization through which professionals practiced new town planning as scientific discipline. INTA produced a steady stream of gatherings and publications as "a forum for debating and resolving the problems confronting all who are involved in the conception, planning, building, financing, management and development of new towns at both national and local levels."[5] For example, at INTA's first international congress, How to Build a New Town, held in Tehran in December 1977, planners from Poland and Hungary debated new town strategies with members of the United States New Communities Administration, with American new town developer

James Rouse, and with the British planning personalities from the new town of Milton Keynes.

British New Town Systems: Experiments in Freedom

Although it could show off some of the most celebrated new towns of the reconstruction era, by the late 1950s Britain had fallen into a mood of general apathy toward utopian projects. Serious doubts arose about the dizzying array of schemes that had flown off the drawing boards in the initial postwar enthusiasm for building towns from scratch. The logic and financing of those massive infrastructure drives had drawn immediate skepticism from political opponents, to say nothing of public misgivings. New town ambitions had generated fevered scientific research, a flood of planning studies, and the inevitable seminars and conferences. New university programs appeared, new planning courses developed. Yet the valiant policies of the reconstruction years that built the first generation of new towns were pronounced disappointing and outdated. Despite the heroic efforts at redistributing industry, economic growth still gravitated toward the southeast and the powerhouse of London, while the north languished—and as a result, the postwar new towns were waved off as failures in stimulating economic decentralization and solving the problem of widening regional disparities. By the late 1950s, eminent regional geographer Gordon Cherry found British planning to be in the "doldrums . . . the planner himself became unpopular and his main work was unspectacular . . . the inspiration of the 1940s had gone."[6]

Then suddenly, in 1961, the British government began designating a second wave of new towns, and then a third wave starting in 1967. The climate of political opinion swung once again in favor of comprehensive planning for the modernization of British society. Both sides of the political spectrum, from Conservative prime minister Harold Macmillan to Labour prime minister Harold Wilson (both of whom led the government in the 1960s), understood the need to strengthen economic development and promote technological innovation throughout England. Particularly for the Labour Party, stimulating new economic activities in the country's declining regions went to the heart of their ideology. Wilson called for regions to "embrace the white heat of technology" and switch production to rising sectors such as plastics, electronics, and automotive engineering.[7]

The second factor prompting more new towns was Britain's increasing population and the baby boom generation's arrival to adulthood. As teenagers and young adults, their demand for goods and services was voracious.

The rationing and soberness of the reconstruction years had given way to material consumption. Access to an array of urban amenities, from shopping centers, movie theaters, and restaurants to recreational and sports facilities, had become normal expectations. The information technology revolution was in its infancy, but the enormity of its impact on daily life was already tangible in the rush to buy telephones, televisions, and household electronics.

It was simply expedient for government to rely on new towns for managing both economic growth and the urban lifestyle demanded by a new generation. These changes forced renewed attention on experimental urban form. Eventually, fifteen new towns were designated throughout England, Scotland, and Wales during the 1960s. Other mint-condition towns were also constructed, though not given official new town status. Together, they amounted to an entirely new urban landscape.

However, the most systemic changes in the land-use panorama resulted from the automobile. Unlike Italy or Germany, Britain was late in developing a national highway system. Fleets of cars lumbered along antiquated roads in a perpetual state of traffic congestion. Highway interchanges around Birmingham and Glasgow were notorious logjams. The automobile age altered everyday urban life and pitted pedestrians against motorists for control of the streets. It made enormous demands on road infrastructure, a reality brought out in a major national conference sponsored by the British Road Federation on Urban Motorways in 1955. The gathering featured a cavalcade of traffic and planning experts from western Europe and the United States, including Robert Moses, who argued the benefits of motorways for business and property values as well as for public safety. A short stretch of the M1 Highway was finally opened in 1959, but at best it was a measly attempt to improve the generally miserable conditions.

The situation was then laid bare by the 1963 Buchanan Report, known formally as *Traffic in Towns: A Study of the Long Term Problems of Traffic in Urban Areas*. It was initially commissioned by Ernest Marples, Transport Minister in the Conservative Macmillan government, and then produced by a working group inspired by leading town planner Colin Buchanan. The report was grabbed up so quickly that it was immediately republished in an abridged version as *Traffic in Towns*.[8] It was instant headline news and had almost talismanic standing among planners worldwide. It was acknowledged immediately as one of the most significant planning texts of the late twentieth century.

The publication of *Traffic in Towns* coincided with a media campaign by the Ministry of Transport to convince the country to invest in highway infrastructure. A BBC broadcast of "Our Strangled Cities" on the popular *Panorama* television show in November 1963 warned of urban breakdown and environmental ruin if something was not done.[9] Buchanan himself had traveled to Germany to study the Autobahn and also to the United States, where he witnessed the automobile's deplorable impact on Los Angeles, and warned of the car culture's insidious threat to the fabric of British life. His way out of these calamities was straightforward. He believed that traffic should be rationalized through highway design and land-use planning. *Traffic in Towns* made the case for a clear-cut separation of people from cars. It was the only way to survive in the automobile age.

The *Traffic in Towns* publication was filled with stunning illustrations of traffic architecture drawn by Kenneth Browne, a well-known artist on the staff of the *Architectural Review*. They created a visual drama of multilevel cities, pedestrian decks hovering over motorways, and bold thoroughfares hurtling out across the countryside. Such a rapturous embrace of highway experimentalism was also evident in architect Geoffrey Jellicoe's plan for the new town of Motopia in Middlesex outside London. It was a dreamscape of elevated highways and spiral roundabouts, with traffic circulation piped directly into the buildings, like drainage and water. Car driving would be on rooftops in the "glass city of the future." Ground-level moving sidewalks separated pedestrians from the automobile traffic overhead. Jellicoe created a three-dimensional model of his city that was captured in a film short by British Pathé and shown in theaters.[10] The imagery flashed across the Atlantic, where Americans were given a glimpse of Motopia by illustrator Arthur Radebaugh. He depicted his own Space-Age version of the sensational city in his Sunday comic strip *Closer Than We Think*.

British planners seized the moment. The Buchanan Report and systems analysis became a breathtaking scheme for carrying out the second generation of new towns. The city became an invention of networks and systems, speed and fluid circulation. The British use of systems analysis was supported by the Ford Foundation and by regular collaboration between American and British planners and policy makers. American planners such as Britton Harris and Melvin Webber were widely read and regularly consulted. With a five-year grant from the Ford Foundation, British politician Richard Crossman (Minister for Town and Country Planning), along with David Donnison and Richard Llewelyn-Davies, established the Center for Environmental Studies (CES) in 1966 at University College London.[11] It was a pivotal gathering place for urban and environmental research in Britain. Planning for the second generation of new towns was among its major goals.

Llewelyn-Davies, who was the main force behind the CES, was also one of the most celebrated architects in Britain. As a leading midcentury urban reformer, he shared in the conviction that social science was the means to solve long-standing urban problems. Although a modernist, he rebelled against much of the architectural avant-garde and focused instead on comprehensive town planning as the way forward.[12] Created a Labour peer in 1964, Llewelyn-Davies was also head of the Bartlett School of Architecture at University College London, and ran a thriving architectural firm with an enviable track record of successful international projects. In 1965, he was president of the World Society for Ekistics. Throughout the 1960s, he exerted profound influence over new town ideology, and was one of the leading proponents of systems planning.

Llewelyn-Davies also emerged as a key figure in the close American-British partnership on new town planning during the 1960s and 1970s. He traveled extensively through the United States, spent a year in residence at the Institute for Advanced Studies at Princeton University, and had close professional and personal ties. Llewelyn-Davis Associates consulted with the New York State Urban Development Corporation on its new community project at Amherst, Massachusetts, and from time to time consulted with James Rouse on his new town of Columbia, Maryland. Llewelyn-Davies's affair with American socialite Marietta Tree also gave him direct access to the liberal Democratic Party establishment. In addition, she helped promote and finance many of his US projects, and accompanied him on the sought-after Delos Symposia cruises.[13] The result was that Llewelyn-Davies was instrumental in cementing an array of British-American partnerships on new towns.

Melvin Webber spent a year at the CES and participated in a working group with Richard Llewelyn-Davies for studying patterns of future urbanization in Britain. American sociologist Herbert Gans, author of *The Levittowners*, the celebrated study of American suburbia, was invited to the CES, and appealed for social science methods in urban planning. These professional networks were not limited to the CES, however; they were also honed at gatherings such as the 1968 Anglo-American Conference on New Towns. Agreements were signed between the British Department of the Environment (DOE) and the Ministry of Housing, and the US Department of Housing and Urban Development on the exchange of ideas and personnel. The DOE also sponsored a United Nations Seminar on New Towns, held in London in June 1973,[14] in which Llewelyn-Davies played a prominent role as interlocutor.

The first generation of postwar British new towns was imagined essentially as individual communities, even when they were to jump-start regional revitalization. By contrast, the new towns of the 1960s were regarded as powerful growth points in a web of regional infrastructure and economic production. They were merged into comprehensive land-use plans, and staged along the expanding system of British motorways. To take advantage of economies of scale, the new towns of the 1960s were also much larger than those of the postwar generation, and were referred to as "new cities."

This more ambitious vision was made possible through systems analysis, which was enthusiastically adopted by British planners. They produced a remarkable series of regional planning studies, beginning with the *South East Study 1961–1981* (1964) for the London metropolitan region, produced by the CES for the Ministry of Housing and Local Government. The study estimated that by 1981, the population of metropolitan London would grow by another 3.5 million to 4.5 million, and recommended another generation of new towns to house some 1 million to 1.5 million people. The CES also completed planning studies for the new towns of Milton Keynes and Stevenage, and a study analyzing the impact of a proposed third airport in the London metropolitan region. Then came *West Midlands, a Regional Study* (1965) and the Scottish *Lothian Regional Survey and Plan* for the Glasgow-Edinburgh metropolitan area (1966). These achievements pushed the government of Northern Ireland to enact its own new town program in 1965, and eventually the *Northern Ireland Development Program* (1970–75).

Planners sought to legitimize these comprehensive land-use planning reports by characterizing them as carrying on the tradition of Patrick Geddes. But these new towns were made for car ownership, so highway engineers played a significant role in how these metropolitan areas were construed. Planners assumed one car for every two persons, far more than the estimate of one car for ten in the early days of new towns. Systems of cities and towns were particularly well suited to mapping out a modern network of national, regional, and local roads.

For example, the report made by Llewelyn-Davies on the Newbury-Swindon area outside London recommended a cohesive network of new communities designed around local roads and the new motorways. Each community would have its own town center, but all would be dependent on one another for a full range of urban amenities.[15] Llewelyn-Davies's plan for the new town of Washington in northeastern England followed the same logic. The design was developed around local roads and motorways, with traffic loads and trip forecasts worked out through mathematical modeling.[16] In 1971, Llewelyn-Davies completed a far-ranging study for the British Road Federation on the environmental impact of London's new motorways. He made a case for knitting transportation directly into urban design by applying a new computer model known as queuing theory.[17] The old idea of intimate, self-contained neighborhoods was swapped for urban systems centered on car ownership. Llewelyn-Davies took Los Angeles as his typology for an automobile city. The result, in his plan for Washington, was a spread of villages and local centers laced around roadways, with a shopping and a civic center, and a sports stadium at its heart.

In Scotland, the *Lothian Regional Survey and Plan* extended almost one hundred miles around the Glasgow-Edinburgh metropolitan area and encompassed motorways, an airport and a maritime port, and railroads as well as industrial incentives and four new university campuses. Both the scale and the comprehensiveness of all these regional land-use plans distinguished them from their predecessors. They were phantasms of wholly observable space from a bird's-eye view, and a mechanics of power that rendered vast areas legible, comprehensible, and controllable.

The *Lothian Regional Survey and Plan* designated eight regional growth points for the Glasgow-Edinburgh area, five of which were to be new towns. Among them, the new town of Livingston in the Almond Valley west of Edinburgh would be "the jewel in the crown," something the Scottish "could be proud of in all sorts of ways."[18] And indeed, Livingston became a much-anticipated stop on the annual international Study Tour of British New Towns, and was visited by hundreds of groups from around the world. In July 1972, six coachloads of passengers from a Soviet cruise ship descended on Livingston to see the futuristic city in the making. The 1976 promotional film *Livingston—a Plan for Living* opens with shots of young planners at their desks poring over materials, creating a utopia where "nothing is left to chance." The town along the Almond River offers everything imaginable to young families. Adult classes in a myriad of leisure-time activities offer self-improvement and personal enrichment. Hobbies and recreation fill time and space. Urban planning becomes social management, and determines what takes place in people's lives. "This is what community is about," the film's narrator confides to a prospective family looking over the ideal urban world. People come together "nowhere more happily than here."[19]

As portrayed in the film, Livingston is not some nostalgic reverie in the Scottish countryside. It is a buzzing regional capital with hundreds of new industries. It is productive and mobile. With its web of transportation, "all of Britain is just down the road." It is a growth pole ready to shape the future of Scotland. The urban design of Livingston was a decisive move away from the old ideal of insular neighborhoods. In its place, the town was spatially opened up around its walkways, underpasses, and road layout, which included the "most modern motorway in Scotland."[20]

A series of smaller communities were then spread out around Livingston's metropolitan area. The *Lothian Regional Survey and Plan* anticipated this kind of widespread growth throughout the Glasgow-Edinburgh region, which would unfold as a polynucleated urban web dependent on the automobile. Writing in 1972, planner Derek Diamond, of the University of Glasgow and a member of the Lothian Survey team, saw the concept as "not fundamentally very different from Ebenezer Howard's idea of 'social cities.'" The result of new town development would be that "the whole pattern of life in Central Scotland by 1980 will be radically different from that in the period 1945 to 1965."[21]

The exaggerated predictions about Britain's 1960s new towns matched those about systems analysis—and they echoed the idealism and aspirations of modernization. The most acclaimed of the British urban idols of that decade was Milton Keynes, England. It was *the* media darling—one of the most written-about, visually devoured, and celebrated places in the broad landscape of twentieth-century urban planning. Milton Keynes was the future: a bright, new, limitless world. The new town was conceived as part of the *South East Study* by the planning team at the CES. It would welcome some 250,000 new town pioneers, who would live in a paradise of cars and consumerism. Located in north Buckinghamshire,[22] it was by far the largest of the British new towns, and was to play the role of regional capital in the zone between Birmingham and London, and Oxford and Cambridge.

Milton Keynes was at the crossroads, literally and figuratively, of southeast England. It would be an incubator for startup electronics and IT companies, particularly those linked to the proposed third London airport. Even further, its concept plan was explicitly focused on social goals and the alleviation of poverty. The city would be committed to the well-being of all its members, including racial minorities. Residents' social needs and aspirations would be carefully monitored and evaluated through scientific measurement. Every dream, every ambition was attached to Milton Keynes. It was seen as the supreme utopian adventure, and clearly one of the most provocative.

In the long process of deciding how to design a city laden with these kinds of magical powers, Richard Llewelyn-Davies insisted that the object was to consider "some of the likely form giving elements of future urban situations." A parade of experts were brought in to advise on concepts for the city and to help convince the public of their worth. Among the experts was Melvin Webber of the University of California at Berkeley, who mesmerized audiences with his vision of the changes close at hand. The knowledge industries, including information handling and systems analysis, would be the key to economic growth. Webber gushed, "Milton Keynes will be, in a sense, a spearhead of this changing phase in urban civilization."[23]

Not everyone was convinced of this quixotic imagery or even understood it. Recounting a meeting on the plans for Milton Keynes, a reporter from the *Observer* quipped that "the language was of such high-flown political, statistical and sociological stuff that it left many of the so-called experts baffled." Systems analysis was not for the uninitiated. Nonetheless, "behind the huge smoke screen thrown up by mysterious terminology lay the basis of what may just possibly be, one day, an interesting place."[24]

The original design for the town's central area was cutting-edge modernism, in keeping with the 1960s ideal of the urban megastructure (see chapter 6). A spectacular multilevel edifice would separate traffic into horizontal planes, provide every service for residents, and look out over a picturesque lake. Four futuristic monorail loops whipped through the central zone. Frederick Pooley, the county architect of Buckinghamshire, designed this first dreamscape of Milton Keynes. Its enthusiasts declared the town would be "newer than Brasília, better planned than New York, more convenient than Paris, would do more for British prestige than a score of misguided missiles or a dozen failures to reach the moon."[25] It was a bold statement of nationalist pride and, unfortunately for Pooley, also laid bare the struggles that accompany utopian production. Pooley's plan prioritized public transportation and subordinated car use, while the goal of the master plan for Milton Keynes was the unrestrained embrace of the automobile.

Pooley's daring design was upstaged by Llewelyn-Davies and Walter Bor's idea for the city as a loose network, an undulating grid of highways following the contours of the natural landscape (fig. 5.1). It was an image of "geometrical rigour" that was "astonishing in its purity of form," wrote British architect Robert Maxwell, who likened "the beautiful city" to French urban tradition or to Frank Lloyd Wright's Broadacre City concept.[26] Rather than being centralized in one place or one megastructure, the urban functions of the new town were spread out indeterminately across a lattice of American-style highways in a deliberately low-density vision of the non-city of the future. It was a loose-fitting, extendable design that would grow

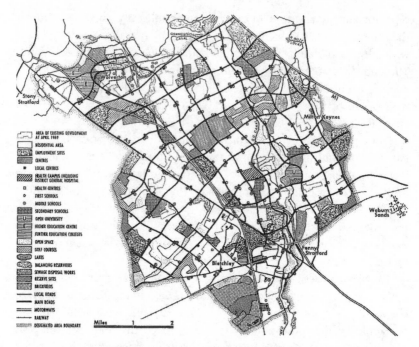

5.1. The master plan for Milton Keynes, England, by Richard Llewelyn-Davies and Walter Bor. Commissioned by the Milton Keynes Development Corporation. 1970. Courtesy of Llewelyn Davies, London, and Llewelyn-Davies Sahni, Houston.

through building envelopes, or modules, clipped onto the matrix.[27] Milton Keynes would be a city for the Pop age.

Traffic circulation patterns were woven directly into the urban form. Residents would act as a mobile network, circulating between various points in the city, speeding their cars like cybernetic impulses at seventy miles per hour along four-lane highways, selecting their commerce and services in complete freedom of movement and choice. Planners could pinpoint development at specific crossroads along the grid as Milton Keynes grew in response to needs. Underground, the city was wired with an extensive cable network for the computer age. In other words, Milton Keynes was saturated with hypermodernity and a modish vision of consumerism.

The city center was undefined and low density, legible by its openness and greenery. The shopping buildings in steel and granite with wide expanses of glass were intertwined with open courtyards and arcades (by Derek Walker, Stuart Mosscrop, and Christopher Woodward; fig. 5.2). It was a climate-controlled pedestrian zone, and at the time the largest shopping center in Britain. Overall, Milton Keynes's design was nonhierarchical and flexible to accommodate an individualized constituency of resident-consumers.

It was the antithesis of the earlier new town creed that by the 1960s was soundly condemned. "The whole concept of planning . . . has gone cock-eyed" was the verdict of the seminal 1969 article "Non-Plan: An Experiment in Freedom" by the modernist rebels Reyner Banham, Paul Barker, Peter Hall, and Cedric Price. The article took direct aim at the early new towns: "Stevenage has become a focus for junkies. . . . Harlow has been parasitical on surrounding communities." Instead of the professional class deploying the latest planning craze, the article called for the people themselves to shape the environment they lived and worked in. This could be done with the new American social science techniques and by imagining the physical environment through the lens of cybernetics as flows of people, goods, and information. "Non-Plan" waved the flag of cybernetics as "a major revolution in our ways of thought" that made traditional planning obsolete.[28] Machine logic would construct urban society. Although "Non-Plan" argued for no planning at all, there was a clear tension between its desire for urban vitality and spontaneity, and its advocacy of the formulas plugged into computers as somehow offering self-determination. The focus remained on planning *itself* as the medium through which social,

5.2. Central Milton Keynes Shopping Center, Buckinghamshire, England. Photograph by John Donat. © John Donat / RIBA Library Photographs Collection.

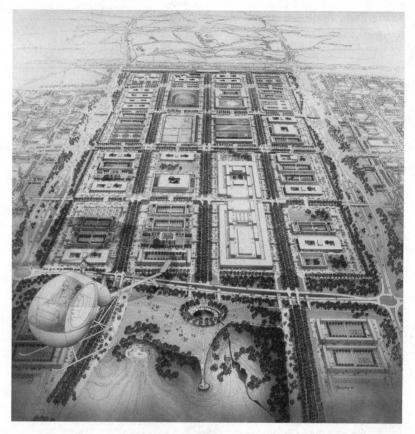

5.3. Central Milton Keynes, England, as visualized by architectural illustrator Helmut Jacoby, 1974. Commissioned by the Milton Keynes Development Corporation. Design by Derek Walker, Stuart Mosscrop, and the Central Milton Keynes Team. © Derek Walker and Homes and Communities Agency. Image courtesy of Milton Keynes Discovery Centre.

economic, and urban reform took place. Was it possible, Llewelyn-Davies queried, "to design a town which can grow outwards in all directions" and to design a new town to accept change?[29] Milton Keynes was the response.

The quixotic imagery produced of Milton Keynes by architectural illustrator Helmut Jacoby (figs. 5.3, 5.4) became a visual semiotics of the city of the future. His drawings remain one of the most significant material artifacts of twentieth-century utopia. Born in Germany, Jacoby immigrated to the United States and attended Harvard's Graduate School of Design, where he specialized in graphics. His illustrations were visual registers of 1960s modernism. Jacoby's renderings of Milton Keynes shape it as urban dreamscape. His sweeping aerial vision captures the city's geometric spatiality as open, mobile, exciting. It is a seductive, thrilling future world. A system of highways and vast interchanges extend the city outward, networked with the surrounding region. The hottest cars and nifty personal helicopters are seamlessly drawn into daily life. The people of Milton Keynes are citizens of the Space Age. Its natural environment is interpreted as a landscape frame into which towns could be positioned.

Architect Norman Foster designed an automobile complex of glass and steel in nearby Linford Wood that was a combination car dealership, exposition hall, and leisure center. Although Foster's project was never

5.4. Queen's Court in the Milton Keynes, England, shopping center as visualized by architectural illustrator Helmut Jacoby, ca. 1970s. Commissioned by the Milton Keynes Development Corporation. Design by Derek Walker, Stuart Mosscrop, and the Central Milton Keynes Team. © Derek Walker and Homes and Communities Agency. Image courtesy of Milton Keynes Discovery Centre.

built, Jacoby's illustrations of it present the wooded setting in which Milton Keynes's high-tech modernism would be staged. Greening was an organizing mechanism that naturalized modernist construction. Planned entertainment facilities for Milton Keynes such as Wolverton Agora and Granada Limited were high-impact modernist structures with grand-scale advertising and supergraphics. This techno-utopianism was the stuff of the future.

The media blitz and image-building on behalf of Milton Keynes extended to *Architectural Design*, Britain's premier avant-garde design magazine, which by the 1970s was also one of the leading voices of the systems approach. It devoted two entire issues to Milton Keynes in 1973. The articles were written by the city's chief architect and planner, Derek Walker, and his hip staff of young designers, all of them under forty years of age. Walker called them a "maverick quota" that would cogitate over their projects at Monday evening design meetings and at seminars with personalities such as Buckminster Fuller.[30] It was a heady time. Milton Keynes was brain food for young professionals of the Swinging Sixties. They cut their professional teeth working on new town projects.

The captivating drawings by Helmut Jacoby and by urban designer Andrew Mahaddie, along with the photographs of Milton Keynes by John Donat, appeared throughout the international architectural press. As had been the case for Tapiola and Vällingby, flashy marketing campaigns, big-budget television advertising, and promotional films made Milton Keynes a household name. "Wouldn't it be nice if all cities were like Milton Keynes?" a TV commercial crooned after a heartwarming two-minute journey with a young boy finding his way home in the new city. The catchy jingle, "You've never seen anything like it: never been anywhere like it," advertised the town's swank shopping mall. Milton Keynes was branded as "good for your health." It was "the kind of city you'll want your family to grow up in."[31]

Naysayers criticized the design of Milton Keynes as an attempt to import California-style suburban car culture to England, and it came to be called Los Angeles in Buckinghamshire. Censors wagged their fingers at Melvin Webber and his influence. Historian Mark Clapson argues that it "came to be the most Californian and therefore the most American city in Britain," and that it was an early "edge city."[32] But the goal for Milton Keynes was something far more sweeping. Planning elites attempted to invent a radically new and different way of life in a non-place urban realm driven by technology, transportation, and communications systems.

French New Towns: "The Beautiful Adventure"

The new towns of France have been studied perhaps even more than those in Britain. Beyond their histories, both official and unofficial, is the surprising number of oral histories and memoirs produced by the individuals who designed and constructed the model cities laid out around Paris. These clearly indicate the prestige enjoyed by participants in "the beautiful adventure"[33] to create an urban ideal for France. The new town endeavor was led by Paul Delouvrier (fig. 5.5) and the new administrative territory of the District de Paris, and was perhaps the ultimate application of systems theory.

A long-standing associate of President Charles de Gaulle, Delouvrier was appointed the French state's chief policy maker in the capital. He followed de Gaulle's directive to clean up the mess in the Paris region with the landmark Schéma directeur d'aménagement et d'urbanisme de la région parisienne of 1965 (known as the Schéma directeur or SDAU). Although it

5.5. Meeting on the Schéma directeur for metropolitan Paris, with Paul Delouvrier (*center*), prefect of Paris; Maurice Doublet, prefect for the Seine-et-Marne; Jean Verdier (*left*), prefect of police; Maurice Grimaud, and the prefect of the Seine-et-Oise (*right*), 1967. Photographer unknown. Courtesy of IAU îdF.

came under withering criticism from the Paris City Council and local politicians, the Schéma directeur remains one of the most grandiose regional land-use programs ever attempted.[34] It anticipated that the capital's population would increase from 8.5 million in 1962 to 14 million in 2000, and that Paris would take its rightful place as the premier capital of the new European Common Market. Even more, Paris would become a global city. The ambitions were lavish and the population forecast overdramatic, as was the sense of immediacy in planning.

Instead of chaotic suburban sprawl, the region's development would be concentrated in five major satellite new towns founded on two main axes in the Seine and Marne river valleys: Marne-la-Vallée, Melun-Sénart, Evry, Saint-Quentin-en-Yvelines, and Cergy-Pontoise.[35] The designation of each town was accompanied by a media blitz. The construction process was recorded and their meteoric rise acclaimed. They were celebrated and talked about in scores of television documentaries and in film. Each of the towns was to accommodate from three hundred thousand to upwards of a million people, so that the Paris region would become polycentric. The towns were linked with the capital and with one another by a complex system of highways and a new 160-mile regional rapid transit system (RER) as well as by train. Each would have its own jobs and industry. The scale of each was enormous by new town standards. In addition, a massive project at La Défense would be an extension of central Paris and create a world-class business center. It was the ultimate land-use plan.

To carry out this extravagant vision, Delouvrier gathered a young generation of architects and planners, engineers, economists, and geographers in the regional planning think tank known as the Institut d'aménagement et d'urbanisme de la région parisienne (IAURP). Theirs was the generation of Team 10, the architects and planners who eschewed a rigid approach to urbanism and mutinied against the modern orthodoxy of CIAM, especially as it had been carried out in the 1950s and early 1960s in the notorious *grands ensembles* public housing projects. As young professionals, they railed against the outmoded French architectural culture in which an omnipotent architect divined and controlled a project based on his taste alone. Instead, "the concepts borrowed from the social sciences and the new ways of approaching the city were to become almost official rule."[36]

The IAURP group worked in over a dozen interdisciplinary teams, carrying out surveys and feasibility studies on new town projects and analyzing every aspect of the Paris region. The hope was to attract high-tech firms and subsidiaries of multinational companies to the constellation of new towns. Each of the towns was treated as a cohesive urban system, and then the entire Paris metropolitan area would be brought together as an interconnected regional system. As a measure of French interest in new town ideology and design, the pages of the major professional journals *Techniques et Architecture, L'Architecture d'aujourd'hui*, and the *Revue Urbanisme* were filled with articles on Tapiola, Vällingby, and Milton Keynes. The 1963 British Ministry of Transport publication *Traffic in Towns* was translated and published in France in 1965, and referred to continuously by French planners anxious to set out high-flying new town schemes within a comprehensive regional plan for Paris.

Paul Delouvrier and his technocratic team set off on a grand tour of the new towns around London, Copenhagen, Stockholm, and Helsinki. They also traveled to Tokyo, Brasilia, and New York. In one of the innumerable oral histories given by Delouvrier, he recounted the helicopter ride over the famous London greenbelt out to the new town of Stevenage. It was from the air that he realized the errors of the earlier British new town program: the towns were too far from London, too small in size, too lacking in jobs. The French were skeptical about the British new towns, which they often compared with the *grands ensembles* in the suburbs of Paris. Planners pointed to the danger of competition between the British towns, questioned whether residents actually could find work, and bemoaned their monotonous architecture and spaces along with the overly technocratic planning that deadened them as urban places.[37] The new towns in the Paris region were formulated in reaction to these international comparisons. Their scale was much larger than their British counterparts, they were carefully integrated into a regional nexus, and a diversified economic base for each town became the priority.

The study trips were documented in a series of publications edited by IAURP's first scientific director, Pierre Merlin. An eclectic urbanist, Merlin was schooled at diverse French institutions ranging from the Ecole Polytechnique to those specializing in demography and statistics to the Sorbonne. He eventually became a founder and president of the Institut français d'urbanisme. Merlin played an outsized international role in the new town movement, editing or writing some sixty books and hundreds of articles and reports that acted as a conduit for the transmission of theoretical models, information, and ideas. He traveled along with the IAURP team throughout Europe, the United States, and Asia in the 1960s, and his research was a vital resource on new town projects in each country. Each of his reports is filled with sophisticated planning diagrams, regional maps, center city schema, tables of socioeconomic data, and explanations of the administrative, financial, planning, and operational aspects of new town development. They were veritable manuals for building a new town.[38] Especially in the Eastern Bloc countries during the Cold War, Merlin's research was a vital knowledge channel.[39]

Merlin located the origins of regional science and the new geography in Germany, the United States, and Sweden, and particularly in the work of Walter Christaller, Walter Isard, and Torsten Hägerstrand (who developed statistical techniques for understanding innovation in the Swedish spatial economy). French regionalists overcame their anti-Americanism to discover the quantitative revolution in geography. Location theory and systems analysis had been introduced in France as early as 1958 with the work of economist Claude Ponsard,[40] and French engagement with the new models continued through the 1960s and 1970s. Merlin helped introduce French planners to the RAND Corporation's research on systems analysis, the Penn-Jersey Study, and computer modeling at the University of North Carolina.

But many in France, including Merlin, remained hesitant. As the plans produced by the IAURP filled up with complicated mathematical equations and initial forays into computer-generated simulations, Merlin along with geographer Phillip Pinchemel signaled the need for caution. As was often the case, the limited resources of local planning agencies meant a fragmentary use of computers and reliance on some problematic combination of statistical analysis and simplified modeling technique. Mistakes and omissions proliferated. Critics such as Merlin and Pinchemel counseled that understanding the authentic *realities* of local geography and the qualitative as well as quantitative totality of place would avoid the dangers of relying purely on abstract reasoning.[41] To achieve this, the French set off on semiparticipatory planning experiments in which the young staff from IAURP were matched with local administrators and elected officials in on-the-ground fact-finding missions. Planning was somehow to be democratic. But the lofty claims of local empowerment were pretense. In reality, the new towns often faced intense local opposition. Decisions were made in the technocratic halls of the IAURP and often in secret, with the justification being to avoid real estate speculation. The new town program was branded by critics as another case of French state authoritarianism.

The Schéma directeur employed a visual semiotics of cybernetics and systems theory, turning the Paris region into a complex matrix of activities, flows, and dynamic connectivity (fig. 5.6). The new towns were conceived as nodes or nuclei within an abstract regional geography of urban places. They would end the physical exhaustion of commuting and the rat race of daily life. They would offer families of even modest means the opportunity

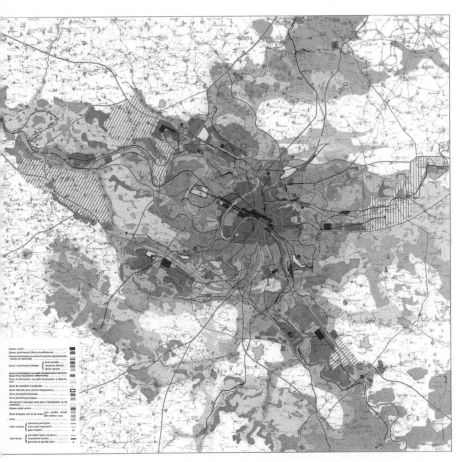

5.6. Schéma directeur for the Paris region created by the Institut d'aménagement et d'urbanisme de la région parisienne, 1965. Courtesy of IAU îdF.

to live with all the amenities of paradise, in nature, free from the aggravations of the big city. The built environment would be a source of pleasure.

In comparison with the much-maligned, ill-equipped, and underprivileged Paris suburbs, the new towns were miraculous objects. Although they fell victim to politics and were only partially realized over the next twenty years, the documentation, charts, maps, and logic diagrams inventing this fantasy were extraordinary. Slews of propaganda material were churned out from state-run presses. Planning was produced as expert science replete with its own vocabulary and ciphers. The visual cryptograms in these official texts evidenced the flattening out, the abstraction, and the suppression of the figural urban environment. They also represented a convergence of avant-garde utopianism with French state modernist strategies for the urban future (see chapter 6). The planning diagrams of the Paris metropolitan region took on an uncanny resemblance to avant-garde fantasies of networked or computer cities.

Abstraction meant estrangement from the actual materiality of everyday urban life. Yet at the same time, the IAURP teams completed site surveys for each settlement that highlighted the local landscape and environment. The distinctiveness of place would give each new town its personality and legibility. Ironically, however, the locations were still largely imagined as blank topography on which the city as a utopian project could be written. An exasperated planner for the new town of Marne-la-Vallée described the urban systems' "utopianists who think it possible and desirable to design the project globally, defining the economic, political, social and cultural objectives, the plan, even the detailed treatment of space, and then leaving to the future any adjustments in policy."[42]

The prototypes for this systems ideal were the new towns of Cergy-Pontoise and Evry. They shared the overriding attention to urban centrality and sociability that French planners believed distinguished a real town from a simple suburban subdivision. Both were made seats of political power as prefectures of their departments. Cergy-Pontoise, thirty kilometers northwest of Paris along a bend in the Oise River, was designed to blend into the local topography. But the tension between the city and its environment was unmistakable. The site was imagined as a continuation of the linear axis from the Eiffel Tower through La Défense. It was given all the accoutrements associated with a "place of exchange, its influence radiating out and integrated into a system of far-flung relations with other cities in a general urban system."[43] Cergy was laid out in a web of three highways, two train lines, and a new regional rapid transit (RER) line that linked it to Paris, with plans for a direct train to the new Charles de Gaulle airport. The transportation arteries came together in the city's center under a massive three-level pedestrian deck built by the well-known Bouygues construction conglomerate. The deck, or *dalle*, provided a formal, unifying coherence to the city's design. It projected, in the words of the proud local planning agency, "a strong central-city image for those who live here as well as for those who visit. It symbolizes the birth of collective life."[44] State buildings and city hall, office towers, a regional shopping center, a university, industrial zones that included two electronics factories, even a small port on the Oise River complemented a variety of housing, leisure, and recreation options.

As an elaborate system of visual signs, Cergy was a model new town to its core. Its public buildings were constructed in a profusion of self-consciously ultramodern styles designed by a galaxy of French architects. They broadcast the city's official posture as a beacon of the future. If the center city of Milton Keynes was entirely indeterminate, Cergy's was unashamedly staged. The prefecture building, designed by architect Henri Bernard as a monumental inverted pyramid in bare concrete, protruded on the landscape as a symbol of state power (fig. 5.7). The city hall was an unembellished glass-and-steel structure of rigorous symmetry by Dominique Armand and Thierry Melot. The municipal building by Claude Vasconi and Georges Pencéac'h was slick 1960s modernism covered in blue and green enamel panels. The steel-and-glass courthouse was by Henri Ciriani, while Antoine Grumbach designed the market hall in iron, brick, and glass. These architects were the favored elites for state new town projects. The center of Cergy-Pontoise was made into an architectural performance of French modernist proclivities, lifted up and joined together as a complete system on a platform stage, which itself was the focal point for the town's transportation web.

5.7. The place de la Préfecture at Cergy-Pontoise, France, with the pedestrian deck, the prefecture building by architect Henri Bernard, and the EDF tower by Renzo Moro, 1977. Courtesy of J. Bruchet / IAU îdF.

Local elected politicians were left completely in the dark about the new town, and so remained on the sidelines as the realization of utopia shifted and changed. "[Bernard] Hirsch [who led the Cergy-Pontoise planning team] said white one day and black the next," one of them quipped. "We had no say. It was a fait accompli."[45] In fact, Hirsch was forced to tack his plans for Cergy-Pontoise in different directions, depending on the political winds in Paris. The communist mayors in the suburban districts were bitterly hostile to what they saw as a technocratic takeover of decision making. Local elected officials organized against the Schéma directeur and the IAURP and began making their own plans. They railed against the redrawing of local administrative boundaries and eventually forced a reform (the Boscher law of 1970) allowing local officials more say in decision making.[46] Others resigned themselves to the scheme, while still others tried to get what they could from the projects, or continued to fight.

The official inauguration of the center city was greeted with protests by local labor unions. The new town program was denounced by acclaimed urban critic Thierry Paquot in the pages of the left-leaning journal *Espaces et Sociétés* as an instrument to technocratize society in alliance with capitalism.[47] Indeed, seducing companies to move to Cergy-Pontoise was one of the main reasons for marketing the city as a model community. "Image was essential," according to a former head of the city's Development Corporation. "Image is much more difficult to nurture than quality of life."[48]

And the imagery was indeed arresting. French New Wave filmmaker Eric Rohmer directed a four-part television series on the new towns in 1975. Rohmer was captivated not only by the urban landscape as cinematic backdrop but by the idea of the city as a living entity that emerges, evolves, and ages. His films depicted the saga of urban modernity. In the first installment, *Birth of a New Town*, Cergy is seen under construction, rising like a specter on a vacant landscape. Bernard Hirsch stands before a vast planning map, explaining his ideas. Shots inside the Development Corporation offices capture planners, architects, and sociologists hunched over their desks, conjuring up the town with slide rules, compasses, diagrams, and models.[49] The result of their work is an urban idyll.

In his full-length feature film *L'Amie de Mon Amie* (1987), Rohmer presented four young people living and working, playing out their love lives in the urban paradise of Cergy-Pontoise. They enjoy a privileged, happy existence, a perfected social order. The education and good jobs, the shopping and leisure activities, the conviviality associated with the magic of modernity, are at their fingertips. Blanche and Fabien, Léa and Alexandre are members of a mod young generation, released from the past. Their lives are circumscribed by the town's unabashed modernist cityscape. The film is a monument to the mammon of state planning, and the ultimate cinematic backdrop for a new town and its new people.

Systems Logic in Navi Mumbai (New Bombay)

Systems analysis seemed to underlie a foolproof, comprehensive land-use strategy. Applied to developing nations, it was a political instrument of modernization, one that could fall back ingenuously on expert science and abstract models. Vast stretches of politically volatile territory were fictionalized as a clean canvas ready to be filled in, organized, and interconnected in a coherent pattern of development. Successful development depended on devising functional typologies and hierarchical arrangements of towns and cities. The terminology of classification and typology, along with the ideal of rank-size, played a key role in this conceptual framework. It started at the national level, where each country was categorized by economic system. Cities were then ranked by population and economic function into pyramidal hierarchies. This process set out the chain of command and filtered control and surveillance from capital city all the way down to the village level. Classification and ordered arrangement were then applied to income groups and social structure. The reasoning was that scientific categories originating in the professional planning culture of the West could be read outward to societies on the periphery.

Towns in developing countries were sources of local data that was then tailored to fit the classifications.[50] Both the UN and the Ford Foundation espoused this kind of logic and devoted enormous resources to field research that could be incorporated into the systems structure of analysis. Rather than imagining the structures of social power and the evolution of cities as organic, it was assumed that urban morphologies and typologies were fixed and could be operated by centralized planning. In fact, the organic growth of cities was despised as flawed, the cause of so much chaotic overdevelopment and social mayhem. It produced what were referred to in planning parlance as structural imbalances.

Urban growth was instead to be shaped and regulated by professionals with the scientific tools to fully understand complex socioeconomic processes and their spatial components. The turbulent metropolitan areas of the Third World would be contained and their population siphoned off into small and medium-sized places that would be given official blessing as new towns. The correct population size for these freshly minted places obsessed planners; wanton overpopulation and congestion were the ultimate sins. Each of the towns was arrayed by functional component: industrial estates, residential areas, town center, recreation areas. Transportation arteries would blend all these elements together.

The typologies of new towns abounded: ring towns, satellite towns, industrial towns, service towns, and so on. The remarkable specificity of these definitions, size classes, and attributes, and the detailed examination of every new town facet, was a mechanics of power and control. These conventions were ordering devices that operated as professional knowledge. Exported to the developing world, they were a mechanism for alleviating what were condemned as social pathologies caused by hyperurbanization or overurbanization, terms used regularly by international development agencies.

Indian professor of planning L. R. Vagale was one of the UN's chief technical advisers on urban development in the 1960s. He produced some fifty publications about urban and regional planning in India and Nigeria, and was instrumental in promoting technical education in developing countries. Through experts such as Vagale, the UN developed an entire narrative about harmonious cities for the emerging nations of the Third World. The term *harmonious* evoked a moral language and was used liberally to connote interactions between planning experts, the social relations that would alleviate ethnic conflict and extreme social disparities, and the interactions between cities and towns across a metropolitan territory. For Vagale, a new town offered this perfect harmonious world:

> A new town is not merely an exercise in building houses, factories, schools and shops; it is a deliberate and planned effort to create the best possible physical and human environment, conducive to a healthy and satisfying life for the individual, the family and the community, and its success should be assessed in terms of improved life and living. The form and spirit of the community, its character and creativeness, the happiness of people at work, in their homes, in their leisure-time pursuits, and in their social relations are a valid test of a town's success.[51]

This picture of new town utopia was already a long-standing one. "If planners are not allowed to build a few utopias," Vagale went on, "they will be merely making half-hearted attempts and producing new towns which have neither dignity nor grace."[52] He then explained an "integrated synoptic approach" for achieving harmony, in which every aspect of urban life was taken into account, from land-use and economic activities to bicycle tracks, street naming, and house numbering.

Paradoxically, Vagale was one of many experts working with international agencies who began attacking Western new town strategies as woefully inadequate and anachronistic to this ideal. Historian Gyan Prakash has argued that Indian planners such as Vagale both internalized Western visions of the good life and critiqued it as inadaptable to India.[53] By the 1960s, they had certainly ceased being awestruck by Western planning strategies.

The Delhi Master Plan of the early postwar years (sponsored by the Ford Foundation and dominated by American and British consultants) had come under withering attack as unworkable. Its vast greenbelt was at best impractical, at worst an unequivocal symbol of Western sanctimony. The satellite towns did nothing to stem the tide of urban sprawl engulfing Delhi. The garden city and neighborhood unit concepts were old fashioned, too costly in terms of land and money, and inappropriate to Asian urban experience. Even if the form was diluted to fit local conditions, the costs of maintaining the garden city fantasy were exorbitant. They were no more than castles in the air, as was the crusade for self-help of the early postwar years. The outcome could often be tiny houses scattered haphazardly across a terrain strewn with garbage and in a permanent state of disrepair. The fact that these forlorn landscapes had been built on so many utopian aspirations made them seem all the worse. In Our Cities, a 1966 Indian symposium, one participant flatly declared that planners for Bombay were not interested in "bhawans, showpieces, expensive plate glass, steel, concrete, aluminum, decorative louvres and frills . . . streets too wide for any earthly use, and parks which cannot be enjoyed, it being too dangerous to cross over to them."[54]

Whereas the first generation of Indian planners readily assimilated the neighborhood unit in terms of the local *mohalla*, or traditional Indian residential area, the second generation began to view it as a foreign concept. They also shared with their Western counterparts a growing condemnation of the neighborhood mantra as standing in the way of a clearly defined, vibrant town center. In 1971, N. S. Lamba, the chief town planner for the Punjab, argued that radical changes in planning were necessary for the "new way of life, scientific advancements, leisure and Gadget oriented living." Lamba looked to the new town of Cumbernauld, Scotland (see chapter 6), as a guide: "No one goes to the local centre twice a day as in the past and one wants to go to the Town Centre for the variety, choice and gaiety." Rather than trusting one architect to conjure up a town design, he recommended that interdisciplinary teams use forecasting methods to determine planning priorities over the long term and "project the image of the people living in a state or territory." With this approach, "it will not be necessary to appoint foreign experts on high salaries who cannot understand the local requirements in such a short period."[55]

The paradoxes implicit in relying on Western planning techniques were increasingly evident. Indian planners looked to Western models of modern town planning, but grappled with their adaptability to their own national context. The Ford Foundation sponsored a major field study of seventeen Indian new towns based on systems analysis, and the simulation models pioneered by American planning experts Britton Harris at the University of Pennsylvania and Lowdon Wingo at the University of California, Berkeley.[56] And in general, the Indian government continued to employ the full toolbox of planning methodologies and extravagant standards recommended by the United States and England. It incorporated cost-benefit analysis into land-use planning. But Indian planners also questioned the cultural authority of Western new town models and their viability as too small and scattered, and the expense too extravagant. The magnitude of the urban problem in India was at a decidedly different scale than in the West.

Even so, 112 settlements were already claiming the title of new town in India by the late 1960s. Their design and construction had preoccupied an entire generation of Indian planners and architects, engineers and administrators. Migration into the new settlements was ongoing and extensive. In some cases, their populations ballooned far beyond the garden city idyll that was initially imagined. Roukela, Bhilai, and Bokaro all had about 2 million inhabitants, a good many of whom lived in slums and squatter settlements in and around the original planned central areas. But most other new towns remained small. At the end of the 1950s, the population living in India's new towns was estimated at about 1.5 million, or less than 1 percent of the country's urban population. The projected population in new towns by the 1970s was only 2.5 million to 3 million.[57] This was a trifle compared with the populations of sprawling slum districts on the peripheries of India's great cities.

Clearly, the new towns could not keep pace with population growth; nor were they equipped to absorb the massive relocation of people, their hopes in hand, into the urban world. In perhaps the most alarming prognosis of the "population bomb," the number of Indians was predicted to double by the turn of the twenty-first century, with 70 percent of them (some 630 million) living in urban areas. It was clear to public authorities that new settlements had to be developed at an entirely different scale to have any real impact. As a result, India's Third Five-Year Plan (1961–66) laid out a comprehensive policy for the balanced and orderly development of towns and cities. Master planning became the order of the day.

No city exemplified this accumulation of problems more than Bombay. With a population of over 4 million in Greater Bombay by the 1960s, it was stretched to its limits and had become, in the minds of planners, simply intolerable. Thousands upon thousands of migrants flocked there each year, ready to take their chances in the brash Bitch City. It was surrounded by vast, informal squatter colonies that had taken on a productive life of their own, but remained unauthorized, illegal, and outside the regulatory power of the official planning apparatus. Bombay was the symbol of modern hyperstimulation, an out-of-control urban omnivore. Official reports churned out the litany of evils: overwhelming density, traffic-snarled streets, monstrous buildings and slums, filth and pollution, disease, deteriorated infrastructure, organized crime. Bombay was sick. It was a city of perpetual crisis.

No end of government committees and official reports laid out recommendations for controlling the growth of Bombay, but these were never put into practice. Powerful entrenched interests systematically blocked any regulatory control despite repeated calls for reform. In 1956, strikes and deadly riots broke out in a round of violence against efforts to make Bombay a separate federal district under government authority. A wide spectrum of popular opinion resisted the reform and forced its withdrawal. Mass protests and violence between Maharashtri and Gujarati ethnic groups continued for political influence in both Bombay and the territorial and linguistic makeup of the new state of Maharashtra. Maharashtrians comprised the majority of the city's population, while trade and industry were controlled by the Gujaratis. By the mid-1960s, the populist right-wing Shiv Sena party emerged to defend Maharashtri power in Bombay, newly renamed as Mumbai in Marathi usage. Mumbai's inclusion in the new state of Maharashtra (which was made official in 1960) was thus born out of brutal local conflict. Navi Mumbai (or New Bombay) was imagined as a solution to these vicious dynamics of caste, class, and corruption in the old city. It was meant for the common man: an emblem of social reform.

Plans for a Greater Bombay region had been drawn up as early as 1948 with the Master Plan in Outline developed by American architect Albert Mayer and engineer N. V. Modak. It was a classic example of land-use planning in the immediate postwar years: a confined urban center surrounded by greenbelts and an array of garden city satellite towns. Although it was never implemented, the vision continued to inspire Bombay's planners throughout the 1950s and early 1960s as the dream of overcoming India's backwardness through modernization. However, the Bombay development acts of 1954 and 1964 provided the opportunity to reformulate the metropolitan region's future. Then in 1965, three of India's leading urban reformers—Charles Correa, Pravina Mehta, and Shirish Patel—laid down a manifesto for the city in "Bombay: Planning and Dreaming," a special issue of the influential design journal *MARG*. Educated in the United States at the University of Michigan and MIT, Correa was a major figure in contemporary Indian architecture. Patel was a civil engineer educated at Cambridge, while Mehta, the third member of the team, was a US-educated architect. All three men represented a young generation of modernists allied with the trailblazing trends of the 1960s.

The *MARG* issue was a watershed moment in shaping a countermagnet to old Bombay. In its luxuriously illustrated pages, Ebenezer Howard's tried-and-true diagram of the Three Magnets appeared alongside lavish land-use maps and glossy photographs of the "splendors and miseries" of Bombay. But this time, the garden city ideal was thrown out as unsuitable: "Where a great metropolitan centre exists—as in Bombay—small satellite towns located haphazardly . . . cannot possibly fill the role for which they are designed."[58] They were too small to balance the hypnotic pull of Bombay. They were exorbitantly expensive to build, and their infrastructure was redundant. Instead, the manifesto in *MARG* proposed a radical alternative in design and scope: one entirely new, self-sufficient city situated on the other side of Thane Creek that would be large enough to draw population, commerce, and industry away from the overcrowded island of old Bombay. It would rebalance Bombay's north–south trajectory with a new east–west axis. Although the low-lying marshland was already occupied by some one hundred thousand agricultural workers and fishermen, it was treated as empty space to be filled with modern fantasies. The new town would "permit the residents . . . to live fuller and richer lives in so far as this is possible, free from the physical and social tensions, which are commonly associated with urban living."[59]

The new Bombay would reflect India's changing economy. Although old Bombay continued to be the heart of the country's banking and financial sector, by the mid-1960s its historic textile mills and traditional industries were shutting down. The city's economic base was shifting toward light engineering and petrochemicals, and its growing commercial film industry. The new city would welcome these forward-looking enterprises along with a new seaport as well as a new airport.

The main *MARG* proposal was for a prestigious central business district across Thane Creek from the old city, featuring cutting-edge financial and commercial services and high-tech firms. Urbane, well-educated white collar workers materialized in this dream setting. New Bombay was meant as a glamour zone of high-rise offices and high-end amenities. It was built for the new business class and would make Bombay a global city. The *MARG* issue's full-page photo spread of Rockefeller Center staged the fantasy of a midtown Manhattan for India. The city would then stretch out in three spokes from this central business district. Skywalks were envisioned, along with underground railways, monorail systems, and fast ferries.

Gyan Prakash has called the New Bombay plan (fig. 5.8) "a richly embroidered dream text. A visual feast of beautifully produced maps, charts, and graphs."[60] It was a phantasmagoria of the future, replete with an array of deluxe planning maps, full-color overlays, and quotes from Lewis Mumford and Lyndon Johnson. This staging made Bombay's future legible as a disciplined spatial production. Planning as visual performance and spectacle was the lifeblood of new towns as seductive fantasy. Whatever qualms Indian planners may have had about foreign influences, the vision of the Twin City, as New Bombay was called, was 1960s modernism. Yet it was contextualized by Correa's sensitivity to natural environment and local place. Rather than a simple transfer of Western utopian aspiration, New Bombay would be shaped by the indigenous landscape. This hybridity was framed through local ecology, the experience of the harbor, and the vision of a "city on the sea." A new skyline would drape over the hills and continue down to the waterfront. People would experience the great estuary and Thane Creek as part of their everyday lives.[61]

New Bombay was one of the most ambitious urban schemes undertaken anywhere. Its protagonists imbibed the thrill and excitement of creating a brand-new city from scratch: Charles Correa described "the *brave new world* idea, the great enthusiasm from the beginning."[62] The official project was almost equal in size to Greater Bombay, and planned for a meteoric influx of some 2 million people. But utopian fantasies usually diverge from their original intent, and this was the case with the proposals on *MARG*'s pages. By 1966, the government's Gadgil Committee of planning experts rearranged the land-use plan to include some twelve to fifteen townships for the new city, each with 50,000 to 200,000 residents. Together they would act as a countermagnet for the multitudes packed into the congested districts of old Bombay. They were a schema of Possible Worlds.

The townships were strung like pearls along mass transit and highway routes. Each would have its own landmark railway station and its own distinctive town center with schools, shopping, and services. Feeder roads led

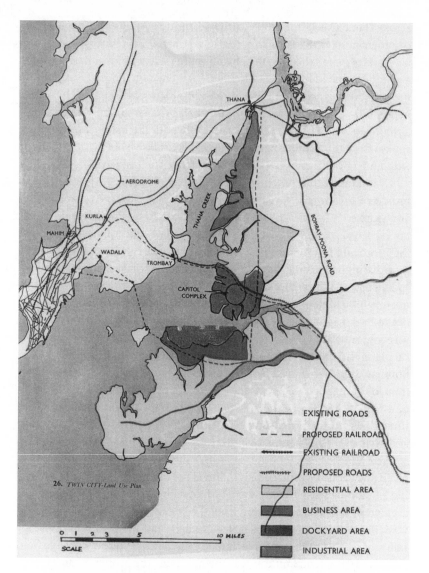

Inside the map (labels):

THANA

AERODROME

KURLA

MAHIM

WADALA

TROMBAY

THANA CREEK

CAPITOL COMPLEX

BOMBAY-POONA ROAD

26. *TWIN CITY-Land Use Plan*

— EXISTING ROADS

--- PROPOSED RAILROAD

EXISTING RAILROAD

PROPOSED ROADS

RESIDENTIAL AREA

BUSINESS AREA

DOCKYARD AREA

INDUSTRIAL AREA

0 1 2 3   5   10 MILES
SCALE

5.8. Twin City Land Use Plan; from "Planning for Bombay," by Charles M. Correa, Pravina Mehta, and Shirish B. Patel, special issue, *Marg* 18, no. 3 (June 1965). Reprinted in *Bombay to Mumbai: Changing Perspectives*, edited by Pauline Rohatgi, Pheroza Godrej, and Rahul Mehrotra (Mumbai: Marg Publications, 1997). © Charles Correa Associates.

into individual neighborhood villages for 2,500 to 10,000 inhabitants. The obsession with urban typology, with ranked classification by population and function, held sway. The townships and their neighborhoods, the connecting network of transportation were all arranged as an integrated, fluid system. As growth took place, new townships could be clipped onto the city's matrix.

Control over the massive project was given to the public-private City and Industrial Development Corporation of Maharashtra (CIDCO), which owned 95 percent of the property at the site. A multidisciplinary planning team was assembled and forecast models developed. Thirteen townships were actually constructed. Vashi was the first and initially the largest of the townships with its market center. Belapur was the central business district, Airoli and Kopar-Khairane the city's industrial heart, and Nhava-Sheva housed the new container port. Housing options ran the gamut from highrise apartments and row housing for the wealthy and middle classes to self-help huts for the poorest families. The New Bombay would be an open city without social and ethnic enclaves and without slums. It was a reform city that staged India's future.

Correa believed that mythic urban images from "outside" (meaning the West) had to be internalized and reinvented: "This is the crucial difference between a superficial *transfer* and that something far more basic which I call a *transformation*." His examples were the houses in Balapur township (fig. 5.9). The plot sizes were more or less equitable. Shelter for working families was simple and low rise: designed and built so they could be improved by their owners incrementally using local materials. This concept accommodated high densities, yet homes were clustered around intimate courtyards in the pattern of Indian tradition. These clusters then formed "modules," which in turn formed "community spaces." Urban hierarchy would generate the "open-to-sky spaces so essential to the lifestyle of Asians."[63]

In invoking the ideal of self-help, Correa was relying on a well-established development discourse for Third World nations. It was part of his larger vision of New Bombay as a self-help city. This narrative began to merge into the idea of sustainability. In a speech at the United Nations Conference on Population, Resources and the Environment in Stockholm in 1974, Correa argued that cities were organic, almost biological, entities: "Once you start a city, you are starting a process." They should be founded on "the natural life style of the people." More than just providing a house, the new town would take into account the "complex system of a family's spatial requirements." The traditional domestic skills of reuse and recycling would be transmitted to the entire city for "healthy and sus-

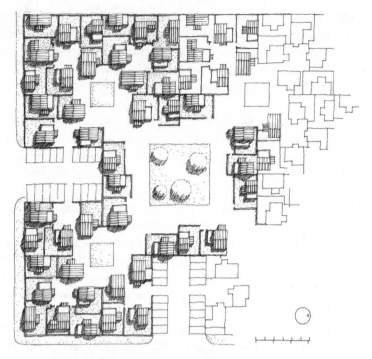

5.9. Charles Correa, site plan showing hierarchy of shared community spaces, Belapur Housing, New Bombay. © Charles Correa Collection / RIBA Library Drawings and Archives Collections.

tained growth." Flexible rather than frozen master planning would provide multiple options for the future and allow towns to evolve and endure as a productive environment. Correa believed these ideals were the "DNA of a new city" that would motivate people to participate in and create a new civic culture. New towns were the chance for an "enormous quantum jump" and "the opportunity to—in Buckminster Fuller's phase—rearrange the scenery."[64]

These dreams were bound to fade when faced with the realities of Bombay's socioeconomic whirlwind. In reality, the new city suffered long periods of slow growth and stagnation. Poor transportation links with old Bombay was the main reason. Few government or commercial offices were willing to cross Thane Creek and make New Bombay into India's version of midtown Manhattan. By 1995, its population hovered at only around seven hundred thousand. To make matters worse, it battled for relevancy against the massive land reclamation and office district project at Nariman Point in Bombay's Back Bay, which soon grabbed the status as the

"Manhattan of Mumbai."[65] As priorities shifted, CIDCO privatized land at New Bombay to pay for infrastructure in the half-built city. Inspiration was drawn less from planners and more from private developers.

In the end, the brand-new city actually did little to cure Bombay's so-called illnesses. Its impact on the greater Bombay metropolitan area was negligible: its presence did nothing to alleviate the infamous slums, and in fact added to them. Floods of poor set themselves up in New Bombay in vast informal colonies around Nhava-Sheva port and even next to the city's swank railway station. What New Bombay did offer was prized middle-class housing—although city residents choked on smog and pollution from the local petrochemical industry, despite having paid a great deal for living in utopia. India's endemic social and caste divisions remained as intractable as ever. According to critics, Charles Correa's plan was little more than "a redistribution of wealth for the upper strata of society."[66] New Bombay gradually became more independent of the old city, although it lacked the excitement and buzz of old Bombay's urban atmosphere. Designing a *locus amoenus* for every man resulted in a drab non-place, a mere shadow of the *locus terribilis* across the creek.

American New Communities

The likelihood of ever realizing a project on the scale of New Bombay in the United States was always remote. The American attitude toward the ideal of new towns was at best ambivalent, at worst hostile. Although he had been instrumental in foreseeing a systems model for planning in his *The Post-City Age*, the most urban theorist Melvin Webber could hope for was that "there may be a national consensus that would permit us to pursue some common objectives in a directed and deliberate fashion."[67] There was no doubt that the country had a long and esteemed legacy of regionalism and garden city thinking. America's urban intellectuals—Lewis Mumford, Clarence Stein, Clarence Perry, and Catherine Bauer—were well known and admired. They traveled to Washington, DC, and used their notoriety to press Congress for town-building initiatives. Stein voiced his belief that "the cost of building new communities is less than that of rebuilding old and obsolete cities. The total cost of carrying on industry and business in the United States would be greatly decreased by a more scientific distribution . . . of goods and people."[68] But discussions of urban planning at this scale fell on deaf ears. Cold War politicians viewed it as a form of creeping socialism. James Clapp, professor of urban planning and a notable figure in the American new town movement, opened his 1971 book on new towns by remarking that mentioning the term was the quickest way to end a conversation. The new town idea was a "dead horse."[69]

Even so, there was a groundswell of enthusiasm in the 1950s and 1960s for metropolitan land-use planning and the possibility of new communities in the United States as well as in Canada. Urban reformers kept up the crusade. In the pages of the American Institute of Architects' *AIA Journal*, noted architect and planner Carl Feiss scolded colleagues: "There is nothing complex about the concept of the New Town. The only complex problem is to find the reason why the New Town idea is not acceptable to the American public." Feiss headed the AIA's New Towns Task Force and was chief planner for the new towns of Telico in Tennessee and Lucaya in the Grand Bahamas. In his estimation, a series of "community coagulants"—large-scale suburban shopping mall, highway interchanges, and large public schools—could all act as anchors for new communities.[70] Similarly, architectural critic Ada Louise Huxtable introduced the European new town movement to a broad reading public in the pages of the *New York Times*. Her articles of the early 1960s on Tapiola, Vällingby, and Cumbernauld described the new communities "designed and planned with a sophistication, economy and coordination, and frequently with an architectural grace and elegance, that critics feel put the American affluent society to shame." Readers drank in jaw-dropping photographs of paradise accompanying her praise. Tapiola was "a beautifully clear, handsome and unequivocal demonstration of the virtues of the planned community" and "the antithesis of current practice in the United States."[71]

Much emphasis has been placed on the Americanization of planning techniques in the second half of the twentieth century and their use in the modernizing process across the globe. But the international diffusion of planning cut both ways: ideas and information arrived from other shores. In setting forth American new communities, government officials from the Department of Housing and Urban Develpment, private developers, university researchers, and the Ford Foundation and RAND Corporation all looked to international examples for inspiration. As planning schools cropped up in major American universities, a new generation of professionals campaigned for a regional perspective on urban growth and explored European models of what could be accomplished. As was the case virtually everywhere, the British new town movement was the baseline for observation, and the Americans and Canadians regularly paid homage to Ebenezer Howard in their discussions. But shared experience had come a long way from the movement's taproot of Letchworth Garden City. By the

1960s and 1970s, an entire postwar generation of new towns worldwide blatantly exhibited both the successes and the failures of urban utopia.

In a series of articles in the *Architectural Record* in 1964 and 1965, Clarence Stein and Albert Mayer continued the campaign for comprehensive land-use planning and regional constellations of new towns. Their articles formed the core of *The Urgent Future* (1967), which Mayer wrote as a definitive statement of their accumulated wisdom. Like many of the urban surveys produced in this period, it was a richly textured, dramatic exposé. It opened with an image of Ossip Zadkine's public sculpture *The City in Mortal Agony*, which was created over the remains of bombed-out Rotterdam and spoke "with equal poignancy to the world-wide twentieth century urban crisis." Fantasy diagrams of American metropolitan areas followed, with new towns set as nodes along networked highways. References to European new towns abounded. Mayer placed a photograph of "Anywhere U.S.A." (taken of the suburban sprawl at Massapequa, Long Island) alongside that of the "Imagination and Control" of ideal Tapiola, rhetorically asking readers, "Which Do You Prefer?"[72] The dynamics of self-help community work and the social science techniques that Mayer used in India were presented as embryonic forces for urban renewal.

Spurring the American movement into action was the overwhelming sense of crisis in US cities. The tragedy of the inner cities, the racism and social pathologies, mounting municipal deficits, white flight and suburban sprawl, were all specters of urban decline. Already in 1956, a series of articles in "By 1976 What City Pattern?," a special issue of *Architectural Forum*, began with a sense of urgency:

> The last ten years have given us an unholy mess of land use, land coverage, congestion and ugliness. This is nothing to what the next twenty promise. Barring annihilation, deep depression, or a more tractable invention supplanting the automobile, we have no way to avert this crisis of growth, no choice but to face it and try to civilize it. Somehow. And not much time to do it.[73]

In her contributed article, Catherine Bauer staged the hair-raising nightmare of sprawl twenty years hence: 55 million more people living in metropolitan areas, with the overwhelming majority headed out toward the fringes of development. The suburban population would double. These figures were a "staggering prospect."[74] The 1950s and 1960s would be the decades of the greatest suburban growth in US history. Development spreading amoeba-like without rhyme or reason, engulfing land, swallowing up rural areas; thousands upon thousands of little houses with cars parked outside their doors, popping up in fringe towns and road towns across America—for Bauer, they offered nothing but social isolation and deadening conformity. Making matters worse, the map of each metropolis was a crazy quilt of political jurisdictions and special-purpose districts that thwarted decision making. New towns were irresistible as a solution to these calamities. Bauer offered the hypothesis of new cities of half a million to a million people built in the countryside, to "re-establish some of the traditional cosmopolitan virtues of urban life which are now lost in the stupid village ideology and class-race exclusionism practiced in suburbia."[75]

While white families fled to the suburbs, poor black communities were left to the slums. The race riots that ripped through American cities in 1964 and 1965 were ample evidence of the toxic atmosphere. Harlem in New York City, Rochester in New York State, North Philadelphia, and Watts in Los Angeles erupted in violence. American cities were literally burning to the ground. The Detroit and Newark riots of 1967 were among the most explosive and violent in US history. The total abandonment and devastation in the South Bronx were searing testimony of failed public policy. The "specter of two societies" could prove fatal, according to architectural critic Donald Canty.[76]

Planning experts with a taste for urban systems and computer simulations were also driven by indignation at this stark social injustice. "We have been tempted to apply city-building instruments to correct social disorders, and we have been surprised to find that they do not work," wrote Melvin Webber in his *The Post-City Age* (1968). "The next generation of Americans is destined to enjoy the unprecedentedly rich life that the post-industrial, national urban society will offer. Our central domestic task now is deliberately to invent ways of extending those opportunities to those groups that future history threatens to exclude."[77] For many reformers, the new town became an instrument for solving racism and social segregation. It would end the schism between the city of the poor and the suburbs of the middle and upper classes. It would constitute a new world of social equitability and balance. The utopian impulse was defined by this tension between the pessimistic perceptions of reality and the wish for a completely different life.[78]

By the mid-1960s, the sense of urgency and the chorus of criticisms against American urban areas reached Washington, DC. President Lyndon Johnson announced the federal government's involvement in two speeches outlining its policy: the first to Congress in 1965, "Problems and Future of the Central City and Its Suburbs," and his 1968 message, "The Crisis of Our Cities." In 1967, the US Department of Agriculture held a prominent event, National Growth Policy and Its Distribution: A Symposium on Communities of Tomorrow. It was a gathering of Cabinet secretaries, city mayors, university experts, and new town enthusiasts headlined by vice president Hubert Humphrey.

The symposium was billed as a historic occasion in which finally, in the words of agriculture secretary Orville Freeman, the vision for an American landscape dotted with hundreds of new communities—from small cities and new towns to growing villages—could be realized. There were impassioned discussions about a national system of cities within comprehensively planned regions, how it might look, and how it might be implemented by government and the private sector. These new towns would not be bedroom communities but solidly built towns with their own commerce and industry.

To the symposium's eager participants, technological innovation and the computer seemed to hold untold possibilities. New town enthusiast Barbara Ward voiced the gathering's "excitement" that with "the full use of computers, with the probably almost unlimited unleashing of energy, which will come from the breeder-reactor [i.e., nuclear energy], with the education explosion following our new methods of information, we are planning for a city whose outlines we can barely see." In the worst of ironies, Ward pointed to the use of computer modeling in the city of Detroit, which earlier that year had been torn apart by riots, looting and destructive fires, and police brutality against black youth. Nonetheless, she waxed eloquent that the computer was "an instrument for the liberation of urban choice, a means, if you like, whereby men achieve the power to 'invent' their urban future, and to do so even in an already congested society with a rapidly growing population."[79] She went on to describe the computer forecast of a Detroit region made rational with a system of new towns settled along transportation and communications lines and bordered by green areas and lakes.

Vice President Humphrey imagined "wholly new urban centers accommodating a large part of our anticipated population increase" that would be settled between the Allegheny Mountains and the West Coast, "where land is cheap."[80] As evidence, Humphrey offered Athelstan Spilhaus's Minnesota Experimental City (MXC) Project (see chapter 6). It was one of the first projects of its kind to receive aid from the federal government's new communities program, and it flaunted a total systems approach to city building: all the parameters and various subsystems intertwined at critical interfaces. The nation needed to focus, Humphrey believed, "on the future relationships of people in urban environments. How can we best accommodate the young and old, the rich and poor, the black and white, the swinger and the square in one community? An experimental city could be a testing ground for just such a living situation."[81] It was precisely this kind of idealistic urban laboratory that the authorities gathered at the Communities of Tomorrow symposium were looking for.

In 1969, the newly formed ad hoc National Committee on Urban Growth Policy produced *The New City*, one of the most influential urban texts of the 1960s. The committee itself was a stellar assemblage of new town advocates from all levels of government, with the backing of the National League of Cities, the US Conference of Mayors, and the Washington, DC, nonprofit organization Urban America Inc. *The New City* volume was edited and written largely by Donald Canty, an architectural critic and editor of *Architectural Forum* in the 1960s who also created the short-lived *City* magazine.[82] He was one of America's leading activists for a socially conscious modernism, and a tireless crusader for new towns.

The New City was meant to educate members of a broad reading public about their advantages. The volume was richly illustrated with black and white photographs of both the worst urban conditions in the United States and the best solutions in new towns. Canty's text blasted the American metropolis as "monumentally ugly." Spontaneous urbanization was "wasteful and destructive of the urban environment."[83] Photographic spreads evidenced the sprawl, the poverty, the pollution. To undo all these horrors, images of Tapiola, Cumbernauld, and Märkisches Viertel outside Berlin appeared as inspirational beacons.

The New City made the case for a dramatic shift in public policy toward comprehensively planned communities, and it offered the new towns of Reston, Virginia, and Columbia, Maryland, as examples of what America was capable of producing. It took a bold step in proposing an entirely new scale of urban development: the population boom demanded not just new towns but new cities. William Finley (who led the planning of Columbia, Maryland) wrote the National Committee on Urban Growth's contribution to the volume. In it, he said that the best response to the population boom would be "the building of 20 cities of one million people each and 200 new towns of 100,000 each."[84] The committee recommended federal funding for 100 new communities averaging 100,000 population and also 10 new cities, each of which would contain at least 1 million people. Together, these would accommodate 20 percent of the anticipated population growth in the United States. This building program was necessary, because "the continuation of current trends will bring the country a succession of one urban crisis after another which will tear at the very fabric of society."[85]

The population problem was one of the strongest and most heavily media-driven issues fueling this atmosphere of crisis. Urban reformers and pundits alike wrung their collective hands over the "population explosion" or "population bomb." Canty's *The New City* began with the threatening population bomb and a crowded future of 200 million Americans that by the end of the century would be 300 million, with the additional 100 million living in urban areas. Paul Ehrlich's best seller *The Population Bomb* (1968) predicted massive famine as population outstripped food supply. American media giants *Time* magazine and *U.S. News and World Report* identified the population bomb as the world's number one problem.[86]

A tsunami of statistics evidenced the coming apocalypse and stoked public fears. At Urban Renewal in America, a 1971 symposium featuring some of the country's best architects and planners, the participants grappled with the fact that "some people predict that we're going to have twice as many people in the year 2000. In other words, we'll have to rebuild the country entirely, and either have twice as many cities or expand the ones we have."[87] The histrionics and sense of impending doom mirrored Cold War apprehensions. They also worked hand in hand with reveries of a new town paradise without care, without these calamities.

The year 1968 was the high point of this unprecedented public debate about the future of cities. A flood of reports spilled from the desks of urban activists. They appeared amid the student protests, the civil rights marches, the riots and violence, and the political upheaval over the Vietnam War. There was the National Commission on Urban Problems' *Building the American City*, the Advisory Commission on Intergovernmental Relations' *Urban and Rural America: Policies for Future Growth*, and the American Institute of Planners' *New Communities: Challenge for Today*. RAND and the Ford Foundation jointly sponsored a 1968 workshop to address urban problems with a star-studded cast of some fifty-six urban reformers from academia, government, and the media. The Urban Institute was created by President Johnson as a military-style think tank to study America's cities. Yet another research foundation, the Urban Land Institute, created a New Communities Council. The Ford Foundation and Urban America Inc. sponsored a trip for New York City municipal officials to the superstars in the new town galaxy: Cumbernauld, Amstelveen near Amsterdam, Vällingby, Tapiola, and Gropiusstadt near West Berlin. American new town celebrities from James Rouse to Albert Mayer took part in the Anglo-American Conference on New Towns in London.

CBS produced *The Cities*, a three-part special report that was beamed into living rooms across the United States in June 1968. After the first two segments gave a searing account of the urban catastrophe, part 3, "To Build the Future,"[88] featured veteran news reporters Walter Cronkite and Mike Wallace interviewing Lewis Mumford, Constantinos Doxiadis, Gunnar Myrdal, and Athelstan Spilhaus. Systems engineering and land-use models were introduced to television viewers as state-of-the-art techniques, although they "were more complex than a rendezvous in space." Doxiadis introduced his multicellular *dynapolis* concept, while Spilhaus depicted his MXC Experimental City. A televised tour of the new town of Columbia, Maryland, was led by James Rouse; the new town of Irvine, California, was described by chief planner Walter Pereira as a "series of ideas." In the concluding segment of the CBS report, Mumford, Myrdal, and New York City mayor John Lindsay all predicted increasing urban chaos, disaster, and the "deterioration of Western civilization" unless comprehensive planning and the dispersion of people into new towns was begun.

Not to be outdone, ABC produced the January 1969 special *Cosmopolis—the Big City*,[89] which featured the new town of Stevenage outside London as well as the fantastic foursome of Cumbernauld, Tapiola, Reston, and Columbia. Interviews with modernist architects Philip Johnson and Moshe Safdie were interspersed with shots of Buckminster Fuller's floating city and a discussion with Paul Douglas, chairman of the National Commission on Urban Problems. Futurism of the 1960s merged with the formulation of government policy. The program hammered home the message of comprehensive regional planning by interviewing a Stevenage housewife who was "passionate about new towns" and the "good life they have created." Children were free and uninhibited. Nature was close by. The "former slum dwellers" of London had found happiness.

An outpouring of books, articles, comprehensive reports, conferences, and meetings campaigned for a new town future. It was the magnitude and utopian quality of this production that marked the new town movement of the late twentieth century and gave the new town such mythic power. Praeger published an entire series of books about new towns that provided an international focus.[90] In 1971, the American Institute of Architects held a new communities conference in Washington, DC. The next year, the University of California at Los Angeles organized an international conference on new communities with representatives from Britain and Israel.[91] In the mid-1970s, the US Department of Housing and Urban Development (HUD) sponsored tours of the new towns in the Paris region and in the Soviet Union. Jack Underhill, a research analyst for HUD and author of innumerable reports about new towns, was a tireless campaigner for open-ing "our minds to fresh ways of looking at our problems and suggesting new kinds of solutions." Underhill worked closely with Pierre Merlin in France as well as with A. O. Koudriavtsev from the Soviet Union's Com-mittee on Construction Affairs, themselves members of the cosmopolitan planning elite formed around new town methodology. Their reports about comparative new town policies[92] produced massive data on every aspect of urban development.

References to the stellar American new towns of Reston, Columbia, and Irvine appeared incessantly in planning documents as signifiers of the fu-ture. Hundreds of articles in the popular press focused national attention on these three communities. In his 1977 overview of the American new town movement, historian Irving Allen referred to this outpouring of am-bitions as "the pop nature of public and much professional thinking on the social goals of New Towns."[93] Planning had broadened out from physi-cal land use to systematized construction of perfect societies.

The utopian quality of the American new town movement was ulti-mately its undoing. New towns became an all-purpose panacea for every social problem, for every evil in American urban life. The federal Task Force on New Communities began its report with the ringing statement, "1968 can be the year when a systems approach is successfully applied to our Na-tion's cities." It continued, "This is why America needs new cities—places where the 'new start' can begin; places where people can gain a new sense of community; places that do not produce rootless and isolated citizens."[94] The new town program would be made scientific through economic mod-eling and feasibility studies. America's new towns would be socially diverse communities where people lived in harmony, without conflict, crime, or violence. HUD eventually declared its priorities as reducing segregation and encouraging the upward mobility of lower-income and minority fami-lies. New towns would improve the self-image of the underprivileged. They would offer social relatedness and provide young people with stimulation and adventure.[95] Poor families would step into a wonderland of personal well-being and community cohesiveness.

This litany of exalted hopes provoked eventual exhaustion. But for a time, Americans joined the chorus of believers. The new town movement in the United States was actually implemented by a combination of pub-lic and private initiatives. The government-sponsored new town program, which was euphemistically called New Communities, was a product of president Lyndon Johnson's Great Society initiative of the 1960s. Federal support for new town construction was initially taken up in the 1964

Housing Bill. Ada Louise Huxtable reported in the *New York Times* that the United States was "on the verge of a New Town boom."⁹⁶ But the stigma associated with European-style socialist planning killed the initiative. Two years later, a revised and still controversial version of the New Communities bill was enacted by Congress as part of the Demonstration Cities and Metropolitan Development Act of 1966. The legislation established federal loan guarantees to private developers who would construct Great Society Model Cities. But it was hedged in by too many restrictions to be effective. The New Communities program was then expanded as Title IV of the 1968 Housing and Urban Development Act, and then expanded further in 1970 to include loan guarantees for state and local government agencies as possible new town developers.

By 1973, fifteen New Community projects had received federal funding. HUD secretary James Lynn enthusiastically predicted that in twenty years, their population would approach 1 million. At least 20 percent of the black population would be evacuated from inner city slums to the new settlements.⁹⁷ These sweeping clearance programs would then open up development of upscale central business districts in America's premier cities. The displaced could start over, from scratch, in the New Community satellites strategically set in place throughout the metropolitan regions. Yet only one—Soul City, North Carolina, built specifically to attract black minorities—was actually a brand-new freestanding town. And the only public agency able to surmount the HUD requirements and receive federal funding was the New York State Urban Development Corporation, which initiated two projects: Roosevelt Island in New York City and Radisson in northern New York State. The rest were initiated by private developers with federal loan guarantees.

The federal government imagined the new towns to be pacesetters in mass transportation, and in the provision of education and health care. In them, it planned to test broadband cable television and ways to turn solid waste into fuel. The new town of Jonathan, Minnesota, proposed turning television screens into visual phone calls. In their application for federal assistance, the town of Pontchartrain, Louisiana, envisioned a "massportation" conveyor belt system that delivered goods and people at forty miles per hour, while the proposed community of New Franconia, New Hampshire, would include a monorail as "the spine and aorta" of a series of villages "that resemble a string of beads."⁹⁸ Shenandoah, Georgia, experimented with solar energy collectors on its buildings. Riverton, New York, wanted to carry out a community medical program.

But the New Communities program remained controversial; political

support was at best halfhearted, sometimes amounting simply to an in-
sider agreement negotiated between the HUD secretary and a developer.[99]
Dozens of applications were blocked by the Nixon administration, which
had little interest in promoting urban reform. The novel projects proposed
by new towns that did begin construction were often quickly scaled back
as too costly or too unpopular. None achieved the growth they originally
expected, and only one was able to repay its federal loan. The schemes
withered in the face of the 1970s oil recession. Then Nixon's 1973 mora-
torium on federal urban renewal spending spelled the end of the Great
Society programs. The government-sponsored New Communities program
collapsed.[100]

Nonetheless, there were some achievements. In a HUD study made in
the late 1970s, the New Communities were found to have mitigated urban
sprawl and have had a clearly positive impact on their local areas' environmental quality. They also provided more housing opportunities than other
development projects, especially for low-income households.[101]

The Private Development Paradise

Perhaps the greatest success of the New Communities legislation was
its immediate embrace by private developers. It stirred enthusiasm even
among those who saw government incursions into land-use planning as a
suspicious form of creeping socialism. The role of the private sector in the
American new town movement is usually pointed to as its defining feature
and its distinguishing characteristic from its European counterpart. Private development corporations dove into new town projects and brought
American-style free market capitalism to the forefront of the debate about
city building. Real estate developers were drawn to the mesmerizing imagery of new towns because of their scale, the appeal of control over land
and investments, and the potential of reaping enormous profits. Indeed, it
was assumed that profit was essential to achieving a better way of life. Of
course, with dreams of riches came the risk of failure. New towns were an
immense gamble, even when capital investments were backed by federal
loan guarantees. Many of these across-the-board development schemes
floundered on the reefs of overspeculation and financial insolvency.

Around 150 privately financed new communities were publicized in
the 1960s and 1970s, although some never made if off the drawing board.
Most were in rapidly growing areas of the country, mainly California, Florida, Arizona, New York, and Maryland. California in particular was the
dreamland of new town development. A rush into large-scale real estate

speculation was tied to the state's population boom and the construction of its famous freeway system (see chapter 4). A majority of these places were no more than glorified subdivisions with some glossy design elements and a shopping center. There was no intention of offering jobs, a downtown, or a sense of urbanity, and the new communities had little relationship with the surrounding area. Some, such as El Dorado Hills north of Sacramento, were designed as complete new towns, but ended up as residential enclaves. Others, like the colossal new town of Irvine Ranch, were celebrated as triumphs at master planning. In 1967, the Ford Foundation financed a study of new towns in California, written by home builder Edward Eichler and planner Marshall Kaplan. Positioning their report as a hardheaded economic feasibility analysis that was largely critical of new towns, Eichler and Kaplan concluded that the return on profit for the immense capital investments required for new towns was too low for the high risks involved. But that did not stop a "new breed of land barons"[102] who saw dollar signs in the quest for urban utopia.

Economist William Alonso at the University of California at Berkeley, whose studies of location analysis had been instrumental to American regional science, offered a broader critique of the California new town phenomenon. As alluring as they were, "it would be misleading to assign to them a central place in the future development of California's urbanization." They provided, in Alonso's estimation, "brave visions of starting afresh without the baggage of previous history and mistakes. They appear Camelots of the future, proving grounds for aesthetic, social, economic and technological breakthroughs. On the whole, the argumentation for new towns resembles more a Rorschach test for men of good will than a rational evaluation of pros and cons."[103] Indeed, certain developers were not immune from these utopian reveries or from the temptation of becoming impassioned city builders. First among them was James Rouse, who despite staggering costs put together the new town of Columbia, Maryland, that was immediately acclaimed as the future of America. The most successful new towns in the United States were all associated with visionary businessmen who represented the best in a social-minded American capitalism. This was the case with Robert Simon's ideal town of Reston, Virginia, and Philip Klutznick's Park Forest, Illinois. The ranks of social entrepreneurs included Richard Mellon in Pittsburgh, and William Zeckendorf and David Rockefeller in New York City.

A self-made millionaire and commercial real estate mogul, James Rouse was the epitome of the American success story. He spent much of his long and productive career as a crusader for better cities, finally making the cover of *Time* in 1981 as a master planner. From his three dozen shopping malls to Faneuil Hall Marketplace in Boston (1978) and Festival Harborplace in Baltimore (1980), Rouse saw his projects not just as profitable real estate ventures but as effective forms of urban renewal. Part entrepreneur, part evangelist for saving the American city, his methods were enthusiastically imitated. Rouse's Columbia, Maryland, was (and still is) considered the most complete and successful new town in the United States, and one of the most admired in the world. Urban scholar Ann Forsyth dubbed the first stages of Columbia's development in the 1960s as "Early Camelot" and "High Camelot," and argued that its atmosphere of innovation was unparalleled in the American new community movement.[104]

Rouse's initial forays into urban reform began in the 1950s and 1960s by using his contacts in Baltimore and in Washington, DC, to campaign for better cities. He was the organizer and guiding force for one of the first national nonprofit organizations dedicated to improving cities, the business-oriented American Council to Improve Our Neighborhoods (ACTION). Supported by the Ford Foundation, Rouse crisscrossed the country delivering impassioned speeches on the need for urban renewal. He organized a stump tour of seminars with urban experts to help guide decision making in the country's metropolitan centers.[105] He became adviser to mayors, governors, and presidents, and a leading international voice in American urbanism. A lifelong Democrat and deeply Christian, Rouse espoused a social welfare agenda combined with urban planning derived from people's needs. For example, he proposed an exhibit entitled "New Image for the American City" for the 1964 World's Fair in New York that would showcase the most advanced planning techniques and could be used as a catalyst for comprehensive redevelopment projects in cities throughout the country. As he argued, it was "easier and quicker and more profitable to build new cities that will work than it is to continue to build scattered, fractured, senseless sprawl as we now do across America. Everyone knows this is absurd. No one could defend the way our cities grow in America today."[106]

Rouse's early ideas for an ideal community were worked out in his design for the Rockefeller property in Pocantico Hills, in the Hudson Valley north of New York City. Assigned the task by the Rockefeller family, Rouse assembled a team of planning experts and social scientists to examine "the needs of people and how they might be best fulfilled in a well planned community." Although the plan was never implemented, the picturesque village quality of Pocantico Hills, much in the garden city tradition, would form part of his vision of the future. So would early suburban corporate campuses being constructed in the late 1950s and early 1960s. Rouse saw the Connecticut General life Insurance headquarters in Bloomfield, Connecticut (designed by Skidmore, Owings and Merrill, with landscaping by Isamu Noguchi), as a transformational form of living drawn from the best qualities of the American small town.

His early shopping malls had this same potential. Following concepts pioneered by planner Victor Gruen (see chapter 6), he imagined the shopping mall as reinventing the sociability and community atmosphere of small town America for the 1960s. Both Rouse and Gruen ruminated endlessly about how design could be used for constructive social ends, and how shopping centers could transform the suburban environment. Rouse believed that "the well-planned, well-managed shopping center is more than simply a new plan for retail expansion. It represents the massive reorganization of the urban community."[107]

Many of Rouse's speeches and writings have been passed down in planning documents and in scholarly texts precisely because he articulated future thinking about cities with such flair. They are worth considering, especially from an international perspective, because they played such an instrumental role in the articulation of utopian imagery. In his speech "Utopia: Limited or Unlimited," Rouse described his expansive dream:

> To hypothesize the working city, the city where people were employed, had incomes capable of coping, lived in communities where there was a delivery of health care, where people were well educated, moved about in reasonable systems of mass transportation, where there was beauty and community health and a city full of spirit: now, that's not beyond our capacity.[108]

This is what the new town of Columbia. Maryland, symbolized. It represented an amalgamation of Rouse's ideals. It was a pristine suburban vision based on the virtues of community life, the neighborhood and garden city ideal, and the need for improving the lives of the poor trapped in the inner city. From the beginning, Columbia was racially integrated and maintained a black population of around 20 percent.[109] It was to be a therapeutic solution to the nation's social and racial discord. Nonetheless, making a profit on the colossal investments required for new town development was a priority. Economic forecasting models and financial feasibility and marketing studies were part of the systems techniques put in place. Columbia was an intriguing real estate package that would sell because of its quality of life. It would be profitable enough for other developers to copy as a better way. In a word, Columbia would be "contagious."

Rouse's commitment to social science techniques in building his vision of the future was considered the most groundbreaking aspect of the Columbia project. He assembled a multidisciplinary work group headed by urban planners William Finley and Morton Hoppenfeld. Like many of the architects and planners recruited to develop new towns, they worked in both public and private sectors as professional elites. The project team was perceived as liberal, hip, and freethinking. Some of them were members of the Space Cadets, a network of people interested in urbanization from the mental health and behavioral points of view. Team director Donald Michael was a psychologist and an expert in systems analysis. He noted that instead of asking "'Where should the high schools be located?' . . . we began with the question 'What role can a city play in the growth of persons?'"[110] The project was wrapped in systems thinking. According to Rouse, "We had to get away from the separate cells with which we regard urban life and problems. Each area . . . must be seen as systems that reinforce each other."[111] Rouse called the strategy the "system of life" and emphasized it in his keynote speech at the 1967 Communities of Tomorrow symposium. He underscored it again ten years later, in his address to the International New Towns Congress held in Tehran:

> We pulled together a group including doctors, ministers, teachers, psychologists, psychiatrists, social scientists. . . . We drew many shares of life in those discussions that influenced the detail of the physical plan but of most importance was the consensus that evolved about the need for community among people, the need for place, for easy relationships that brought a feeling of support, for the need to belong, to be important, to have a base from which man, woman, and child could move out into the world.[112]

The new town imaginary had shifted from providing housing, jobs, and a stable neighborhood to a holistic, psychological fantasy of utopian living. In designing Columbia, Rouse "wanted to look at the optimums as we could conceive them. What would be the best educational system we could conceive in a city of 110,000 people—the best health system—what is the best communication system—what are the causes of loneliness and delinquency—what are the causes of happiness?"[113]

Despite its American capitalist pedigree, Columbia's genesis was in part international. Like most new town enthusiasts, Rouse and members of his team made the obligatory pilgrimage to the European new towns and participated in various European new town conferences. Hoppenfeld in particular was inspired by Tapiola, which became one of the models for Columbia, as did the British new towns. But Rouse and his team put a decidedly American spin on the new town intellectual corpus. In his 1964 presentation of the new town plan to Howard County in Virginia, Rouse described a mosaic of villages of 2,500 to 3,500 families surrounding a central core area. The villages were designed in a cloverleaf pattern, with each petal a cul-de-sac residential neighborhood with its own commerce and services arranged around a central square and loop road (fig. 5.10). In the tradition of Clarence Perry's neighborhood unit, the focal point of community life was the primary school and community center. This was a performance of planning ritual. Columbia exuded a sentimental desire for a "scale of life reminiscent of the small towns which form such a rich heritage of America." The entire system would be connected by roads, by public bus routes as well as walking and bike trails. Every acre of the site was analyzed for trees, slopes, streams, and natural features "which needed attention."[114] As in Tapiola, the natural environment was interpreted as scenic and recreational.

Columbia offered housing options for a range of incomes, from high-density apartments to single-family homes, although they were segregated by neighborhood. Housing for low-income families was subsidized by the federal government. Columbia was launched with a business development program that attracted global firms such as General Electric, Toyota, Shell,

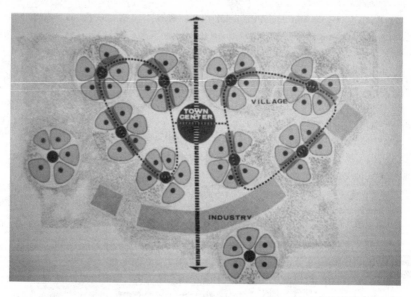

5.10. Diagram of village "community pods" for the new town of Columbia, Maryland. Community Research and Development, Inc. Photograph by Robert Tennenbaum. Courtesy of Columbia Archives.

and Pfizer. Rouse envisioned a dense urban center for the new town packed with restaurants, hotels and offices, theaters and concert halls, a college and hospital. The town center bordered an artificial lake, where "in the evenings, the lights and sounds of the restaurants and cheerful cafes along the lakefront will welcome people out for an evening." At its heart was an enclosed "completely air-conditioned" shopping mall with over a hundred stores.[115] The Mall in Columbia, opened in 1971, was the largest shopping project Rouse had built. It was featured on the cover of the *Architectural Record* and described as creating "a city-like world." The exposed steel and glass were the mall's "heraldic symbol."[116] Its decorative lights and greenery, splashing fountains, and glamorous design made shopping into a tantalizing wonder-world experience.

New towns promised much. They were imagined as large-scale, blank canvases on which dreamscapes could be built from scratch. Although Columbia certainly was not without its problems, particularly financial ones, it came to represent a specifically American success story. As James Rouse stepped into the international spotlight of the new town movement, Columbia became transportable imagery and joined the lexicon of utopian urban places. Much like Milton Keynes, it was a Pop icon of the future, a pilgrimage site around which the new town faithful congregated. Writing in *The Architectural Forum*, Donald Canty described the project as "unprecedented . . . almost daily setting new precedents."[117] The Rouse Company ran bus tours of the first neighborhoods and opened an Exhibit Center. The planners and residents of Columbia during these early Camelot years understood themselves as pacesetters—part of a unique experiment in living. As one of the early residents recounted in an interview, "Most of us felt that we were in a very important drama, that this was something that had, if not worldwide, certainly a national stage. We had visitors, and there were people who were doing TV shows and people who were doing articles about Columbia and so on. It makes you feel different."[118] This enthusiasm and eagerness about new towns, the sense of being an explorer of the future suffused the genre's ethos. Stepping onto new town turf was a statement of confidence and audacity.

SIX

Architecture for the Space Age

As icons of the future, new towns were also an imaginative playground for the architectural avant-garde of the 1950s and 1960s. Inventing Space-Age worlds—with the possibility of actually seeing them built—captured the creative energies of these trailblazers and sparked an outpouring of visionary production. Creating new towns was an indulgence born of the spirit of playful optimism. It was a fusion of the utopian imagination with systems thinking, space vehicles, and space colonies, the prospect of a world controlled by computers. Cybernetics was a sophisticated tool for making space travel possible. It was essential to the rocket systems and space capsules, the complex, cutting-edge technology that the Space Age represented. But this science was also a philosophy and, in the hands of the architectural avant-garde, an art form.

This chapter focuses on architecture and design in new towns. The scope is narrowed to the built environment and the architects, planners, and futurists who conjured up these brave new worlds. They shared an unwavering belief that the physical structure of cities could transform society. The first part of the chapter examines the impact of the Space Age on their design theories. The city was reinvented as a megastructural cybernetic machine that could support life and launch out into the cosmos. And this idea was far more than just a space dream. With the support of the state and its technocratic corps, new towns were built that attempted to carry out these utopian fantasies on earth.

The second part of the chapter examines the architecture and design of these actual places. New towns were deliberate and highly symbolic acts. Whether these places were just dreamscapes on paper or were actually built, cybernetics and the Space Age produced a profusion of futuristic urban imagery. It is the magnitude and diversity of this discourse about new towns that made it into a distinctive movement. This chapter highlights this abundance of utopian practice, and ends with its eventual demise.

Dreaming along with Technocracy

The unconventional performances of avant-garde groups such as Archigram, Archizoom, the Metabolists, the Situationists, and GIAP are traditionally interpreted by scholars as provocations and subversive tactics for generating freedom outside the iron cages of technocracy and capitalist consumerism. These vanguards are understood as hostile to or at the very least ambivalent toward the modernization that took place in the 1950s. The avant-garde camp unleashed a scathing critique of the first generation of new towns that had veered, in their minds, so dreadfully far from their original ideal. In 1959, Situationist artist Constant Nieuwenhuys vilified new towns as "cemeteries in reinforced concrete . . . where great masses of the population are condemned to die of boredom."[1] British architect Peter Smithson of Team 10 belittled the dead-end urbanism of the English new towns.[2] Even the favored child of Tapiola came under withering criticism. A young generation of Finnish architects branded it a failed forest city and insisted that no more be built.

The baiting shenanigans by the rebels were meant to shock modernist orthodoxy and incite the profession to think boldly.[3] Using formats ranging from fanzines and comic strips to collages and illustration, they produced rabble-rousing versions of an urban future infused with Pop culture. Their designs had a needling spirit of jest, a ridicule of outdated practices as inappropriate to the modern metropolis. Static space and architecture had to be replaced by mobility, flows of people, transportation and communication, adaptability to change.

The avant-garde was "in the image business," as architectural critic Reyner Banham put it. Self-promotion and publicity were integral parts of their production. Architectural illustration functioned in the genre of pulp novels and science fiction, and shifted the discourse to the mischievous and the fantastic, meant for journalistic propagation. By disseminating their projects in the media, their visual discourse on future cities was instrumental in expanding the sense of what was possible and shifting popular awareness toward a starry-eyed future. A world of ethereal urban vistas materialized on their drawing boards. The dreamscapes discussed in this chapter—colossal multilevel cities, massive concrete megastructures raised on platforms, weird cybernetic modules—were mesmeric in their intellectual and emotional force. Utopia in the mass cultural age became a series of whimsical designs, fantastic forms and feats of engineering, elaborate electronic gadgetry. Avant-garde production can be construed as hedonistic, kitschy, tongue-in-cheek parody and a tactic of social and political agitation, which indeed it was. It turned architecture and space into exaggeration and spectacle. Visual drama was part of utopian expression, and this quality permeated the entire new town movement.

Yet ironically, no matter how antiestablishment and outlandish these fantastic cities of the far future were, they worked in conjunction with state power. When the imaginaries of the avant-garde were made tangible, when they crystallized into actual places, it was under the auspices of the state. Utopian architecture, along with the cutting-edge technologies on which it was based, conformed to political authority.[4] Cybernetics and the computer revolution, and their vital link to space exploration, created a state-supported scientific technocracy to carry out fast-paced technological innovation. The planning and architectural community joined forces with this regime of expertise as a way to achieve the utopian dream. It was the chance to actually make the future they imagined *real*. The wild fantasies of a cybernetic, Space Age urban realm had only traces of radical social content and actually paralleled technocratic thinking. Although utopian production defined itself against state plans, the avant-garde community never rejected the underlying architectural and spatial logic. It was based on the same fixation with technology, rational organization, fluidity, and mobility as the official discourse.

Rather quickly, the architectural subculture found itself in an ambivalent sociopolitical position. Their urban designs ended up somewhere between dissident counterspace and a handy pool of resources for *dirigiste* states and real estate developers eager to cash in on projects. Even more, the space exploration that acted as muse for their designs was deeply embedded in the military-industrial complex and Cold War that the avant-garde despised. A conflict of purpose, a fundamental disjuncture, existed between the ideals of freedom and provocation and the modernizing framework in which they were enacted. It was a dialectic tension that promoted extraordinary creativity. The plans and diagrams produced by state technocrats themselves exuded a spearheading aesthetic. Experimental modernism and fantasy cities were internalized by new town planners and the political elites who commissioned the projects, and even the developers and construction companies who carried them out. Technocratic functionaries saw themselves as part of the avant-garde.

They all were members of the young Sixties generation—a hip, trailblazing elite. They had steadfast faith in the capacity of the state to enact progressive measures for the good of society. As a result, the similarities between the visual artifacts produced both by avant-garde architects and by technocratic elites were remarkable. The critique of the status quo, the hopes for a better way of living passed into the hands of government administration and new town development corporations that had the capacity to put even uncanny utopian concepts into practice. The impact was ironically to reinforce the very regimes the architectural avant-garde hoped to undermine.

New towns of the 1960s tended to be the places where this creative collision took place. Innovative architectural forms, megastructures, and shopping malls could be found in a host of urban places. But in the case of new towns, architects and planners had the tempting opportunity to design a city ex nihilo as the fullest expression of their utopian caprice. It is in conceptions of new town *centers* as all-inclusive, multilayered megastructures that we particularly find the frictions implicit in futuristic urban fantasies. In the 1960s, the megastructure was a significant arena of innovative design production. Each was produced as a revolutionary interpretation of the new urban whirl of leisure and consumerism, as a solution to the conflict between cars and people, as a renewal of community life. The visual and promotional pageant of these avant-garde central-city structures, their excitement and magic were unprecedented.

Yet the hubris in attempting to incorporate the whole urban environment into a gigantic architectural assemblage was apparent almost immediately. When Reyner Banham published his book *Megastructure* in 1976, he had already dubbed the megastructures the "dinosaurs of the Modern movement" and laid bare the reality that architects were "forced to recognize that the homogeneously designed 'total architecture' . . . would be as dead, as culturally thin, as any other perfect machine."[5] The spectacular futuristic behemoths were quickly appropriated by commercial interests, and downgraded into consumer palaces in which any activity other than buying was excluded. Utopian aspiration collapsed with amazing speed into the hypnotic act of shopping that the avant-garde had attempted to break through, and this time in massive cement assemblages accused of destroying all hope of urban life.

The Playground of the Avant-Garde

Heterogeneous and at times contradictory factors were behind the avant-garde concepts that upended the old-style new town of the 1950s. First and foremost were cybernetics and the mesmerizing possibility of space travel and space colonies—and the space comics version of what that might mean. The dawn of the Space Age produced an enthralling utopian dream of spreading humanity out into the cosmos. We would glide between the stars and vacation on the moon. There was a tremendous public appetite for computers and high-tech gadgetry, rocket ships and spacecraft, and visions of intergalactic travel. Films such as Stanley Kubrick's *2001: A Space Odyssey* (1968), with its enthralling views of the cosmos, its space station, and its unstable HAL 9000 computer, were enormous popular successes internationally. Urban dreamers basked in the glow of this futurism and concern for the greater good of Spaceship Earth, the term coined by Buckminster Fuller and then adapted as the title of the 1966 book by Barbara Ward.

This kind of cosmic vision found its way into fantasies about urban form and function. The architects connected with British Archigram were obsessed with Cape Kennedy. They were convinced that cities could be designed that would take off and become independent living entities. They played off NASA's futuristic visions of orbiting space stations, lunar bases, and robotic missions to Mars. Archigram was also the first to connect the word *capsule* with home. The pod was separating from the urban mother ship, or urban frame. It was an autonomous, self-regulating organism dependent on electronic information streams and onboard computing technology. Clearly, the influence of space exploration was not limited to flights of imagination. Scientists and engineers from NASA and private aerospace companies were involved in a variety of projects to bring aerospace technologies and systems management directly to bear on urban problems.

Two urban theorists in particular exemplify these influences: Paolo Soleri and Athelstan Spilhaus. Their megaprojects display the flamboyance and provocation of avant-garde creative production in the 1960s. Yet both received official sanction by state interests—utopian fantasies had genuine meaning as a pathway to the future. Italian visionary architect Paolo Soleri, who trained under Frank Lloyd Wright at Taliesin West in Arizona, was perhaps the foremost practitioner of Space Age environmentalism. Soleri designed entire cities as hyperstructures of extreme population density. His two publications, *Arcology: City in the Image of Man* (1969) and *The Sketchbooks of Paolo Soleri* (1971), are a stunning set of illustrations in which architecture and ecology are fused.

Soleri called for an ecological radicalization of urban systems. Cities, in his estimation, must move toward more concentrated, efficient forms, and be multilayered and multidimensional rather than wastefully scattered and sprawled. The ideal of Arcology was a self-contained, self-sustaining community inside a giant megastructure (fig. 6.1). It was a holistic environment

6.1. Paolo Soleri, BABEL IIB Arcology, population 520,000. Page 113 in *Arcology: City in the Image of Man* by Paolo Soleri. Original publication 1969 by MIT Press. © Cosanti Foundation.

to support life—essentially a space station. This new urban form could be achieved through enlightened technology and the computer revolution.

For Soleri, cybernetics provided a choice between *nemesis* or *genesis*, and a new topography of the human world.[6] He took aim at Constantinos Doxiadis's sprawling vision of *ecumenopolis* as entirely wasteful and the result of an erroneous computer forecasting model. The only way to rescue humankind from this "map of despair"[7]—which Soleri visually depicted as a map of the United States overwhelmed by "Ecumenopoly"—was to build up into a third dimension, above Earth. From the outset, his theory was linked with the goal of human habitation in outer space: "Asteromo is an asteroid for a population of about 70,000 people. It is basically a double-skinned cylinder kept inflated by pressurization and rotation of the main axis. . . . He will be able to fly without the need of any power devices. Man, standing head toward the axis of rotation, will be enveloped in a solid ecology."[8]

Soleri's idiosyncratic schematics were meant to jolt the senses and perform an ecological breakthrough. He envisioned orbital colonies, urban rivers sculpting the surface of the planet and joined to hovering megastructures for hundreds of thousands of people. In his drawings, the colonies are weightless, floating over the landscape. This was not a way to live with nature so much as separating mankind from it. This was the only way to deal with the sheer magnitude of the population bomb and allow the continents to return to their virginal state. Human survival required a new measure of collective solidarity. Loyalty to the planet was a moral obligation.

Soleri's first attempt at this superdense environment was the settlement of Arcosanti near Phoenix, Arizona, begun in 1970. It was designed as a colossal, precast concrete megaform for five thousand people, reliant on solar energy and entirely computerized. It would rise twenty-five stories above the desert as a vaulted structure honeycombed with greenways, solar corridors, and air spaces. The critics were damning and accused Soleri of indulging in science fiction. But there were shards of utopian value in these dreamscapes, and his schemes had the support of both the academic and the political establishments. A 1970 exhibition of his work at the Corcoran Gallery in Washington, DC, was sponsored by the US Department of Housing and Urban Development and the Prudential Insurance Company.

A second utopian project, the Minnesota Experimental City (MXC), further evidenced the merging of cybernetics and futuristic urban form. It received the blessing of three different federal agencies as an official new community and also became a pet project of vice president Hubert Humphrey and the State of Minnesota. MXC was endorsed by a host of business interests, including the University of Minnesota and newspaper mogul Otto Silha of the Minnesota Star and Tribune Company.[9] From their perspective, cities built from scratch meant jobs and big profits. The project's steering committee was headlined by Buckminster Fuller, Paul Ylvisaker from the Ford Foundation, and William Wheaton from the Institute of Urban and Regional Development at the University of California, Berkeley. General Bernard Schriever, creator of the US missile system and former chief of Air Force Systems Command, was also a member—an indication of the ongoing relationship between systems thinking, urban utopianism, and the Cold War military establishment.[10] The basic concepts for the city were hashed out in a series of fourteen workshops at the University of Minnesota. It would be built for a population of 250,000 in the wilderness 120 miles north of Minneapolis-St. Paul.

The guiding light of MXC was futurist and environmental scientist Athelstan Spilhaus, dean of the University of Minnesota's Institute of Technology. Among his many roles was creator of the enormously popular cartoon strip *Our New Age*, syndicated in some 110 Sunday newspapers worldwide from 1958 to 1975 (fig. 6.2). The strip was populist techno-utopianism replete with the latest news on space travel, lasers, people movers, and experimental cities. Its sensational scenes inspired the MXC project. For Spilhaus, the basic "problem in the world is that there are too many people," and "too many people drift into too few cities."[11] The solution to accom-

6.2. Athelstan Spilhaus, Experimental City from *Our New Age* Sunday comic strip, March 5, 1966. Spilhaus (Athelstan) Papers. Courtesy of the Dolph Briscoe Center for American History, University of Texas at Austin.

modating this population bomb of 3 million more souls added each year to the world's population was optimally planned new towns.

MXC was to be an entirely enclosed community covered by a geodesic dome of massive volumetric proportions. Spilhaus designed it as a cybernetic machine. The interior environment would be programmed by computers and adjusted by sensors. Electrical wiring was woven directly into the dome's exoskeleton. People, food and goods, and recycled waste would travel through a warren of tubes and tunnels. The concept was similar to Peter Cook's "Plug-In City," with its podlike units clipped onto a tubular framework containing the city's electrical, water, and sewer systems (fig. 6.3). Towns became science fiction; MXC was a glimpse of the city as a Space Age living container, a superintelligent cyborg with its own volition. It was an aggregate of networks and flows, a total experiment in social science, human ecology, and environmental engineering without slums, traffic jams, or pollution.[12] It was eccentric, but the utopian impulse exhibited in projects such as this attracted corporate backers, because it responded to the problems and opportunities of the present. Ford Motor Company devised MXC's transportation system of automated roads, computerized minibuses, and moving sidewalks. Residents would shop, bank, and conduct meetings virtually through electronic networks.

MXC was one in a profusion of ideal communities conjured up in the 1960s. These ranged from back-to-the-land hippie communes to grandscale and highly lucrative real estate deals such as James Rouse's Columbia new town in Maryland. Utopianism had legitimate, sincere meaning. It was a philosophy of hope. For many of those within the countercultural movement who wandered off to Arcosanti or model communities like it, Spilhaus's *Our New Age*, Buckminster Fuller's *Operating Manual for Spaceship Earth* (1969), and Stewart Brand's *Whole Earth Catalog* (1968–72) were their inspiration. Arcosanti and MXC belonged to a utopian genre that merged the ideal of a Space Age urban vessel with cybernetics and the

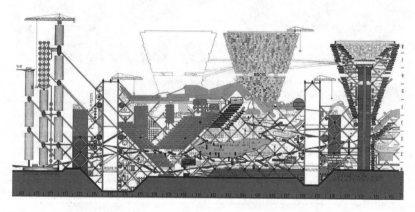

6.3. Peter Cook, Plug-In City, section, Max Pressure Area. © Archigram 1964.

nascent environmental movement. Cybernetics and computer-controlled megastructures were envisioned as the path to collective freedom and fulfillment, to ecological balance and ultimately to the earth's survival.

These fantasies were a pointed intervention in the present. For all their provocative features, they also displayed the fears of the atomic age. Lifting off from the earth's surface was a tactic for survival after a nuclear holocaust, as was the hope of human existence in a sealed, self-sustaining structure. As utopian archetypes, they embraced the very same technological politics as the era's military-industrial complex and Space Age exploration. The absence of any real subversive or seditious content is evidenced not only by their official sanction but by the lack of any options one could legitimately hope for except in the remotest of futures. They were simulations in the same sense as computer programs, war games, and cybernetics itself.[13]

Arcosanti came to fruition, but shortages of money, materials, and people made progress painfully slow. It became a semifinished, semiruined landscape of Sixties optimism. The local opposition to the MXC project as offbeat and bizarre was intense. Although Spilhaus worked on the plans from the mid-1960s onward, it was dead within a few months of its public announcement in 1973. Both prototypes had fallen victim to the 1973 economic crisis and the political shift toward the new conservatism.

As evidenced by both Soleri's and Spilhaus's elaborate theoretical designs, cybernetics had a profound influence on 1960s architectural avant-gardism and visionary futurology.[14] The term *architecture* also entered into common parlance to describe the internal organization of a computer system. It was not simply a question of architecture's influence on the engineering and design of computers, or cybernetics' influence on architecture, but rather the common adherence to a new vocabulary of perception and new imperatives[15] that ultimately became the framework for designing entire new cities.

In 1960, Reyner Banham curated a special issue of the British *Architectural Review*, "The Science Side," about the future of architecture. In a telling contribution on "Computers" for the issue, M. E. Drummond of the IBM Corporation claimed that systems simulation, linear programming, and queuing theory enabled architects to formulate and optimally solve multidimensional planning challenges. The mathematical modeling of street plans, highways, and traffic flows was his example.[16]

Banham eventually became leery of the architectural profession's eagerness to gulp down the latest fashions in cybernetics and computers, just as he had with megastructures. But the possibilities were spellbinding. The

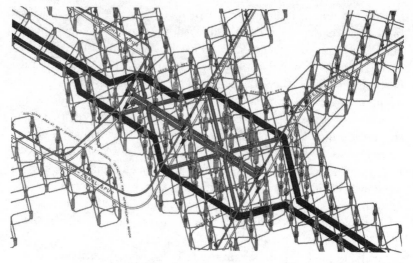

6.4. Dennis Crompton, Computer City, Axonometric. © Archigram 1964.

built environment not only could become entirely quantifiable, but could be imagined as a living machine intelligence. For their part, members of Archigram reinvented the city as a megastructural container of naturally regenerative circulation and information systems based on computational diagrams, such as Dennis Crompton's rendition of "Computer City" (fig. 6.4). Constantinos Doxiadis as well moved toward an increased engagement with electronic data collection. He analyzed the evolution of cities through electromagnetic maps and computerized cartographatrons that displayed hidden networks and organic force fields.

Other sources of inspiration for this outpouring of architectural avantgardism were the behavioral and social sciences as well as the media. In his contribution to Banham's curated *Architectural Review* issue, Richard Llewelyn-Davies argued that the social sciences "provided us with an understanding of what people need in buildings. They offer the hope that we may break through the rigid, nineteenth century convention of the architect's 'programme' and get back to a broader, more imaginative comprehension of our task." Group dynamics in particular involved building up a picture of social life and how it functioned.[17] Collective identity, belonging, and the ideals of a humane social democracy were integral to urban aspirations in the 1960s. This perspective was far less nostalgic than the old neighborhood unit had been; it was far more irreverent, permissive, in line with the 1960s creed of liberation.

Architects explored the worlds of advertising and communication, consumerism, and automobile culture. These did not just offer new tools but became urban environmental models. Built forms were meant to make people happy and take advantage of commercialism and mass culture. Pleasure and fun became permissible in architecture. Flexibility in design was in line with the quest for nonconformity, free-spiritedness, and creativity. The design of social systems, architecture as an intuitive form of collective communication, and liberated spaces of public life became avant-garde manifestos. Architects were gripped by the challenge of interconnecting, clustering, lumping together the various components of the city into an amalgamated built form. It gave shape to the idea of community and urbanity in the age of mass society.

The Megastructural Dreamscape

The combination of these influences found form in a new scale and monumentality. There was a popular preoccupation with anything colossal, gigantic, super: superman, supermarkets, superhighways. Collective urban life was visualized in monumental architectural gestures that would encourage social contact across mass society. The New Brutalism, which lay at the heart of the megastructural form, was an examination of "the whole problem of human associations and the relationship that building and community has to them."[18] Brutalism built up colossal structures that joined the functions of urban society into a unified whole. These structures represented a group form—"One World," in Sixties jargon. They were monumental, fortresslike masses with their structural elements, functions, and flows exposed to reveal community life. But the term *megastructure* did not just mean gigantic size. Megastructures exploited and unabashedly displayed the aesthetics of exposed precast concrete as construction material. They were rawboned, angular, and multiform in shape. Concrete columns and steel crossbeams lifted the massive structures into the air.

Their whole makeup simulated a Space Age vessel composed of membrane, spines, tubes and plates, benevolent forces of power. They were knotted with circulatory ducts, stairs, ramps, elevators, and escalators. Like the space station, the cybernetically controlled megastructure contained and supported life. It was epitomized as "organic" and "natural," in a biomorphic vocabulary that became omnipresent in megastructural descriptions. The organic embodied the plasticity and mutability that preoccupied the avant-garde and their engagement with mass society. The form could be rearticulated, or could physically grow through flexible joints and seg-

mented components clipped onto the frame. The megastructure was nothing less than a vehicle for the revolutionary reconfiguration of the urban
world in totality. There was nothing puny about it. It was brutalist, colossal. The impact on anyone confronting such a structure was either awe inspiring or gut-wrenching revulsion.

An inexhaustible series of monumental musings paraded off architectural drawing boards. They pushed the edge of what was possible. Their
scale responded to the mass of people, the numbers and density that Japan
in particular was coming to grips with. Already in 1960, the Tokyo metropolitan region, with nearly 13 million inhabitants, was one of the two
largest in the world. Congestion and urban sprawl were choking daily life
in Japan's capital. In keeping with frightening prophesies about the "population explosion," demographers predicted that Tokyo's inhabitants would
double in thirty years. This criticial situation, along with Japan's housing
crisis, supplied the avant-garde Metabolists with an unprecedented experimental terrain.[19] A rebel band of Japanese architects and designers, the
Metabolists experimented with plug-in megastructures and prefab modular units that could expand to accommodate mass society. Architect Kisho
Kurokawa, for example, conceived his new towns of Hishino and Fujisawa
as networks, or clusters, of settlement modules linked in elaborate highway systems. Dismayed by the authorities' inability to cope with Tokyo's
growth, the Metabolists then proposed a radical new vision for the capital,
one in keeping with Japan's postwar economic boom and the city's global
recognition as the site of the 1964 Olympic Games. For the World Design
Congress in 1960, architect Kenzo Tange created an elaborate illustration
of a gigantic megastructural city perched on a vast platform over Tokyo Bay
(fig. 6.5). It would house 10 million people, and was designed around its
swirling highway system and communications channels. The megastructure was an engine of economic growth and prosperity that would grow
metabolically.

Tange and his fellow Metabolists were deeply influenced by European
modernism. Tange attended CIAM meetings and was involved with Team
10. He joined Constantinos Doxiadis's famous Delos cruises and debated
the merits of a cybernetic approach to human society with the coterie of
celebrity voyagers. His concept for the Tokaido megalopolis (comprising
Tokyo, Osaka, and Nagoya) was based on the work of both Jean Gottmann
and Doxiadis's *ecumenopolis*. Likewise, Tokyo was a key case study in the
ekistics movement and figured prominently in Doxiadis's City of the Future project for the new Pakistani capital of Islamabad (see chapter 3).

The Japanese Metabolists and Tange's Tokyo Bay project were intro-

6.5. Kenzo Tange, A Plan for Tokyo, 1960. Photograph by Akio Kawasumi.
© Akio Kawasumi. Image courtesy of Akio Kawasumi and Tange Associates.

duced to the West in 1961 in a ten-page spread in *Ekistics*, then again in a 1964 special issue of the British journal *Architectural Design* devoted to the "dream planning" of Tokyo. By 1967–68, these megastructural performances had reached their pinnacle. However, it was Reyner Banham's *Megastructure* (1976)[20] that framed interpretations of the form. In his text, Banham described megastructuralism as an international movement derived from Le Corbusier and the Russian futurists as much as by the Japanese avant-garde. One could also harken back to the drawings of La Città Nuova by Italian futurist Antonio Sant'Elia to find its roots. What distinguished the 1960s version of the form was its relationship to the Space Age, and then its actual execution in new towns. While the megastructure may have achieved extravagant virtuosity on paper, it was this actual appearance as utopian artifice that created the most hullabaloo.

Pop Idols and New Town Form

New towns were themselves megaproductions of infrastructure as well as modernist theater. They lent themselves to making megastructural dreamscapes a reality. A raft of the huge brutalist structures appeared in new town projects of the 1960s. Nordweststadt in the suburbs of Frankfurt in West Germany was a case in point (fig. 6.6). Publicized as New Frankfurt, it was located near the old experimental suburb of Römerstadt, built by Ernst May during the 1920s (see chapter 1). The planning and design of the town was

6.6. Nordweststadt, outside Frankfurt, West Germany, 1968. Photograph by Klaus Meier-Ude. © Institut für Stadtgeschichte Frankfurt am Main, S7C1998/37252.

led by architect Walter Schwagenscheidt, who was steeped in prewar social-ist housing reform efforts. He had taken part in the original Neue Frankfurt project with May and also accompanied him to Russia. Schwagenscheidt's book *Raumstadt* (1949) spoke of his commitment to rectify the homeless-ness and feeling of loss in Germany in the aftermath of the Second Word War. The new town of Nordweststadt was to heal these injuries. In his 1964 book *Die Nordweststadt: Idee und Gestaltung*, intricate, playful line drawings portray a dreamscape of happy families.[21] An enchanted urban world cel-ebrates collective life and the ordinary, everyday world of living.

The new town was planned with every advantage for its population of sixty thousand—from playgrounds and schools to the angle of the houses to allow for sunlight. In Schwagenscheidt's sketches, families ram-ble through cheerful neighborhoods free of car traffic. They enjoy leisure time together in courtyards, parks, and public spaces woven around the residential buildings. Hans Kampffmeyer, the longtime socialist city coun-cilor in Frankfurt who was the political champion of the Nordweststadt project, called it "a milestone in the living culture of our day." It was "a gleaming white town,"[22] a textbook study in social betterment and utopian aspiration.

But it was the megastructure at the town's heart that created the most international buzz. In a competition won by architects Otto Apel and Hansgeorg and Gilbert Beckert in 1961, the city center was designed as a huge concrete mass lifted up into the air (fig. 6.7). Three levels of interior pedestrian space were designed with a cosmopolitan ambiance: it would be "full of people all the time."[23] The megastructure became an unlocking mechanism, liberating the residents of Nordweststadt to enjoy the ameni-ties of modern life with carefree joy. No doubt the first shopping malls introduced in new towns were a response to the increasing affluence of the 1960s and the cornucopia of consumer goods becoming available to a mass public. But as an urban environment, the Nordweststadt mega-structure was conceived as fulfilling the diverse functions of a downtown for the contemporary age. It encompassed and sustained all of collective life. Along with commercial shops were cafés, movie theaters, offices, an indoor pool, day-care and elderly centers, library, police and fire stations, a trade school with dormitories, and outdoor theater.[24] It was a lively lei-sure and recreation zone. A two-level community center comprised assem-bly halls and gymnasium, club rooms and workshops. The structure was a densely packed geometric stronghold with a hypermodern interior of glass and steel. Upper-level apartment towers completed the complex. There was

6.7. Nordwestzentrum by architects Otto Apel and Hansgeorg and Gilbert Beckert, Nordweststadt, West Germany, 1968. Photograph by Tadeusz Dabrowski. © Institut für Stadtgeschichte Frankfurt am Main, S7C1998/37639.

an absolute visual power in the sheer scale and cubic massing of the gray concrete colossus, an overwhelming grandeur to its vertical and horizontal planes. Thousands attended its opening in 1968.

Perhaps the most famous (or infamous, depending on the point of view) megastructural feat overlooked the new town of Cumbernauld in Scotland. As with all Scottish new towns, Cumbernauld was the answer to the housing crisis in Glasgow, which remained as grim as ever despite the construction of thousands of dwellings. The old city was a blemish in the minds of planners. Its worst slum areas had been torn down, but there was little space left in the capital, or any desire to build there. In the original *Clyde Valley Plan of 1946* for the Glasgow region, legendary planner Patrick Abercrombie set out a huge greenbelt of thirteen thousand acres enveloping the city. Slum dwellers would find new lives in entirely new towns outside this *cordon sanitaire*: East Kilbride, Livingston, Irvine, and Cumbernauld. Cumbernauld (Gaelic for "meeting of the waters") was located on a two-mile-long windswept hill fifteen miles northeast of Glasgow. It was to be home for seventy thousand newcomers, 80 percent of whom arrived in the early 1960s from the dilapidated neighborhoods of Glasgow. This was a far higher number than had been the case for the earlier Mark I British new towns. Cumbernauld represented the new scale of town building as growth points and new cities. In the 1970 promotional film *Cumbernauld: Town for Tomorrow* directed by Robin Crichton, it is touted as "a town for the future," and "a new concept of community living."[25]

This new version of utopia was designed by urban planner Hugh Wilson, who worked with squads of international experts, from physicists to economists and sociologists. It was a reflection of the technique and teamwork of systems engineering imported to solve complex urban design problems. The master plan featured seven compact, high-density neighborhoods with cottage homes and apartments in vernacular Scottish style. Although each neighborhood was outfitted with the requisite parks, schools, and community center and corner shops, none of them were meant to look inward in the older garden city tradition of neighborhood units. Instead, they were clustered like supplicants around the Central Area. Drawing on American highway design, the roads wound leisurely through the town. There were no grids, no intersections or traffic lights. Residents were protected from traffic by a series of pedestrian pathways under and over the streets, which then spread out to the surrounding countryside, where satellite villages would house another twenty thousand people along with high-tech and light-manufacturing zones.

What made Cumbernauld so trailblazing and drew the wild excitement of architects, planners, and journalists worldwide was the multilevel modular megastructure at its center (fig. 6.8). Designed by Geoffrey Copcutt, known as one of Britain's flamboyant young architects, Cumberland Town Centre was a milestone in urban design. The monumental concrete structure was perched atop Cumbernauld's highest elevation, making the new town a "city on the hill." Its opening was a major propaganda coup for the government-sponsored new town program. The state had successfully responded to criticism of the postwar new towns by offering a revolutionary alternative. On May 18, 1967, Princess Margaret officially inaugurated the showpiece with great fanfare, pronouncing it "fabulous."[26] Although the megastructural concept was mesmeric for architects working in the 1960s, it was mostly resigned to dreamscapes rendered on paper. Cumbernauld was the exception. It was the experimental fantasy come to life. Writing about Cumbernauld for the American *Harper's Magazine*, architectural critic Wolf Von Eckardt waxed eloquent that "Leonardo da Vinci, nearly five hundred years ago, envisioned a city where all the vehicles move underground, leaving man to move freely in the sun. Leonardo might also have sketched Cumbernauld's town center, a soaring citadel surrounded by meadow."[27]

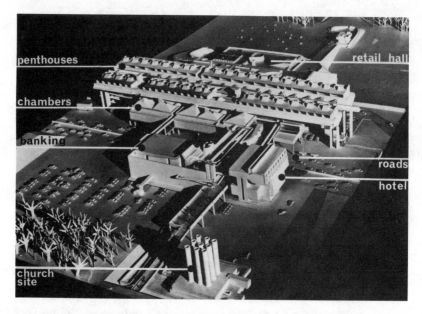

6.8. Model of the Cumbernauld Town Centre, Cumbernauld, Scotland, ca. 1963. Photograph by Bryan and Shear Ltd. Courtesy of North Lanarkshire Archives. © North Lanarkshire Archives / CultureNL.

Cumbernauld Town Centre was part architecture, part designed environment. It was an object building that demonstrated the possibilities of steel and precast concrete. The terraced, cubiform structure was meant to elicit new codes of community behavior. It was this nourishing of collective life that lay at the heart of megastructural aspiration. As well as shops, the structure contained a health center, hotel, ice rink and bowling alley, library, police, fire, and ambulance services, and a technical college, as well as penthouse apartments. It was a colossal living vessel, a terrestrial version of a space station. Inside, people streamed through the structure's eight levels in a maze of escalators, elevators, ramps, and stairways. They melded into the totalized environment. Soon after it was opened, the Town Centre was applauded as an exemplary model of civic architecture. It won the American Institute of Architects' R. S. Reynolds Award for Community Architecture in a competition against the new towns of Tapiola, Finland, and Vällingby, Sweden.

The Town Centre became the darling of the architectural press around the world. It was hip, young, a phenomenon of the Swinging Sixties (fig. 6.9). Copcutt imagined it as a vast terminal facility decorated with a kaleidoscope of advertising and Pop art. *Cumbernauld: Town for Tomorrow,*

Robin Crichton's promotional film from 1970, showcased a consumer and leisure mecca of self-service supermarkets, chic boutiques and restaurants, go-go dancers at the red-hot discothèque. Cumbernauld's Town Centre was a "happening," a social performance where amazing things took place. It was a tangible version of the Fun Palace imagined by Archigram follower Cedric Price. The structure was a source of play, a subversive breaking free from the ordinary through color, space, and design. Cool young couples could relish the fun and entertainment of the city's Central Area and then descend to their convertibles in the parking lot below and speed out onto the motorways.

Yet if they were a sweeping experimental gesture, megastructures such as Cumbernauld Town Centre were also decidedly operational for state technocrats and capitalist interests. Among the many honors for his design, Geoffrey Copcutt received a merit award from the US Department of Housing and Urban Development—an indication of how easily cutting-edge design could be appropriated by the status quo that the avant-garde was bent on undermining. That technocracy could recognize such a trailblazing project was evidence of its own measure of cool. Real estate developers

6.9. Drawing of Geoffrey Copcutt's design for Cumbernauld Town Centre, Cumbernauld, Scotland, by Michael Evans, ca. 1963. Courtesy of The Royal Incorporation of Architects in Scotland © RCAHMS (RIAS Collection). Licensor www.rcahms.gov.uk.

in the 1960s also approached their schemes from the viewpoint of floor space, and the megastructure had plenty of it. It attracted a broad cross section of users and consumers, and maximized profits. The Cumbernauld Development Corporation had toured retail centers in the United States and Britain for ideas.[28] This commercial logic gave Cumberland's Town Centre its status as the first shopping mall in Britain. But from a conceptual point of view, it was imagined as far more. The megastructure was an explicitly 1960s architectural expression. It was Space Age as both architectural style and biospheric living container. And it integrated car culture directly into cutting-edge urban design. The city became a pulsating stream of movement, connectivity, and flow. Reyner Banham acclaimed Cumbernauld Town Centre as "the nearest thing yet to a canonical megastructure that one can actually visit and inhabit."[29]

The cachet of the megastructure filtered down into a variety of futuristic townscape schemes. The magnitude and radical nature of this design production marked it as a school of thought. This was a sweeping transformation of the utopian impulse toward a Space Age frame, an avant-gardist zeal for collective urbanity and automobile mobility. Brutalist megastructures and spaghetti-like highway systems were not for the faint of heart. But an entire design philosophy grew up around them. It sprang from the rebel avant-garde scene, and also from city offices such as the London County Council. For example, working for the LCC, architect Graeme Shankland and his team produced the "Hook Book"[30] for a new town ruled by a sensational multilevel megastructure. The bold imagery made the "Hook Book" an instant primer, already in its sixth printing in 1965 and translated into a host of languages. Publications such as these laid out an entire theoretical corpus for megastructural design and new town ideology.

The gauntlet was laid down yet again, this time in the new towns of Runcorn, England, and Irvine, Scotland. They were the work of Hugh Wilson (chief planner of Cumbernauld) and David Gosling. The two were among the most prominent new town planners in Britain. Gosling in particular was captivated by the possibilities of megastructures as symbolic of new town ideology. He had studied urban planning at Manchester University and then at MIT and Yale University in the United States. His mentors were Kevin Lynch and Lewis Mumford along with Gordon Cullen, and he was heavily influenced by the theoretical writing in the *Architectural Review* of the late 1950s and 1960s. In other words, like many in this midcentury generation, Gosling swam in the stream of the urbanistic avant-garde. His seminal publication, *Design and Planning of Retail Systems* (1976), was a gold mine of information and ideas about commercial forms. He recounted their history and the diversity of designs, and laid out his own visionary theories on the future of retail systems. Gosling saw megastructures as part of the natural evolution of town planning. Although he admitted they could be construed as utopian and futuristic, he insisted they were viable, high-density interpretations of modern city centers.[31]

In 1968, Gosling became chief architect and planner at the new town of Irvine, southwest of Glasgow. It was among the last to receive new town status (in 1966) and was unique as an expansion of the already existing port of Irvine along the Scottish coast at the Firth of Clyde. Like all Scotland's new towns, it was to provide a new life for families escaping the slums of Glasgow—in this case, some 120,000 people. The design for Irvine was based on flexible growth and car ownership. Highway designers and traffic engineers were star players on the planning team. The housing areas (baptized "urban acceleration units"), services, and manufacturing were all situated along a linear highway spine. At the town's center was a gargantuan megastructural deck spanning the thoroughfare and the river Irvine, and spreading out to such an extent that it linked the local rail station with the old harbor area (fig. 6.10). For all intents and purposes, Irvine's megastructure was a built version of Alan Boutwell's proposal for "Continuous City-USA," stretching from New York to San Francisco (see chapter 4). The immensity and bulk of the elevated deck required the controversial demolition of historic sections of the old town of Irvine as well as the old bridge that crossed the river. The new town megastructure was transcendent, floating outside time and place—in a sense, "in the air." The old is arrested, underneath, destroyed, and buried.

Both the vertical and the horizontal organization of the Irvine Town Centre complex became an instrument for inventing public life. The structure's design was based on a meticulous calculation of retail floor space and consumer demand, with multiple supermarkets, department stores, and specialty shops acting as a catalyst for economic growth. The project was financed by private capital (Land Securities PLC and Ravenseft Properties, which were involved in the development of a number of new towns) in partnership with the public Irvine Development Corporation. It contained restaurants, movie theaters, and its own central square.

Gosling's drawings of the imagined interior feature high-impact advertising, Pop art, and a social mix of Mod shoppers enjoying a consumer dream world. It was a playground for the new leisure classes and the coordinates for their desires. They enjoy complete freedom in a consciously designed, climate-controlled environment. His sketches interpret megastructures as upbeat emblems of the Sixties youthful middle classes, a

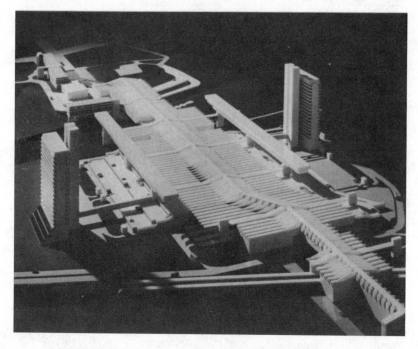

6.10. Model of the Irvine Town Centre from the Irvine, Scotland,
New Town Plan. Published by the Irvine Development Corporation,
1971. Image courtesy of Ross Brown, © North Ayrshire Council.

design strategy not only for enhancing consumerism, transportation, and communication but for countering loneliness and alienation. In yet another influential twist, Gosling had lived in Brazil and was impressed with the recreational facilities available for middle class families there. He imagined the new town center as a Brazilian-style social club open to everyone regardless of income. The result was that Irvine's megastructure was a 1960s social force and became one of the most popular facilities on the west coast of Scotland.

As the design went through multiple iterations, the girth of the shopping floor was increased to become two parallel malls over the widest part of the deck, with branches spreading out on each side of the river. The influence of private developers and their voracious demand for commercial space was palpable. A column-and-beam system supported the exoskeleton packed with wiring, ducts, and circuitry. It was then covered with a state-of-the-art wide-span roof of steel trusses locked together to form vertebral ribs. The result was a thrilling Space Age aesthetic. The terraced pedestrian promenades and commercial floors were stacked over the spine highway running through the town. Ramps led down to the renovated waterfront, pedestrian paths, and a civic and entertainment area. For Hugh Wilson, it was a "coherent system" that was flexible and adaptable to local conditions, and "a climax to the town in the social and architectural senses." It was a symbol of new town promises.[32]

The megastructure's authority was based on its monumentality and sheer spectral force. The structure hovered over Irvine like a colossal space station that had left orbit and docked on Earth, there to colonize the Firth of Clyde and make it modern. Like a space station, it enveloped life. It was bionic: an extensible organism that could grow by clipping on new segments as development expanded outward. Its limbs stretched out and fused with the linear highway. Although it was only partially built and required constant upkeep and a facelift almost immediately, the striking imagery of Irvine's colossal rib-vaulted megastructure epitomized the 1960s' possibilities for the future.

Irvine, Cumbernauld, and Nordweststadt were mercurial places. Their mind-boggling impact embodied a *dirigiste* interpretation of a 1960s countercultural wonderland. The planning, the enormous capital and material resources needed to build megastructures of this scale as the centerpiece for entirely new towns involved an operation of both capitalist and state power. The architectural avant-garde joined forces with this regime in an uneasy but highly creative alliance. The result was that their chimerical designs and fantastic forms actually materialized, but as a normalized configuration used unendingly to signify the forces of modern development.

In a display of showy confidence, the Cumbernauld Development Corporation produced the 1970 film spoof *Cumbernauld Hit*, a James Bond parody about an evil woman's plan to hijack the new town. She arrives by helicopter to disco music. In a stab at the city's uberdesign, she stands before a planning map and proclaims to the city's stereotypically British residents, "Cumbernauld! Tomorrow is already here! A veritable jewel on the Scottish countryside!"[33] A chase scene through the Town Centre celebrates its labyrinthine design. The bad guys pursue an innocent professor through its shopping corridors, up and down escalators, stairways, and ramps, out onto the bridges over the highways. They merge with the infrastructure; they travel its innards. They are a fluid, pulsating circulation in a cybernetic organism. They pierce its membrane and climb out onto the roof. The footage of the wild pursuit is a kaleidoscopic megastructural triumph.

The Cellular Metropolis

These extraordinary fantasies did not last long. The monumental structures that embodied so many Space Age utopian possibilities were rather quickly downgraded into shopping malls. This metamorphosis can best be examined in the work of architect Victor Gruen (fig. 6.11). Having fled Austria in the wake of the Anschluss in 1938, Gruen soon established himself in the United States as a charismatic figure in the architecture and design world with a huge international agency operating out of Los Angeles. Gruen is

6.11. Victor Gruen, date unknown. Photograph by Ivor Protheroe. Courtesy of Victor Gruen Collection, box 56, American Heritage Center, University of Wyoming.

known most famously as the father of the American-style shopping mall, although he vehemently "refused to pay alimony to those bastard developments. They have destroyed our cities."[34] He detested the suburban sprawl that surrounded American cities and the crippling impact of the automobile, which he feared would bring about the "physical and psychological starvation of urbanized man."[35] He saw the shopping center as a remedy for the conformity and anomie of suburbia.

More than simply a commercial outlet, the mall in Gruen's imagination was an ideal community space derived from the tradition of the market squares of Europe or the main street of American cities. Shopping was only part of "the widest possible palate of human experiences and urban expressions" that would take place there. Malls would act as social promenades, incorporate civic and educational facilities, and play the role of the Greek Agora. Influenced by Ebenezer Howard, Lewis Mumford, and eventually by Jane Jacobs, Gruen shared fully in urban reformism and the desire to restore a sense of community belonging. The shopping mall was for him the space of modern public life. In the biomorphism that pervaded systems thinking, the mall would be "the heart of the city." His aim was to counteract "formless suburbanism" by "clustering" development around a pedestrian-oriented urban core.[36] The architectural form that best fit this technique was the megastructure. Gruen pointed to Cumbernauld as the best example of this ideal. With its neighborhoods gathered in homage to the massive Town Centre, Cumbernauld, in Gruen's estimation, was "more principled and sophisticated" than the earlier new towns around London.[37]

Gruen's own first experiments with mall design were at Northland Center outside Detroit, and Southdale Mall in the Minneapolis suburb of Edina. Southdale Mall was a self-contained, controlled environment—a megastructure equipped with the latest technologies. As described in *Architectural Forum*, the two-level mall "uncannily conveys the feeling of a metropolitan downtown: the magical, intangible assurance that here is the big time, this is where things happen, here is the middle of things." Southdale was what a busy, bustling downtown should be if real "downtowns weren't so noisy and dirty and chaotic."[38]

In the Cherry Hill Mall outside Philadelphia, Gruen worked with developer James Rouse to craft a merchandising space that would both embody their community design theories and enrich community life.[39] Access to a theater of consumption was a social act, a release from the monotony of daily routine. More than just commercial fluff, the mall was a fantasy world. Like the neighborhood unit, it would support and nurture both the collective and the individual psyche. It was holistic and comprehensive, socially responsive and profitable. Along with its shops and public courts, Cherry Hill boasted a community hall and a four-hundred-seat auditorium. A variety of civic events were held there during the 1960s, from a local junior prom to civil rights meetings and demonstrations against the Vietnam War.[40]

Victor Gruen was seemingly everywhere in the 1950s and 1960s, appearing on American television, being featured in newspapers and magazines. In 1961, he presented a megastructural project for Roosevelt Island in New York City's East River, which was slated to become a model "new town in town" by the federal New Communities program. Gruen envisioned a vast platform covering the entire island, on top of which eight fifty-story towers and three vast, curving structures would be erected. These extravagant schemes and his notoriety as a "mall maker" made him a media celebrity. In 1956, he consulted for *1976*, an American television documentary for NBC, to explain his visions twenty years into the future. The project was produced by the American Petroleum Institute as a tribute to the automobile's reign in urban design. Then in 1962, Gruen was on the cover of *Fortune* magazine. He also wrote prolifically, sketching out his designs in text and visual diagrams.

Over time, the scale of Gruen's work mushroomed. He proposed urban regions organized into medium-sized "cellular metropoles" separated by green corridors devoted to recreation, parkland, and transportation infrastructure. They would be built up from an organic hierarchy that began with the individual and family, then neighborhood, community center and town, and finally city and polynucleated metropolitan region. In *The Heart of Our Cities: The Urban Crisis; Diagnosis and Cure* (1964), Gruen rendered the Metropolis of Tomorrow in a floret diagram: a metro center surrounded by ten cities, each city center surrounded by ten towns, each town center surrounded by four communities, each community center surrounded by five neighborhoods. He was famously suspected of copying Ebenezer Howard's garden city diagram, an accusation he vehemently denied. Certainly by the 1950s, this kind of formal utopian composition was a ubiquitous part of the planning lexicon. Gruen was one among many who merged it into their concepts. More significant is the quantitative source for these visual productions as both art and science. The Metropolis of Tomorrow was based on precise tabulations of density and population size, and transportation data for each cellular cluster.

Accordingly, *The Heart of Our Cities* is filled with mathematical formulas and graphs depicting statistics. Queuing theory determined land-use and transportation patterns. Gruen argued that Ebenezer Howard's vision "had to be brought up to date." It was the new town system around Stockholm (Vällingby and Farsta) that were the models for "clusterization."[41] He devised a futuristic land-use plan for the entire United States based on a system of planned satellite towns that he called "living cells," linked by expressways. They would combine the benefits of the American small town with the urbanity and cultural institutions of the nation's great cities. But the idea looked much like the Soviet techno-futurism of *The Ideal Communist City* (see chapter 4).

Gruen's project for Tehran, Iran, demonstrates how allegiance to this urbanistic fantasy could be aligned with global geopolitics. His cellular vision became the framework for the first master plan for the metropolitan region of Tehran, developed in the late 1960s in partnership with Iranian architect Abdol-Aziz Farmanfarmaian. Urban planner Fereydoon Ghaffari, who was employed by Gruen Associates, worked as the interlocutor between the partners and the Iranian government. The project was part of the sweeping White Revolution, a package of five-year modernization plans inaugurated by the Muhammad Reza Pahlavi regime. The plans were undertaken with the direct guidance of the United States, development institutions such as the Ford Foundation and the UN, and numerous Western advisers (many of whom were actually French and Soviet). They were paid for with petrodollars and massive loans from the World Bank and the US Development Fund. The United States saw in Pahlavi Iran one of the most promising candidates for the reshaping of national identity around the West's developmental model. The Pahlavi regime could promote US foreign policy interests in the Middle East and counteract the lure of communism.

What this meant was an all-out American-style modernization of Iran and the exporting of utopian urban visions directly to Tehran. The plan for Teheran opened the door to a flood of American engineering companies and planning firms. Gruen competed for projects with the likes of Constantinos Doxiadis and Richard Llewelyn-Davies. It was Gruen's celebrated urban theories and above all his work in the United States that won him the contract to develop the plan for Tehran.

The postwar growth of Tehran had been unprecedented: an unending stream of rural migrants filtered into the city and swelled its population fourfold in the space of twenty years. By the mid-1960s, Tehran had reached a population of nearly 3 million without any real planning to guide growth, and the situation was politically explosive. While the affluent middle classes moved into modern high-rise apartments along tree-lined streets in the northern precincts, the surge of rural migrants found shelter in the city's slums or pieced together vast shantytowns in the old industrial districts along the southern periphery of the capital.[42] Widespread unemployment, poverty, and disease were part of daily life and fueled mounting anger and protests against the Western-backed Pahlavi government. In reaction, the slums of Teheran were condemned as a notorious seedbed for communist terrorists and insurgencies supported by the Soviet Union next door. Iran's government-controlled Construction Bank had already developed small new town projects in the late 1950s and early 1960s based on strict social segregation: Kuy-e Narmak northeast of the capital for middle-income residents, Nazi Abad and Narmak for working-class families, the northern garden city of Tehran Pars for the upper classes. They did little to mitigate social tensions.

Victor Gruen was involved with the initial concepts for the *Tehran Comprehensive Plan* (figs. 6.12, 6.13). They were a bundle of urban renewal ideas and modern dreamscapes imported directly from his work in the United States. Critics argued that the plan looked shockingly like Los Angeles, strewn with highways and dominated by development interests. It was put into practice by a combination of American companies and Western-minded Iranian elites, many of whom (such as Abdol-Aziz Farman-farmaian) had studied in England or at the École des Beaux-Arts in Paris and also worked in the United States (for example Fereydoon Ghaffari).

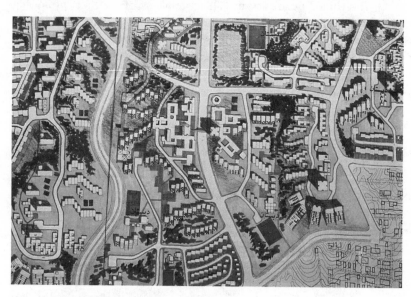

6.12. Diagram from the *Comprehensive Plan for Tehran* prepared by Victor Gruen Associates with Farmanfarmaian Associates, 1966–70. © Gruen Associates.

6.13. Illustration from the *Comprehensive Plan for Tehran* prepared by Victor Gruen Associates with Farmanfarmaian Associates, 1966–70. © Gruen Associates.

This young generation of architects and engineers were secular and nationalist in political stance and Western in taste and outlook. Their short-lived journal *Architecte* (begun in 1946 and edited by Iraj Moshiri) was an avantgarde forum on architecture and planning, and provided vital information on new materials and construction techniques.

The reliance on Western experts was assumed to be essential for giving form to Iran's modernization. Like their counterparts in the West, Iran's new elites were urbanites with little knowledge or experience of the no-man's-land beyond the central city, especially the shantytowns to the south. Even more, they were derisive of its way of life. Their method of dealing with it was abstract and wrapped in systems thinking. Extensive social, economic, and demographic studies and quantitative analysis of traffic flows were undertaken to bring the areas under rational control. These brought the *Tehran Comprehensive Plan* to some five volumes. The growth of the metropolis would be restricted to a population of 5.5 million over a twenty-five-year period, and a new order imposed on its physical fabric. Sweeping slum clearance projects in the congested center of Tehran would open up and modernize the city. A tourist zone would be created around the old bazaar district. These urban renewal programs would force the relocation of upwards of six hundred thousand of Tehran's people out to the suburbs.

A system of ten satellite cities (*mantagheh*) would welcome them. The cities were to be situated along an east–west linear axis and linked by a hundred-mile system of superhighways and rapid transit routes embedded in verdant green corridors.[43] Waxing eloquent on the beauty of the highway system was one of the more noteworthy aspects of the *Tehran Comprehensive Plan*. Illustrations of sleek, sinuous ribbons of highway bordered by high-rise towers grace its many pages. It was the universal image of modernity. The parkland and gardens around them were designed by American landscape architect Ian McHarg as an ecological microcosm of the world. Following Gruen's cluster theory, each satellite city would consist of four residential zones situated around a dense urban core that was dominated by a megastructural shopping mall suspended over layer-cake parking structures. The mall was a sequence of horizontal planes filled with shops and boutiques oozing Western consumer goods. This central area was encircled by its residential zones, each divided into four communities, and then each community separated further into five neighborhoods. Each level in the urban hierarchy would knot around its own central commercial area and its schools.

The northern precincts of Tehran at the foot of the Alborz Mountains became a modernistic dreamscape of new towns. They were designed for Iran's urban elites and inspired by the Western hedonistic lifestyle and consumer spectacle of the 1960s. The megastructure had been converted into a swank shopping mall to create a fledgling base for luxury consumerism. Its role as a civic center and containment of social life vanished. In what were perhaps the most blatant emblems of Westernization, the Royal Hilton Hotel and Chattanooga Restaurant opened their doors in the early 1960s near the new town of Abbasabad. This layout of Western attributes was surrounded by modern apartment towers with swimming pools and tennis courts, the flats outfitted with televisions and refrigerators. These were the consumer cravings of a Westernized Iranian elite. The architectural style for the apartment complexes was an unashamed 1960s modernism. There was not a trace of tropicalism or orientalist fantasies in these new towns to the north of Tehran; everything exuded Western glamour. Meanwhile, the ethnic and rural poor stuck in the slum districts to the south could do little but stare enviously at the glitzy landscape. For them, Victor Gruen's ideal of invigorating community life was a cruel joke. His vision of the Metropolis of Tomorrow was meant for the rich. The new towns further inflamed Tehran's deeply rooted social and spatial divisions and the conflict between a Western secularized Iran and a traditionally Islamic one.

The *Tehran Comprehensive Plan* proved impossible to implement in total. It attempted an archetypal and oversimplified reproduction of the American dream that clashed with Tehran's complex urbanity. The interlacing highway system was carried out, but the commercial centers were never realized. Nonetheless, the plan had a significant impact on the physical form of the metropolitan region, particularly north and west of the city. It guided decisions about not only the capital but all the major cities of Iran over the next ten years. As important as it was, Gruen's vision was only one in a series of utopian illusions for Tehran by the foremost Western new town planners. Despite their promotion of scientific analysis and extensive study of local conditions, each planner arrived ready to apply his theories whole cloth to the capital. They were saturated with the systems logic that dominated urban design theory.

In 1972, totally ignoring the *Tehran Comprehensive Plan* and Gruen's new town clusters, Constantinos Doxiadis appeared before the Pahlavi regime with his ekistics concept and advocated a Western linear extension of the city in the image of his *Dynapolis*. Doxiadis used computer analysis to generate and evaluate data on Tehran and offer a scientific plan. Then in 1975, Richard Llewelyn-Davies and the team from Milton Keynes were hired to design the grandiose civic and ceremonial Shahestan-e Pahlavi complex in northern Tehran at Abbasabad, which had initially been developed as one of Gruen's new towns. The project reflected Reza Shah's obsession with sweeping showcase projects that signified Iran's catching up with the West and arriving at the Great Civilization.

In reality, private real estate speculation and construction booms spearheaded by state modernizing elites had more to do with the form of modern Tehran. The regime's urban policies intentionally encoded the social segregation between the rich Westernized northern districts and the sprawling slums in the south. Ultimately, however, with the Iranian Revolution in 1978–79, all the projects came under ferocious condemnation as a legacy of the hated shah. In the words of historian Bernard Hourcade, "In 1978 the people of Tehran retook their city for themselves."[44] How urban fantasy could communicate political power was crystal clear, especially in the case of high-profile Western firms such as Gruen's. Progressive architecture aligned itself with state interests and showcased the modernization regime. It made practicing utopia more than a mere pipe dream, but the new town could easily embody a vanity project for elites and a treasure trove for foreign investment in a dangerous geopolitical chess game.

The "Heart of the City"

The degree to which avant-gardism could serve political agendas and amplify state power was particularly tangible in the case of France. The French contribution to the decade-long fascination with the Space Age was largely the work of GIAP (Groupe International d'Architecture Prospective), founded in the 1960s by urban theorist Michel Ragon. The French avant-garde met only occasionally with the varied casts from Archizoom or Archigram, or with Paolo Soleri or the Metabolists. Yet the power of invention surrounding cybernetics and the Space Age was so strong that it saturated French utopian imagery regardless of the relative isolation. It was a shared visual semiotics that was interpreted locally and produced an extraordinary array of quixotic urban ideals.

In 1962, Ragon presented "Dossier du Paris futur" in the weekly journal *Arts* in which he surveyed many of the avant-garde ruminations on the Paris region's future.[45] Among them were the drawing studies of Paul Maymont, a member of GIAP known as the architect of the atomic age. His "Spatial Paris" project (fig. 6.14) was intended for the Plaine de Montesson just west of the suburb of La Défense, and was based on monumental cone-like hyperstructures of extreme density hovering over the wetland area, with 15,000 to 30,000 people residing in each tower. The influence of astronautics fueled fantasies for freeing terrestrial dependence and living in mammoth floating cities. Maymont's imagery shared much with Paolo Soleri's Space Age megastructures as well as Soleri's ecological concept of literally lifting techno-settlements off the land. A hollow central column in the gigantic cones would contain wiring and ducts: the cables would unwrap to circulate energy and connectivity for the structure. Each conical city is arranged in layers with suspended plazas, vertically stacked rail and

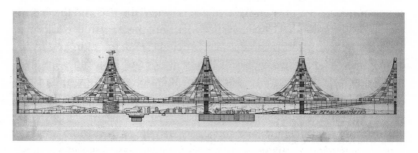

6.14. Paul Maymont, Floating Cities; elevations and sections. Expansion study of Paris, 1962. Photograph by Philippe Migeat. © CNAC / MNAM / Dist. RMN-Grand Palais / Art Resource, NY.

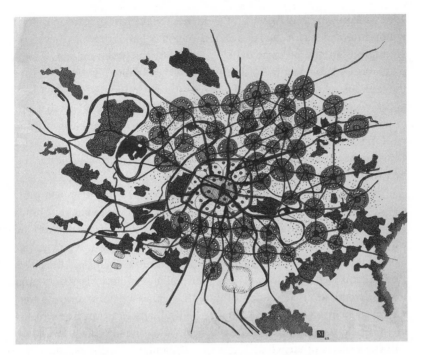

6.15. Paul Maymont, Floating Cities; map of Paris and its surroundings, 1965. Photograph by Philippe Migeat. © CNAC / MNAM / Dist. RMN-Grand Palais / Art Resource, NY.

road systems, and housing and commercial areas. Maymont's conical cities are a floating mosaic of aesthetic objects, linked with one another by suspended motorways in a science fiction map of the Paris region (fig. 6.15). They dot the metropolitan landscape in a panorama of satellite towns.

In order to sustain this scale and monumentality, architecture became a specter of itself. For architectural critic Manfredo Tafuri, Paris became the topography for a supertechnological utopianism and the institution of a championship, a contest of the architectural imagination.[46] Maymont wrote of an "urbanism entirely conceived by spaces that are easy to draw . . . and immediately graspable by the public."[47] The inference was that the new Paris would go from the musings of illustration to actual built environment with ease.

This fantasized urbanism emerged precisely during the *trente glorieuses* of extraordinary modernization in France, which provided unprecedented opportunities for collaboration between the French state and the avant-garde futurists. This alliance is what brought the architectural vanguard into the campaign for new towns and made these places so chimerical. In

1962, Ragon was invited by Paul Delouvrier to take part in a group reflecting on future plans for the Paris region. Ragon believed that GIAP's visionary projects should be immediately adopted by the government as a template for the Paris region. To do this, the group adapted its rhetoric from the language of utopia to that of technocratic revelation. Its "paper cities," as they were called in GIAP's magazine *Utopie: Sociologie de l'urbain* that surfaced in Paris in 1967, were not just visual codes but codes for action. They were the "diagrammatic expression of the spatial imaginary the purpose of which is the communication of master ideas at a very journalistic level, and by virtue of their pregnancy, are destined to be built."[48]

The magnitude of these visual narratives appeared in the French journal *L'Architecture d'aujourd'hui* in a special issue on new towns in 1969.[49] The presentation seamlessly merged actual new town projects with the wild ideas of the rebel experimenters. Paolo Soleri's diagrams for a vast undersea city and for his outer-space city of Asteromo embellished the pages alongside Archigram's "Instant City" and Italian architect Massimo Maria Cotti's "Dendratom" rendering of urban utopia as a giant manmade tree. American architects proposed a soaring vertical hyperstructure as a replacement for Detroit, and a stretched-out horizontal one to replace Brooklyn. Full-color plates of French architect Jean Renaudie's "la ville est une combinatoire" depicted the city as a topography of organic cells linked by a synchronized communications system. Cybernetic logic diagrams laid out the schema for urban form and structure.

We might interpret all these simply as far-fetched utopian provocation. But they were interlaced in the special journal issue with real places such as Milton Keynes and Nordweststadt. The 1965 Schéma directeur's five new towns of the Paris region were splendidly displayed: Marne-la-Vallée, Melun-Sénart, Evry, Saint-Quentin-en-Yvelines, and Cergy-Pontoise. The presentation posited the state's plans for these very real towns as the opportunity to fulfill the radical fantasies just a few pages before. In the case of Renaudie, his "combinatory city" became a proposal for the new town of Val-de-Reuil in Normandy (fig. 6.16).

French avant-garde projects paralleled technocratic thinking far more than scholars have accounted for. France developed a specific spatial culture that was topographical and quasi-magical in orientation. The potentiality of the new age was understood by its architects and planners in terms of visionary sociospatial acts. State architectural and planning competitions and contracts were golden opportunities to carry them out. Architects and urban designers were given free license to play, to be eccentric and inventive with vast urban territories. They behaved as technocrats, invest-

6.16. Jean Renaudie, new city of Val-de-Reuil, Eure, Haute-Normandie, France, 1967–68. Town planning plan study; project for the competition, unrealized. Photograph by Adam Rzepka. © CNAC / MNAM / Dist. RMN-Grand Palais / Art Resource, NY.

ing tremendous energies into producing fantastic urban forms. Theirs was a self-proclaiming architecture in the service of the French state, a uchronic vision of anticipation realized through conscious utopian acts.[50]

Part of this ambition was to create a superstructure at the heart of each new town. Practitioners were determined to avoid the alienation and isolation that plagued the *grands ensembles* housing projects in the Paris suburbs. By the early 1960s, they were already infamous as shameful planning failures. The French were fixated on finding an alternative to these fiascos, and it made their discourse on the urban future particularly grandiloquent. The megastructural *dalle*, or deck, became an inflated formula, a force field within which an idealized collective life would be forged under state guidance. The massive decks were interpreted as open, free, endlessly flexible spaces. They were "a place of public life, where the inhabitants can find possibilities of *flânerie* and spontaneous encounters, information, and can also provide the identity and expression of collectivity." They represented "a kind of social paradise, an idyllic overflowing with optimism and naiveté."[51]

The imagery was arresting: cities raised up on spacious platforms suspended over highways, in a rapturous version of Paul Maymont's floating cities. Working in tandem with state planning agencies, private construction companies responded to this idyll by churning out state-of-the-art construction techniques and materials: ribbed decks in reinforced concrete, massive solid decks, deck flooring with specialty T- and U-shaped steel support systems, watertight ceiling structures. Civil engineering companies such as Barets, Beaupère, Camus, Cauvert, Coignet, Estiot, Fiorio, Sproma, and SSTP became the heroes of modernity. Architecture and building magazines were filled with glossy feature articles about their feats of design and construction.[52] It was precisely this disruptive dynamic between an antiestablishment architectural community, state technocracy, and private capitalism that made carrying out futuristic ambition a real possibility. New towns were the flamboyant register of their alliance. Ironically, the architectural avant-garde worked to lionize the very forces of the modernization regime they had critiqued. The result: a spectacle of utopian expression that permeated the entire new town movement.

Debates about democracy and the right to the city pervaded sociological thinking about urban life from the late 1960s onward, especially among the young architects of Team 10. This engagement with the social sciences was a reaction against the rigid orthodoxy of modernist functionalism and produced a myriad of interpretive design. Shadrach Woods, who was responsible along with fellow Team 10 members Georges Candilis and Alexis Josic for the influential *dalle* at the new town of Le Mirail (fig. 6.17) in the suburbs of Toulouse, argued that contemporary society was an open, non-hierarchical cooperative of total participation: "Today space is total and society is universal . . . the world is one, a continuous surface, surrounded by continuous space. . . . Total space and universal society are interdependent."[53] The mazelike composition of the *dalle* at Le Mirail was a model of Woods's stems concept. It was a way of connecting space in a coherent system of mobility, interaction, and encounter.

It makes sense to differentiate the design theories that proliferated in the 1960s quest for an urban form suitable to mass society. Le Mirail was an indication of the richness of utopian production. Yet in their pursuit of a built environment suitable to collective life, the avant-garde shared a powerful spatial language derived from systems and cybernetic logic. Woods imagined the stem as a flexible organic structure or web. He described it as a system of circulation in which dwellings could be "plugged in" to the "streets in the sky."[54] The result was the city as a cohesive megastructural living environment. For his part, Candilis refused to accept "that cities should only be conceived in terms of His Majesty the car." At Le Mirail, he likened the reinforced concrete *dalle* to a collective "living room" reserved for pedestrians free from the terror of automobiles flying by. The

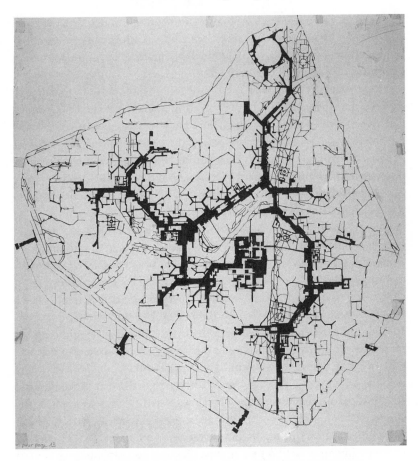

6.17. Georges Candilis, Alexis Josic, and Shadrach Woods, pedestrian traffic map, Toulouse-Le-Mirail, France, 1961. Photography by Philippe Migeat. © CNAC / MNAM / Dist. RMN-Grand Palais / Art Resource, NY.

car was relegated (literally and figuratively) to a lower level. The raised deck would revitalize linearity and "the notion of the street."[55]

The dialogue about megastructural space as the "heart of the city" played out with particular force at the new town of Evry, located about twenty miles southeast of Paris. Evry was anointed as one of the original new towns in Paul Delouvrier's Schéma directeur for metropolitan Paris (see chapter 5). Along with Cergy-Pontoise, it was meant as an ideal urban prototype. Evry's town center became a laboratory of creativity and inventiveness. It was an obsession at France's premier planning think tank, the Institut d'aménagement et d'urbanisme de la région parisienne (IAURP), and at Evry's Development Corporation. Both groups fretted over every detail. The reports on the town center were filled with mathematical equations and systems diagrams, speculations on public space, and preliminary blueprints of a designed wonderland.

The surplus of brainpower behind the town center project immediately exposed the tensions behind the competing visions for Evry. The planners began their work with two contradictory models: the center city of the new capital of Brasilia and the brand-new Parly 2 shopping mall to the west of Paris. To complicate things even further, they also saw a clear distinction between the private sphere of commercial development and the public sphere of civic space. How to combine all these incongruous elements tested the meaning of urban centrality.

The new town projects in the suburbs of Paris took place just as large-scale supermarket and department store chains appeared on the suburban landscape. The quintessential Paris department store Printemps was experimenting with suburban branches, and approached the Evry planning team with a proposal to act as retail anchor for the town center project. Victor Gruen's American shopping mall projects were also well known, and from 1967 to 1971 Gruen consulted with the IAURP, principally on the design for the center of Evry. The proposals elicited a fierce struggle over the purpose of center cities. Evry's planning team toured the United States, where they "suffered from an indigestion of shopping malls that looked exactly alike except for their size."[56] The French skewered Gruen's concepts as the epitome of American crass commercialism. Gruen tried desperately to distance himself from this reputation, swearing, "I am an internationalist . . . I have no intention of blindly imitating the American experience . . . in fact I am warning against imitating the American shopping center."[57] But in the atmosphere of generalized anti-Americanism that pervaded France in the 1960s, the open hostility was enough to push him into the background.

In order to solve the conceptual dilemmas swirling around the design of Evry, planners turned instead to what were deemed European urban models: the Forum at Bellingham, England, and principally the Agora, designed in the mid-1960s by Dutch engineer-architect Frank van Klingeren for the new towns of Dronten and Lelystad on the Zuiderzee in the Netherlands. The terms *forum* and *agora* were used extensively as part of the ideological rhetoric of town center planning from the late 1960s through the 1980s. The terminology shared much with what Gruen had in mind for his multifunctional shopping malls. Van Klingeren's Agora concept was on the cutting edge of spatial design and attracted immediate and widespread enthusiasm. It was, according to Van Klingeren, "where everybody meets

everybody, where you can watch and learn." It would fashion local identity through a mix of recreational and cultural activities, and provide an
open-ended democratic atmosphere where "everyone could see and hear
the same thing."[58] The Forum was laid out beneath a floating roof held up
by nine exposed steel columns and filled in with glass walls. When completed, it was considered one of the earliest and most avant-garde examples
of integrated urbanism used in European new town strategies.

Van Klingeren's Agora concept drew the attention of the team at Evry
almost immediately. It was applied in the 1971–72 design for Evry 1 (the
first of four development zones to comprise Evry Centre). The project was
awarded to Michel Andrault and Pierre Parat, cutting-edge practitioners
known for their arrangement of volumetric spaces and subtle handling
of concrete. Their proposal offered a stacked ziggurat residential complex
similar in design to Moshe Safdie's "Habitat" at Montreal Expo 67, and a
succession of linear pedestrian decks perched above Evry's transportation
hub (fig. 6.18). Eventually, the massive superstructure measured four hundred by five hundred meters, replete with commerce and services, office
space, and a succession of landscaped public promenades on two levels.

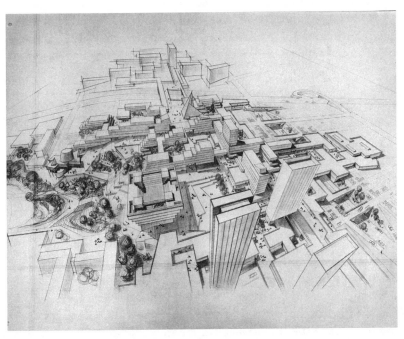

6.18. The platform-*dalle* at the center of Evry, France. Courtesy of J. Bruchet / IAU îdF.

It was a bravura statement characteristic of new town creed. Defined as "the heart of the city," Evry's megastructure was a "privileged space . . . that would reinforce a sense of place and spatial landmark."[59]

The fluid *dalle* surface spread amoeba-like to the surrounding residential buildings, the city hall, the prefecture, the chamber of commerce, Mario Botta's Cathedral of the Resurrection, and the local university campus. The multilevel Agora beneath was equipped with a cornucopia of cultural and recreational activities, from bowling alleys and swimming pool to basketball courts, ice skating rink, cinemas and discos, a vast shopping mall, and a community center. Inveterate French urbanist Pierre Merlin described the platform deck and Agora as embodying the main concept for all the new town projects: "It is 'urban integration,' that is to say a grafting of the different elements of the city onto a space of liaison open to all."[60] For the neglected, ill-served suburbs southeast of Paris, the Agora was a revelation. Its inauguration was attended by thousands, and reported in the press and on television. In the substratum of the monumental complex rested the SNCF railroad station and the RER regional rapid transit system. It was intersected by four main boulevards, bus and bike lanes, and parking garages. Altogether, it was a flamboyant, avant-garde marvel made real under the auspices of the French state.

In his *Megastructure* volume, Reyner Banham argued that despite a wide gulf of intention, real commonalities existed between mid-1960s architectural fantasies and projects such as Cumbernauld, Gruen's plans for Tehran, and places like Le Mirail or Evry.[61] The structures expressed the aesthetics of the Space Age and the ability to construct great horizontal and vertical forms that accommodated all of urban life. These were the possibilities of the future. Yet it was symptomatic of new towns, and all the utopian aspirations they contained, that the dialogue shifted rapidly from superlative to abysmal. Cracks began to appear in the megastructural cast even before the mold was dry. The forms were always a quixotic ideal. Their origins lay in the shock of the uncanny, avant-gardist provocation, the jolt into a new way of seeing community and urbanity. They were imagined as bionic and cybernetic (or cyborg), as an architectural device for freedom and democracy, and as mass-cultural apparatus. But the potential for an innovative theater of public life was fraught with the incongruities of official state sponsorship and with the commercial logic of private development and commercial interests. These required constraint, regulation, and control, a narrowing of the megastructure's use to the act of shopping. The less-than-happy effects were apparent almost immediately.

Perhaps more than any other feature of actual new town construction, colossal megastructures represented Fredric Jameson's swindle of utopia. The leap into the future was short circuited. Fifteen years after opening, the Nordwestcentrum was already out of date and considered a failure. As Cumbernauld's Town Centre literally and symbolically crumbled, the town won the Carbuncle Award for the most dismal place in Scotland. These were not the kinds of accolades avant-garde architects had in mind, but they were an embedded part of the utopian discourse from its beginning.

The prizewinning personality of Cumbernauld cut both ways. The dark side of the futuristic vision was always there to see. Almost as soon as it broke ground, the megastructure with its assorted pedestrian *dalles* and ganglion highways began to lose its avant-garde appeal. By the early 1970s, it carried on as the favored design for state development projects in alliance with private commercial ventures. Its successors are today's omnipresent shopping malls surrounded by layer-cake parking structures.

The Space Age vanguardism practiced by 1960s architects became a repressive monster, a seizure of the city by developers and profiteers. Judged ugly and distasteful in the court of public opinion, the few megastructures that were built symbolized a genuine loss of credibility and trust. Visionary architects found themselves under attack for their technological fetishism and for seeing the built environment as little more than a modular, cybernetic communications system.

The megastructure left modernism with a scathing reputation as alienating, overscaled, and inhumane. The model of a computerized, cybernetic spaceship was ineffective, if not downright loathsome. In the ironic turn symptomatic of utopian aspiration, the megastructure ended by destroying exactly what its planners and designers had hoped to achieve: a vibrant collective spatiality and urban life. Nothing became more ubiquitous than photographs of Cumbernauld's and Evry's decaying town centers. Massive structural faults were found in the Cumbernauld central building and adjoining apartments. Businesses declined and went bankrupt. There was always a trial-and-error quality to these massive projects, both in concept and in execution. Piece by piece, Cumbernauld's Town Centre was demolished until nothing was left but the biggest supermarket in Scotland. By the 1990s, what remained of the structure was sold to a private management company and remade into a banal shopping mall. Shadrach Woods and Georges Candilis's *dalle* at Le Mirail was a forlorn and neglected void, parts of which were demolished in the first decade of the twenty-first century. Reyner Banham's characterization of the New Brutalism as "memorability as an image"[62] found its greatest expression as dystopia.

Marxist philosopher and sociologist Henri Lefebvre's *Notes on the New*

Town (1960), written while he stood on a hill overlooking the new town of Mourenx in the southwest of France, poured unending scorn upon this enslaving, segmented, bourgeois way of living. But he finally came around to considering the potential for emancipation in the new town. Lefebvre wondered about the function of such urban spaces as Evry's Agora, asking, "Will people be compliant and do what the plan expects them to do, shop at the shopping center, asking for advice at the advice bureau, doing everything that civic center officers demand of them like good, reliable citizens?" Everyone is like everyone else. Yet at the end of his meditation, Lefebvre argued that the "boredom is pregnant with desires, frustrated frenzies, unrealized possibilities. A magnificent life is waiting just around the corner, and far, far away. It is waiting like the cake is waiting when there is butter, milk, flour and sugar. This is the realm of freedom."[63] In this, Lefebvre was willing, despite his harsh reservations about the new town, to consider that it might have some potential for human emancipation, in the true tradition of the avant-garde.

New Towns in the Twenty-First Century

In 1970, economist and planner William Alonso penned "The Mirage of New Towns." "For many," he ruminated,

> glimmering images of simpler, future Camelots combine the American nostalgia for the small town with a desire to escape from the biting reality of our complex urban problems. But mostly the idea of new towns has some magic that fires the imagination, stirring some Promethean impulse to create a better place and way of life, a calm and healthy community of crystalline completeness.

Trying to apply new towns as policy was a fool's errand, because even government decision makers, he continued with exasperation, adopt the "vague rhetoric of architectural and utopian writers."[1]

But that was exactly the point. The new town was always the pot of gold at the end of the rainbow. That Alonso was writing his article to counter a popular belief that somehow the majority of population growth in the United States could be channeled to new towns was evidence enough of their mythic power. In the late twentieth century, there was unfettered confidence in utopianism and a belief in its legitimacy as an opening to the future. New towns were a significant part of that conviction. They moved beyond the limitations of the present into a modern urban world that was qualitatively different from what existed. It was magical thinking indeed.

And that has been the argument of this book. The new town movement of the mid- to late twentieth century carried on the legacy of urban utopianism. This engagement with future cities was instrumental in framing the imaginary on how to live and in shifting attitudes toward modernization. The golden age of new towns coincided precisely with what Zygmunt Bauman has termed "heavy modernity." The metaphor of heaviness alludes to the drive to territorialize and homogenize, along with a preoccupation with scale and space.[2] Urban paradise was linked to reconstruction and renewal, to nation building and industrialization, to the welfare state and the invention of a happy citizenry. Capital investment in colossal city-building and infrastructure projects was a leap into a new and better future. Towns and regions were made modern and manageable, and were absorbed under state control. These processes took place in the West and then were exported to Western spheres of influence throughout the world, where they were molded to local circumstance. From this perspective, new towns were not built from a corpus of fixed socialist or capitalist ideals so much as they functioned to define these ideals around modern development. In the most general sense, the new town movement advanced, organized, and reproduced the late twentieth-century model of progress through urbanization and modernization.

Although a skeptic might ask whether this new town movement even registered on the seismometer of late twentieth-century urbanization, the planning, designing, and building of these places were in fact dazzling projects, and their influence was extraordinary. Under the regime of heavy modernity, fulfillment was always somewhere in the future. The new town accelerated progress and the step into this time ahead. This movement to design and build the promised land was carried out by an international community of architects, planners, and urban reformers. Utopian ambition was immensely meaningful for them. They were sailing into the future, tacking back and forth, this way and that, searching for urban paradise.

As a transnational knowledge workforce, they created a phantasmagoria of texts and illustrations around new town praxis. These publications were accompanied by advertising, celebration in film, literature, and song, and images of daring New People living in paradise. The media blitz converted the new town into utopian theater. Consequently, the new town movement was propagated as much through the transmission of these dreamscapes as by actual building. It formed a conceptual machinery steeped in the marvelous. Viewers interpreted what they saw in deeply utopian terms. There was a psychology of perception at work that made new towns into an ongoing commentary about the future. It is this discourse and practice of utopian aspiration that this book attempts to capture.

Perhaps the most quixotic aspect of this vision was that physical design could affect human action and standards of conduct. With the correct urban design and configuration, it was believed, neighborhoods, towns, and regions, and ultimately the nation, would transmute into a modern way of life. In alliance with the modernization regime, the planning of new towns became a laboratory for this social engineering. In the postwar years of reconstruction, the garden city and the neighborhood unit were the guiding doctrines for social stabilization and collective harmony. The mirage of happy families enjoying everyday life in sylvan neighborhoods was replicated endlessly, not only in North America and Europe but most significantly as a form of colonial and postcolonial policy. It was in the deserts of Africa and the Middle East that garden city fantasies met their demise.

By the 1960s, the city as text was written increasingly in the language of cybernetics and systems analysis. The 1960s were the glory years of modernization. Urban fantasies poured from the pens of avant-garde architects and planners. A new Space Age ethic claimed the life of the city. New towns were an experimental zone for ingenious megastructures, futuristic transportation and communications networks, and biomorphic configurations. They were speculations on the future. Utopia became a social freeing up, a racing forward into the realm of science fiction.

Yet all this could form a fatal snare. By the mid-1970s, the new town mania had gone into retreat. The ideal of the clean sweep had lost its enchantment, and utopianism was exhausted. Tipping the scales against new towns was a skepticism that the state could or even *should* be in charge of these ambitions. The quest for the ideal society came under withering criticism.

The concept of utopia has always been subject to a myriad of definitions and interpretation. Architectural historian François Choay defines it as a mechanism to fathom, record, and spell out insoluble social problems. It is a kind of preliminary to political commitment and social transformation. Yet despite the never-ending search to find a built form for changing society, Choay accuses urban theorists of jettisoning the social and political dimensions of utopian production and concentrating solely on a spatial model. This was the devil's due for aligning themselves with state interests and the established power structure. When their fantasies were turned into reality, as in new towns, they did so in oppressive fashion and imposed a sterile technical order.[3] Here, then, was the deep tension in the utopian impulse: new towns were both a critique and a reproduction of mid- to late twentieth-century society.

Despite the growing misgivings, building a city from scratch continued as the most powerful high-tech tool in the professional planning world until 1973, when Douglass Lee published his famous "Requiem for Large-Scale Models" in the *Journal of American Planners*. Not coincidentally, 1973 marked the first global oil crisis and the end of massive state spending on capital- and infrastructure-intensive megaprojects. Lee accused planning models of a variety of sins, from being too large and complex to too expensive, too prone to mathematical error, and too general and wrongheaded in their deductions.

These indictments were eventually leveled at the modernization regime itself. The quantitative geography and location theories of Walter Isard and William Alonso that were so fundamental to new town ideology were upended in a revolt led by David Harvey, Gunnar Olsson, and Doreen Massey. In the pages of the journal *Antipode*, they advocated a critical social geography as the way forward. Planners also came to realize that although quantitative models were well-intentioned instruments for solving intractable urban problems, their results frequently ranged from nonsensical to just plain bad, and could be distorted as well as intensely politicized. Despite the daring efforts to create cities through machine intelligence, to make them operate with the efficiency of a space capsule, computers were not creating better cities. The reaction against systems logic was ferocious. The taste for Space Age avant-gardism likewise faded away.

The objects and artifacts left behind by this new town groundswell are a kind of ontological theater of modernity, the ruins of utopia etched into the landscape. For many, new towns seem without much significance: at best banal, at worst ideological, and in any case ending in dismal failure. And yet the new town *movement* embodied a significant corpus of utopian ideas and influences and held a commanding sway over how cities were envisioned.

New Towns Turned Old

The new towns of the second half of the twentieth century are no longer "new," although some have managed to build outsized reputations at the extremes of good and evil. Beyond those new towns that managed to make it on to the hit parade of triumphs or fiascos are the numerous satellite and new towns that are far more ordinary places. How they weathered time as something less than modernist dreamscape has depended on local conditions, from the gyrations of public policy to mundane investments in maintenance. Modernization and modernism have always been refracted by local prisms.

Eisenhüttenstadt in the former East Germany is a good example. Now well over fifty years old, the town is currently undergoing a facelift to restore its glamour as an ideal communist city. The collapse of the values that built this new town has provoked a growing nostalgia for the utopian good old days. The original 1950s housing blocks are being faithfully renovated. The industrial economy that was Eisenhüttenstadt's reason for existence has long since disappeared. But global steel corporation ArcelorMittal has taken over its famous steel plant, though with a far smaller workforce than in the halcyon days of its unveiling. Eisenhüttenstadt sports its own City Center shopping mall, McDonald's, and Burger King, and prides itself on having received a visit from American actor Tom Hanks in 2011. These are welcomed as signs of the city's endurance and acknowledgment by the outside world.

The celebrated city center in Vällingby, Sweden, is also enjoying a makeover. It now serves a local population of some sixty thousand. The city's sought-after residential neighborhoods continue to be graced with good schools, supermarkets and restaurants, parks and recreation areas. Vällingby is one of those places where the promises of the new town did indeed come true. This is also the case with Tapiola, Finland. It is being reimagined as an eco-city, with an eye to preserving Aarne Ervi's original design concept. The city center will be updated with the largest green pedestrianized commercial zone in Finland.

By the turn of the twenty-first century, Cumbernauld, Scotland, had a population of well over fifty thousand. Although it still suffers a punishing reputation as one of the ugliest places in Britain, it has been recognized by the UN as a heritage site of twentieth-century architecture and won the Scottish Design Award for Best Town. After suffering years of neglect, parts of Cumbernauld's almighty megastructure were demolished. The event was featured on television as a symbolic end to what was considered one of the worst brutalist failures.

Many of the most egregious megastructures, such as Cumbernauld's or the *dalle* at Le Mirail outside Toulouse, France, have fallen to the wrecking ball. But a good number of these oversized creatures of the 1960s remain after being scaled back and refurbished as shopping malls. This was a predictable part of their future from the very beginning—the shadow of dystopia always lurked just below the surface of paradise. And despite their Space Age pedigree, the megastructures were made possible only by commercial investment. Realizing utopian aspiration was a very down-to-earth affair involving an uneasy coalition of lofty avant-gardism and crass financial interests. What remained of Cumbernauld's Town Centre found new life as the Antonine Centre shopping mall, opened in 2007 and anchored by Tesco and Dunnes superstores. After weathering a dismal decline in the 1980s, when most of its stores were shuttered, the Centrum in Nordweststadt, Germany, was also taken over by investors and resurrected as an entertainment and shopping venue.

New Town Encore

Although new towns continued to be built after the mid-1970s, they were far fewer in number. More important, they lost much of their moral authority. For the most part, the deus ex machina of new towns turned in on itself and disconnected from any sort of revolutionary agenda. Instead, most new town projects followed the standard principles of real estate development cloaked in finery appealing to those with the money to buy. The Middle East has continued to be one of the most active arenas of new town development, due largely to its booming and very young population.

In Egypt, a constellation of new towns appeared on the Nile delta around Cairo and Alexandria[4] that were inspired by the Western models described in this book. Ramadan City, Sadat City, 6 October City, and New Cairo are new towns built from scratch, springing up in the desert and containing upwards of five hundred thousand to a million people as well as their own industries, government offices, and carefully planned neighborhoods. They are a mélange of luxury compounds, villas on golf courses, and gated communities, along with a sprinkling of low-income housing. Millions of dollars have been poured into these projects by the Egyptian government and private developers.

Yet despite these attempts at dispersing Cairo's population to overspill towns and solving the acute housing shortage, they have done little more than accentuate social segregation. New towns are beyond ordinary pocketbooks. Some 30 percent of the residents of Cairo (now the largest city in the Middle East and Africa) live in slums and informal settlements. The huge slum of Manshiet Nasser alone houses an estimated 1 million inhabitants, almost as much as all the new towns put together in the Cairo region. Many of the city's slum dwellers appeared on global television screens as hundreds of thousands of young activists flooding into Tahrir Square in the Revolution of 2011. And now the Egyptian government is planning an entirely new capital city for 5 million people. Its scale has all the pomposity of the new town dream: twenty-one residential districts, over 600 hospitals, 1,250 mosques and churches, and a theme park four times the size of Disneyland.

Notwithstanding these fantastic plans and developments, the entire history of new town development in the Middle East has been wracked with political volatility. Israeli settlements continue to be built in the occupied territories of the West Bank and are regularly scenes of violent clashes between Jewish newcomers and Palestinians. Constantinos Doxiadis's extension of Baghdad, now known as Sadr City, is one of the saddest commentaries on new town utopian dreamscapes that one can find. Meant to provide a new life for Baghdad's urban poor, it is convincing proof that urban planning alone cannot shape social transformation. It is a vast slum district of 3 million poor people, and ground zero for the Mehdi army and Shiite radicalism in Iraq. In another example of utopian treachery, this time in South Asia, Doxiadis's Korangi Town outside Karachi, Pakistan, has become a forlorn world with its own slums, its public spaces eaten up by buildings, its wide streets reduced to shadowy gullies, and its original houses buried under garbage and debris.[5] It is a breeding ground for radicalism and gang violence.

Are conditions in Sadr City, Korangi, the satellite cities around Tehran, or any other less-than-perfect new town at least partly to blame for political extremism? Is the situation any different in these peripheral locations than in the vast slums and decrepit housing estates in large cities proper, such as those in the center of Cairo or Karachi? Probably not—but it was the utopian prospects of ideal new towns that made disillusion so unbearable. Yet rather than seeing these places through the lens of unremitting failure, we might also understand them as socially and politically subversive arenas. In discussing these questions with scholars of contemporary Iranian politics, Tehran's satellite towns were pointed to as precisely the zones that escaped surveillance and the all-seeing eye of security forces, where young Iranians could enjoy some measure of personal freedom.

The ways that new towns were instrumentalized by politics and then contributed to political turmoil is demonstrated in plans for New Kabul in Afghanistan, officially called Dehsabz-Barikab City. Financed largely by the Japan International Cooperation Agency, it has all the markings of utopian fantasy. Writing in *Harper's Magazine*, journalist Matthieu Aikins described the new city as geometrically exact, with a large park and a teardrop-shaped lake at its center:

> Wide avenues radiate outward in concentric triangles, separating the business, residential, and industrial districts. Where the old city's streets were crowded and hectic, the houses here are set apart by lawns and tidy little sidewalks; it is a green city, full of grass and trees and powered by renewable energy. On the outskirts, there's a farm belt reclaimed from the desert where tranquil little irrigation canals run under footbridges.[6]

It is a scene taken right out of the annals of any new town of the late twentieth century. Utopia has a rigid spatial form that is intimately connected to social ordering and political authority. New Kabul is likely to form a gated community for Afghanistan's newly rich.

Similarly, Georgian president Mikhail Saakashvili's dream of a new town for half a million people on marshland near the Black Sea had the authoritarian dash implicit in gargantuan megaprojects. The futuristic plans for Lazika included clusters of skyscrapers and a massive cargo port and free trade zone that were to make it, according to its original enthusiasts, a global trading hub, the Manhattan on the Black Sea.[7] Despite headlong construction on what were to be preserved wetlands, the project gave a cold shoulder to environmental impact. The city was conjured up entirely behind closed doors by Saakashvili's coterie. And the grandiose plans fell apart once his party was defeated in the 2012 elections. Whatever their objectives, utopian reveries can quickly evaporate into thin air.

Do these examples mean that new towns were ultimately just idealistic flimflam, the proverbial castles in the air? I don't think so. The features of utopian aspiration may have changed, but cities are still the device for carrying it out. Culture, sustainability, and smart technologies have become the scaffolding for twenty-first-century ideals. As for culture, the search for cultural symbolism and "placemaking" is the new fixation, much of it based on vestiges of memory. The Disney Company's new town of Celebration in Florida famously falls into this category. Inaugurated in 1996 as a model of New Urbanism, it exudes a theme park neotraditionalism. It is less an ideal city of the future than a rose-colored vision of the past, an enclave for white middle-class families. But this does not mean that the new town formula has calcified or simply become slick advertising. In a sense, utopia has always involved simulated spaces and temporalities. One could just as easily argue that the original garden city fell into this same neotraditionalist category.

In another definition of utopia, Canadian literary and social critic Northrop Frye saw it as a kind of contract myth. It is the desire for the restoration of what society has lost, forfeited, or violated.[8] On the other hand, for Ernst Bloch, one of the twentieth century's great theorists of utopia, it is an "anticipatory illumination"; it breaks with the present, and stands on the horizon of a new reality. It is a dramatic otherness, filled with what Bloch would call "a surplus of meaning," an overshooting of ideology[9] and a radical fantasizing of new kinds of spaces and social worlds. It is precisely this tension between a longing for what has been lost and a yearning for the future that makes utopian reveries such dramatic urban theater. The new town stood at this temporal frontier.

The recent new towns popping up across Asia are perhaps the most evocative examples of these utopian features.[10] In 2001, China announced its plans to build twenty new cities each year through 2020 to accommodate the vast migration of its people into the urban whirl in search of jobs. Among these, the new towns around Shanghai have generated the most excitement internationally. In 2001, Shanghai's *One City, Nine Towns* plan envisioned ten new towns: the medium-sized city of Song Jiang plus nine compact smaller towns. Hundreds of thousands of brand-new apartments and a titanic infrastructure have been built within a vast suburban ring. The plan is saturated with fantasies of Western urban design and architecture, in what urbanist Harry den Hartog has called "theme-park urbanism" and a form of self-colonization.[11] Famously constructed to simulate other places and times, Thames Town is a model of Victorian London, while Luodian is Scandinavian, Caoqiao has a Dutch theme, and Anting is a German village planned by Albert Speer and Partners, the firm founded by Hitler's favorite architect. True to its heritage, Anting is near a Volkswagen factory and is planned as a hub for China's automobile industry.

For critics, Shanghai's new town constellation is an imported environment of European chic for China's prosperous new elites. Yet these historicist dreamscapes were not just copies of Western ideas. The very act of translation is a form of metamorphosis and co-optation to local needs, ideals, and fantasies. Rather than simply mimicry, each new town was a negotiated space of possibilities.

A second trajectory that carries the torch of utopian aspiration is the ideal of sustainability and ecotopia. Historians often see a direct line from the garden city of the early twentieth century to the green city of the early twenty-first century. The increasing role of landscape architects and environmentalists in the development of new towns, particularly since the 1970s, is evidence of this genealogy. This was the case for the new town of The Woodlands north of Houston, Texas, which opened in 1974, and by 2000 had a population of over fifty-five thousand. Its developer was oil and gas tycoon George Mitchell, who looked to ecological and environmental planning as the framework for his urban paradise. Famed landscape designer Ian McHarg used land-use suitability analyses and ecological data inventory to develop a master plan.[12]

The quest for ecotopia also follows in the gloried footsteps of the architectural avant-garde. The most recent garden city imaginaries, such as Garden by the Bay outside Singapore, look like video games or the computerized sets of director James Cameron's 2009 film *Avatar*.[13] As well as lush gardens and "super-trees" designed as part of the natural ecosystem, Garden by the Bay features the steel-framed geodesic "bio-domes" that have conjured the future since their introduction by Buckminster Fuller. Sustainable development clearly indicates that utopian ideals have hardly vanished. The Sky City project in Changsha in central China is an attempt at a sustainable vertical city inside a colossal 220-story megastructure made of recycled, prefabricated modules that slot together. As home to some seventeen thousand people, it will have all the requisite urban services, including a hospital and five schools. Masdar City in Abu Dhabi, built with the classic new town formula of massive government financing and design by noted British architect Norman Foster, offers a pioneering model of car-free, energy-efficient urban living. Bankrolled by oil wealth, it is a farsighted experiment in a high tech eco-city. But as in all utopias, freethinking bumps up against reality. Although it originally anticipated personal rapid transit systems as a replacement for cars, the project was too expensive (even for oil wealth). Masdar will instead rely on clean energy and electric vehicles.

The quest for sustainability slides into the third utopian arc, that of "smart" cities that apply cutting-edge digital technologies to save the day. Backed in this case by global corporate giants IBM, Siemens, and Cisco, lightning-fast computer networks, distributed sensors and robotics, and data analytics will create superefficient urban places.[14] According to IBM, smart cities will cultivate "charisma, resiliency, and vitality."[15] The company built its now-famous smart city model online as the portal to a future of safe neighborhoods, quality schools, affordable housing, and smooth traffic flows. In this utopian scenario, cities become platforms of innovation. Urban life will become more transparent and creative: policy makers will be able to shift choices and resources on a dime. The meaning of sustainability has transmuted to that of resiliency and the ability to withstand crisis (especially global warming) and manage risk by building computer models and running future scenarios in real time. Smart city innovations will create a twinkling wireless planet of peace, harmony, and equality. These new technologies have reenergized the dream of building new towns from scratch. A new generation of massive capital investments and megaprojects in intelligent design amounts to a "new city-building industry," according to MIT urban planning guru Michael Joroff.

The smart city of Songdo, built from scratch on the outskirts of Seoul in South Korea, is an entirely networked, automated world beyond the imagination of the 1960s avant-garde enthusiasts. India is planning to build one hundred smart cities for the future. The renderings for the flagship projects of Gujarat International Finance Tec-City and Dholera display the high-tech designs, the sparkling towers and "intelligent, green buildings," the automated transportation corridors that capture the twenty-first-century utopian imagination.

These few examples prove that cities built from scratch, or in the current lingo, pop-up cities, are hardly washed up, nor have they been relegated to the dustbin of history. New towns as ambitious blueprints for the future still fascinate urban practitioners. They are fast-forward places that act as transformational objects, as critique and corrective of the present. Every society deserves its utopia. New towns are the material structure of potentialities and aspiration, and the frontier between two worlds: the present and the future.

NOTES

INTRODUCTION

1. For a broad view of new towns, see Miles Glendinning, "The New Town 'Tradition': Past, Present—and Future?," in *Back from Utopia: The Challenge of the Modern Movement*, ed. Hubert-Jan Henket and Hilde Heynen (Rotterdam: 010 Publishers, 2002), 206–15.

2. See Arnold Bartetzky and Marc Schalenberg, eds., *Urban Planning and the Pursuit of Happiness: European Variations on a Universal Theme (18th–21st Centuries)* (Berlin: Jovis, 2009), and Robert H. Kargon and Arthur P. Molella, *Invented Edens: Techno-Cities of the Twentieth Century* (Cambridge, MA: MIT Press, 2008).

3. This point is made by Fredric Jameson, *Archaeologies of the Future: The Desire Called Utopia and Other Science Fictions* (London: Verso, 2007), xiii. See also Ruth Eaton, *Ideal Cities: Utopianism and the (Un)Built Environment* (Oxford: Thames and Hudson, 2001).

4. David Harvey, "The Right to the City," *New Left Review* 53 (2008): 23.

5. Quoted in Walter Schwagenscheidt, *Die Nordweststadt: Idee und Gestaltung = The Nordweststadt: Conception and Design* (Stuttgart: Karl Krämer Verlag, 1964), 8.

6. This point is made in the introduction of David Pinder, *Visions of the City: Utopianism, Power, and Politics in Twentieth-Century Urbanism* (New York: Routledge, 2005).

7. Margaret Mead, "New Towns to Set New Life Styles," in *New Towns: Why—and for Whom?*, ed. Harvey S. Perloff and Neil C. Sandberg, Praeger Special Studies Program (New York: Praeger, 1972), 120.

8. Karl Mannheim, *Ideology and Utopia: An Introduction to the Sociology of Knowledge* (New York: Harcourt, Brace, 1954), 173.

9. Jameson, *Archaeologies of the Future*, 3.

10. An excellent analysis of the architectural avant-garde can be found in Van Schaik and Otakar Mácel, eds., *Exit Utopia: Architectural Provocations 1956–76* (New York: Prestel, 2005).

11. James C. Scott, *Seeing Like a State: How Certain Schemes to Improve the Human Condition Have Failed* (New Haven, CT: Yale University Press, 1998), 187.

12. See for example the discussion in Arie S. Shachar, "The Role of New Towns in National and Regional Development: A Comparative Study," in Perloff and Sandberg, *New Towns*, 30–47.

13. Michel Foucault, *Security, Territory, Population: Lectures at the Collège de France, 1977–1978*, trans. G. Burchell (Basingstoke, UK: Palgrave Macmillan, 2009).
14. Dilip Parameshwar Gaonkar, ed., *Alternative Modernities* (Durham, NC: Duke University Press, 2001), 18, 21.
15. Interview conducted by Philippe Estèbe and Sophie Gonnard, *Les Villes nouvelles et le système politique en Ile-de-France* (Paris: Ministère de l'Equipement, du Transport et du Logement, 2005), 35.
16. Lewis Mumford, "Utopia, the City and the Machine," *Daedalus* 94, no. 2 (1965): 277.
17. Ernst Bloch, *The Utopian Function of Art and Literature*, trans. Jack Zipes and Frank Mecklenburg (Cambridge, MA: MIT Press, 1988), 12. This quote is given in the exceptional discussion of utopia in Carl Freedman, *Critical Theory and Science Fiction* (Middletown, CT: Wesleyan University Press, 2000), 67.
18. Mannheim, *Ideology and Utopia*, 236.

CHAPTER ONE

1. In this, as in all histories of twentieth-century planning, see Peter Hall, *Cities of Tomorrow: An Intellectual History of Urban Design in the Twentieth Century*, 4th ed. (New York: Wiley-Blackwell, 2014). See as well Stephen V. Ward, *Planning the Twentieth-Century City: The Advanced Capitalist World* (West Sussex, UK: John Wiley and Sons, 2002). On the garden city, see Peter Hall and Colin Ward, *Sociable Cities: The Legacy of Ebenezer Howard*, 3rd ed. (Chichester, UK: John Wiley and Sons, 2002), and Kermit C. Parsons and David Schuyler, eds., *From Garden City to Green City: The Legacy of Ebenezer Howard* (Baltimore, MD.: John Hopkins University Press, 2002). Finally, Robert Fishman, *Urban Utopias in the Twentieth Century: Ebenezer Howard, Frank Lloyd Wright, Le Corbusier* (Cambridge, MA: MIT Press, 1982).
2. Catherine Cooke, "Le Mouvement pour la cité-jardin en Russie," in *URSS 1917–1978: La Ville, l'architecture*, ed. Jean-Louis Cohen, Marco De Michelis, and Manfredo Tafuri (Paris: L'Equerre, 1979), 200–233. See also Stanley Buder, *Visionaries and Planners: The Garden City Movement and the Modern Community* (Oxford: Oxford University Press, 1990).
3. For excellent essays on the international diffusion of garden city concepts, see Robert A. M. Stern, David Fishman, and Jacob Tilove, *Paradise Planned: The Garden Suburb and the Modern City* (New York: Monacelli Press, 2013), as well as Stephen V. Ward, ed., *The Garden City: Past, Present, and Future* (London: Routledge, 1992).
4. Michel Geertse, *Defining the Universal City: The International Federation for Housing and Town Planning and Transnational Planning Dialogue, 1913–1945* (Amsterdam: Vrije Universiteit, 2012). See as well the compendium of information in Ewart G. Culpin, *The Garden City Movement Up-to-Date* (London: Garden Cities and Town Planning Association, 1913).
5. Stephen V. Ward, "What Did the Germans Ever Do for Us? A Century of British Learning about and Imagining Modern Town Planning," *Planning Perspectives* 25, no. 2 (2010): 117–40.
6. John V. Maciuika, *Before the Bauhaus: Architecture, Politics, and the German State, 1890–1920* (Cambridge: Cambridge University Press, 2008), 238. Margaretenhöle was designed by George Metzendorf, and Hellerau under the direction of Richard Reimerschmid and Heinrich Tessenow. Hellerau contained a population of some eight hundred artists, craft workers, and intellectuals by 1913.

7. John Robert Mullin, "Ideology, Planning Theory and the German City in the Inter-War Years: Part I," *Town Planning Review* 53, no. 2 (1982): 118.

8. Ilse Irion and Thomas Sieverts, *Neue Städte: Experimentierfelder der Moderne* (Stuttgart: Deutsche Verlags-Anstalt, 1991).

9. Thank you to professor David Gordon for guidance on Thomas Adams's Canadian projects. Gerald Hodge and David Gordon, eds., *Planning Canadian Communities*, 5th ed. (Toronto: Nelson, 2007).

10. On the extraordinary influence of New York's first master plan, see David A. Johnson, *Planning the Great Metropolis: The 1929 Regional Plan of New York and Its Environs* (New York: Routledge, 2015).

11. Susanna Magri and Christian Topalov, "De la cité-jardin à la ville rationalisée : Un tournant du projet réformateur, 1905–1925; Etude comparative France, Grande-Bretagne, Etats-Unis," *Revue française de sociologie* 28, no. 3 (1987): 417–51.

12. James C. Scott, *Seeing Like a State: How Certain Schemes to Improve the Human Condition Have Failed* (New Haven, CT: Yale University Press, 1998), 97.

13. Mervyn Miller, "Garden Cities and Suburbs: At Home and Abroad," *Journal of Planning History* 1, no. 1 (2002): 17. See also Dennis Hardy, *Utopian England: Community Experiments 1900–1945* (New York: Routledge, 2012).

14. After Howard's death in 1928, Unwin became the president of the international organization and remained so until 1931, when Frederic Osborn took over leadership.

15. See Daniel T. Rodgers, *Atlantic Crossings: Social Politics in a Progressive Age* (Cambridge, MA: Harvard University Press, 2000).

16. On the history of town and regional planning in the Netherlands, see Coen Van der Wal, *In Praise of Common Sense: Planning the Ordinary; A Physical Planning History of the New Towns in the Ijsselmeerpolders* (Rotterdam: 010 Publishers, 1997), as well as Hans Van der Cammen and Len De Klerk, *The Selfmade Land: Culture and Evolution of Urban and Regional Planning in the Netherlands* (Antwerp: Spectrum, 2012).

17. On the Frankfurt Experiment, see Susan R. Henderson, *Building Culture: Ernst May and the New Frankfurt Initiative, 1926–1931* (New York: Peter Lang International Publishers, 2013). See also Barbara M. Miller, "Architects in Power: Politics and Ideology in the Work of Ernst May and Albert Speer," *Journal of Interdisciplinary History* 17 (1986): 287–88.

18. John Robert Mullin, "City Planning in Frankfurt, Germany, 1925–1932," *Jounal of Urban History* 4, no. 1 (1977): 3–28. On Römerstadt, see most recently Susan R. Henderson, "Römerstadt: The Modern Garden City," *Planning Perspectives* 25, no. 3 (2010): 323–46.

19. Miliutin's book was republished by MIT Press: N. A. Miliutin, *Sotsgorod: The Problem of Building Socialist Cities* (Cambridge, MA: MIT Press, 1974).

20. Arthur Korn organized the Proletarian Building Display as a communist response to the Berlin Building Exhibition. Thilo Hilpert, "'Linear Metropolis': The Forgotten Urban Utopia of the 20th Century," in *Megastructure Reloaded : Visionäre Stadtentwürfe der Sechzigerjahre reflektiert von zeitgenössischen Künstlern = Visionary Architecture and Urban Design of the Sixties Reflected by Contemporary Artists*, ed. Sabrina van der Ley and Markus Richter (Ostfildern, Germany: Hatje Cantz Verlag, 2008), 60.

21. Ernst May, "Cities of the Future," in *The Future of Communist Society*, ed. Walter Laqueur and Leopold Labedz (New York: Frederick A. Praeger, 1962), 179. See as well Ernst May, "Villes nouvelles en U.R.S.S.," *La Cité* 9(July 1931): 229–91. Much has been written about urban planning theory in the Soviet Union in the 1920s and

1930s. See most recently Heather DeHaan, *Stalinist City Planning: Professionals, Performance, and Power* (Toronto: University of Toronto Press, 2013).

22. Clarence Stein, quoted in Kristin Larson, "Cities to Come: Clarence Stein's Postwar Regionalism," *Journal of Planning History* 4, no. 1 (2005): 36.

23. For a full discussion of the "fourth migration" concept, see Robert H. Kargon and Arthur P. Molella, *Invented Edens: Techno-Cities of the Twentieth Century* (Cambridge, MA: MIT Press, 2008), 21–24. See as well Kermit C. Parsons, "Collaborative Genius: The Regional Planning Association of America," *Journal of the American Planning Association* 60, no. 4 (1994): 462–82, and Carl Sussman, ed., *Planning the Fourth Migration: The Neglected Vision of the Regional Planning Association of America* (Cambridge, MA: MIT Press, 1976).

24. Benton MacKaye as quoted in Larry Anderson, *Benton MacKaye: Conservationist, Planner, and Creator of the Appalachian Trail* (Baltimore, MD: Johns Hopkins University Press, 2002), 229.

25. See the powerful article by John L. Thomas, "Holding the Middle Ground," in *The American Planning Tradition: Culture and Policy*, ed. Robert Fishman (Washington DC: Woodrow Wilson Center Press, 2000), 33–64.

26. As recounted in Clarence S. Stein, *The Writings of Clarence S. Stein: Architect of the Planned Community*, ed. Kermit C. Parsons (Baltimore: Johns Hopkins University Press, 1998), xxiii.

27. See Lewis Mumford, "Regionalism and Irregionalism," *Sociological Review* 19, no. 4 (1927): 277–88, as well as "The Theory and Practice of Regionalism," *Sociological Review* 20, no. 1 (1928): 18–33.

28. Perry's vision appeared in its most elaborate form as "The Neighborhood Unit, a Scheme of Arrangement for the Family-Life Community," in volume 7 of the 1929 *Regional Survey of New York and Its Environs* sponsored by the Russell Sage Foundation, which also published Perry's reports on "Wider Use of the School Plant" (1911) and "The School as a Factor in Neighborhood Development" (1914) as well as "The Neighborhood Unit" (1929).

29. See James Dahir, *The Neighborhood Unit Plan: Its Spread and Acceptance* (New York: Russell Sage Foundation, 1947).

30. This terminology is used in Geertse, *Defining the Universal City*, 116.

31. For an excellent analysis of *The City*, see chapter 3 of Howard Gillette, *Civitas by Design: Building Better Communities, from the Garden City to the New Urbanism* (Philadelphia: University of Pennsylvania Press, 2010). On Lewis Mumford, see Mark Luccarelli, *Lewis Mumford and the Ecological Region: The Politics of Planning* (New York: Guilford Press, 1995).

32. Federico Caprotti, *Mussolini's Cities: Internal Colonialism in Italy, 1930–1939* (Youngstown, NY: Cambria Press, 2007), 127. See also Diane Ghirardo, *Building New Communities: New Deal America and Fascist Italy* (Princeton, NJ: Princeton University Press, 1989).

33. On Gottfried Feder, see Tilman A. Schenk and Ray Bromley, "Mass-Producing Traditional Small Cities: Gottfried Feder's Vision for a Greater Nazi Germany," *Journal of Planning History* 2, no. 2 (2003): 107–39.

34. On Nazi settlement policy, see Mechtild Rössler, "Applied Geography and Area Research in Nazi Society: Central Place Theory and Planning 1939–1945," *Environment and Planning D: Society and Space* 7 (1989): 419–31. See also John Robert Mullin, "Ideology, Planning Theory and the German City in the Inter-War Years: Part II," *Town Planning Review* 53, no. 3 (1982): 115–30.

35. Klaus-Jörg Siegfried, *Wolfsburg—zwischen Wohnstadt und Erlebnisstadt* (Wolfsburg: Stadt Wolfsburg, 2002), 30. See also Christian Schneider, *Stadtgründung im Dritten Reich: Wolfsburg und Salzgitter* (Berlin: Heinz Moos, 1978). In English, see Kargon and Molella, *Invented Edens*, 38–46.

36. Housing Committee, *A Report of the City Architect and Director of Housing on Speke, a Self-Contained and Protected Community Unit* (Liverpool: City of Liverpool, 1946).

37. See Greg Hise, "The Airplane and the Garden City: Regional Transformations during World War II," in *World War II and the American Dream*, ed. David Albrecht (Washington, DC: National Building Museum; Cambridge, MA: MIT Press, 1995), 150.

38. Quoted in Kargon and Molella, *Invented Edens*, 69. On Willow Run, see the recent work by Sarah Jo Peterson, *Plannning the Home Front: Building Bombers and Communities at Willow Run* (Chicago: University of Chicago Press, 2013).

39. These examples are described in Hise, "The Airplane and the Garden City," 150.

40. Christine M. Boyer, "Aviation and the Aerial View: Le Corbusier's Spatial Transformations in the 1930s and 1940s," *Diacritics* 33, nos. 3–4 (2003): 93–116. See the interesting articles in Mark Dorrian and Frédéric Pousin, eds., *Seeing from Above: The Aerial View in Visual Culture* (London: B. Tauris, 2013), as well as Jeanne Haffner, *The View from Above: The Science of Social Space* (Cambridge, MA: MIT Press, 2013).

41. Mitchell Schwarzer, *Zoomscape: Architecture in Motion and Media* (New York: Princeton Architectural Press, 2004), 158–59.

42. See Tanis Hinchcliffe, "Aerial Photography and the Postwar Urban Planner in London," *London Journal* 35, no. 3 (2010): 277–88.

43. Jennifer S. Light, *From Warfare to Welfare: Defense Intellectuals and Urban Problems in Cold War America* (Baltimore: Johns Hopkins University Press, 2003), 126–27. See also Melville Campbell Branch, *Aerial Photography in Urban Planning and Research* (Boston: Harvard University Press, 1948).

44. Lewis Mumford, "An American Introduction to Sir Ebenezer Howard's 'Garden City of Tomorrow,'" *New Pencil Points* 24(March 1945): 73.

45. Jean-Louis Cohen, *Architecture in Uniform: Designing and Building for the Second World War* (Montreal: Canadian Centre for Architecture; Paris: Hazan, 2011), 86. See also Gerald D. Nash, *The American West Transformed: The Impact of the Second World War* (Bloomington: Indiana University Press, 1985).

46. *Circle* (General Telephone Company magazine), November 1954. These examples are given in Hise, "The Airplane and the Garden City," 171–72.

47. J. Tyrwhitt, J. Sert, and E. N. Rogers, eds., *CIAM 8: The Heart of the City; Towards the Humanisation of Urban Life* (London: Lund Humphries, 1952), 3.

48. Eliel Saarinen, *The City: Its Growth, Its Decay, Its Future* (New York: Reinhold Publishing, 1943), vii–viii, 217.

49. Jean Gottmann, *Essais sur l'aménagement de l'espace habité* (Paris: Mouton, 1966), 164–65.

50. Jacob L. Crane and Edward T. Paxton, "The World-Wide Housing Problem," *Town Planning Review* 22, no. 1 (1951): 17.

51. Foreword by Sir William Beverage, page 5; Madge quote, page 9, in John Madge, *The Rehousing of Britain*, Target for Tomorrow (London: Pilot Press, 1945).

52. Lewis Mumford, *City Development: Studies in Disintegration and Renewal* (New York: Harcourt, Brace, 1945), 157. First printed as *The Social Foundations of Post-War Building*, no. 9 in the Re-building Britain series (London: Farber and Farber, 1943).

53. See Pierre Desrochers and Christine Hoffbauder, "The Post War Intellectual Roots of the Population Bomb: Fairfield Osborn's 'Our Plundered Planet' and William Vogt's

'Road to Survival' in Retrospect," *Electronic Journal of Sustainable Development* 1, no. 3 (2009): 73–98.

54. Saarinen, *The City*, 143.

55. Lewis Mumford, *The Plan of London County*, Re-building Britain series, no. 12 (London: Faber and Faber, 1944), 167–68.

CHAPTER TWO

1. Peter H. Oberlander, "New Towns: An Approach to Urban Reconstruction," *Journal of the Royal Architectural Institute of Canada* 24 (1947): 202.

2. Halas and Batchelor Cartoon Films, *Charley in New Town* (London: Central Office of Information, 1948). YouTube video, 8:13, https://www.youtube.com/watch?v= 6ophEYd4A-Q.

3. See Ernst Bloch's three-volume *The Principle of Hope*, trans. Neville Plaice, Stephen Plaice, and Paul Knight (Oxford: Blackwell, 1986), and the analysis of Bloch in Ruth Levitas, "Educated Hope: Ernst Bloch on Abstract and Concrete Utopia," *Utopian Studies* 1, no. 2 (1990): 13–26. Urban geographer Jennifer Robinson argues that imagining the urban future was a composite endeavor drawn not just from the West and its great capitals but from "ordinary" cities around the world. Jennifer Robinson, *Ordinary Cities between Modernity and Development* (New York: Routledge, 2006).

4. Oberlander, "New Towns," 202, 11.

5. Erkko Kivikoski, dir., *Puutarhakaupunki Tapiola—Garden City Tapiola* (Helsinki: Filmiryhmä Oy, 1967). YouTube video, 7:50, https://www.youtube.com/watch?v= zoH0dQOuya4.

6. See the interesting articles in Clemens Zimmermann, ed., *Industrial Cities: History and Future* (Frankfurt-on-Main: Campus Verlag GmbH, 2013).

7. For Abercrombie's planning concepts, see Michiel Dehaene, "Urban Lessons for the Modern Planner: Patrick Abercrombie and the Study of Urban Development," *Town Planning Review* 75, no. 1 (2004): 1–30, and Peter Hall, "Bringing Abercrombie back from the Shades," *Town Planning Review* 66, no. 3 (1995): 227–41.

8. The two great postwar plans for Scotland were Patrick Abercrombie and R. Matthew, *The Clyde Valley Regional Plan* (Edinburgh: His Majesty's Stationery Office, 1949), and F. C. Mears, *A Regional Survey and Plan for Central and Southeast Scotland* (Edinburgh: Central and Southeast Scotland Regional Planning Advisory Committee, 1948). For a broader perspective on postwar reconstruction, see Nicholas Bullock, *Building the Post-War World: Modern Architecture and Reconstruction in Britain* (London: Routledge, 2002).

9. D. C. D. Pocock, "Some Features of the Population of Corby New Town," *Sociological Review* 8, no. 2 (1960): 218. See also "Making a New Town: Corby Ironworks, and Industrial Revolution," *Guardian* (Manchester), November 28, 1933.

10. "New Town 'to last 100 years,'" *Guardian* (Manchester), November 30, 1949.

11. K. C. Edwards, "Corby—a New Town in the Midlands," *Town Planning Review* 22, no. 2 (1951): 129.

12. James Morris, "Corby: Mirror of the New England," *Guardian* (Manchester), September 30, 1960.

13. On Abercrombie's international projects, see Marco Amati and Robert Freestone, "'Saint Patrick': Sir Patrick Abercrombie's Australian Tour 1948," *Town Planning Review* 80, no. 6 (2009): 597–626, and Lawrence Wai-Chung Lai, "Reflections on the Abercrombie Report 1948: A Strategic Plan for Colonial Hong Kong," *Town Planning Review* 70, no. 1 (1999): 61–87.

14. "Report of Regional Boundaries Committee," New South Wales, Sydney, 1944, as quoted in Macdonald Holmes, "Regional Planning in Australia," *Geographical Journal* 112, no. 1/3 (1948): 79.

15. A. A. Heath and R. N. Hewison, "New Towns for Australia," *Town and Country Planning* 19 (1951): 363–66. See also Robert Freestone, *Urban Nation: Australia's Planning Heritage* (Clayton, Victoria, Australia: Csiro, 2010).

16. See in particular Brian Head, "From Deserts the Profits Come: State and Capital in Western Australia," *Australian Quarterly* 57, no. 4 (1985): 372–83, as well as the publication on Kwinana produced by the Government of Western Australia, *Kwinana* (Perth: Government of Western Australia, October 1969).

17. Margaret Feilman, "Kwinana New Town," *Town and Country Planning* 23 (1955): 385. See the excellent article by Ian MacLachlan and Julia Horsley, "New Town in the Bush: Planning Knowledge Transfer and the Design of Kwinana, Western Australia," *Journal of Planning History* 14, no. 2 (May 2015): 112–34.

18. Government of Western Australia, Department of Housing, "Centenary Flashback: Creating Kwinana, Our First Industrial Township," www.housing.wa.gov.au. Document created August 21, 2012.

19. Norman E. P. Pressman, *Planning New Communities in Canada* (Ottawa, Ontario: Ministry of State, Urban Affairs, August 1975), 9.

20. Norman Pearson, "Elliot Lake: 'The Best-Planned Mining Town,'" *Canadian Architect* 3 (November 1958): 55–57.

21. In the United States, the sites were Los Alamos, New Mexico; Oak Ridge, Tennessee; and Hanford, Washington. In Britain, research was carried out at Billingham and Cambridge. The Chalk River Research Laboratory was established to move McGill University's research facility away from Montreal and closer to the source of uranium as well as a local military base. The uranium used in the first atomic bombs came from the Chalk River site.

22. On the designs for Frobisher Bay, see "Frobisher Bay, N.W.T.: Federal Government Project for a New Town," *Canadian Architect* 3 (November 1958): 44–49. The project has been researched by Rhodri Windsor Liscombe, "Modernist Ultimate Thule," *RACAR: Revue d'Art Canadienne/Canadian Art Review* 31, no. 1/2 (2006): 64–80, as well as by Matthew Farish and P. Whitney Lackenbauer, "High Modernism in the Arctic: Planning Frobisher Bay and Inuvik," *Journal of Historical Geography* 35, no. 3 (2009): 517–44.

23. Peter C. Newman, "The Utopian Town Where Our Atomic Scientists Live and Play . . . ," *Maclean's Magazine*, September 15, 1958. Accessed July 16, 2015, at Scientific Technical Translation: http://www.sttranslation.com/deep-river/almost-perfect-place-live/.

24. See the description in "Boom in the Big Triangle," *Popular Mechanics*, June 1952, 102–7.

25. Clarence S. Stein, "Kitimat: A New City," *Architectural Forum* Special Reprint (July–August 1954). Originally published in *Architectural Forum* 101 (August 1954): 125.

26. Corporation of the District of Kitimat and Pixie Meldrum, "Kitimat: The First Five Years" (Kitimat: Corporation of the District of Kitimat, December 1958), 37.

27. Thank you to Louise Avery, Curator at the Kitimat Museum and Archives, British Columbia, for the introduction to this film. Marlon Brando narrates, with music by the New York Philharmonic Orchestra. United Nations Film Services, *Power among Men*, directed by Thorold Dickinson and J. C. Sheers in collaboration with Alexander Hammid, G. L. Polidoro, and V. R. Sarma (New York: United Nations

Film Services, 1958). The other episodes are the rebuilding of a blitzed village near Monte Cassino in Italy, the modernization of a farming neighborhood in Haiti, and the introduction of an atomic energy reactor in a rural area of Norway.

28. Quoted in B. J. McGuire and Roland Wild, "Kitimat—tomorrow's City Today," *Canadian Geographical Journal* 59, no. 5 (1959): 150.

29. Ira M. Robinson, "New Industrial Towns on Canada's Resource Frontier," in *Program of Education and Research in Planning* (Chicago: University of Chicago, 1962), 87–88.

30. On the public narrative of happiness, see Cor Wagenaar, ed., *Happy Cities and Public Happiness in Post-War Europe* (Rotterdam: NAi Publishers / Architecturalia, 2004).

31. Grundsätze des Stadtebaus, Berlin 1950, quoted in Michel Grésillon, "Les Villes nouvelles en République démocratique Allemande: Problèmes d'intégration," *L'Espace géographique* 7, no. 1 (1978): 32. On American interpretations of socialist cities in the Cold War context, see for example Jack C. Fisher, "Planning the City of Socialist Man," *Journal of the American Planning Association* 28, no. 4 (1962): 251–65.

32. This point is also made by Christoph Bernhardt, "Planning Urbanization and Urban Growth in the Socialist Period: The Case of East German New Towns," *Journal of Urban History* 32, no. 1 (November 2005): 109–14. See as well Ivan Szelenyi, "Urban Development and Regional Management in Eastern Europe," *Theory and Society* 10, no. 2 (1981): 169–205, and David Crowley and Jane Pavitt, eds., *Cold War Modern: Design 1945–1970* (London: V and A, 2008).

33. See the excellent discussion by Jay Rowell, "Du Grand ensemble au 'complexe d'habitation socialiste': Les Enjeux de l'importation d'une forme urbaine en RDA," in *Le Monde des grands ensembles*, ed. Frédéric Dufaux and Annie Fourcaut (Paris: Editions CREAPHIS, 2004), 97–107. See also Andreas Schätzke, "Nach dem Exil: Architekten im Westen und im Osten Deutschlands," in *Grammatik sozialistischer Architekturen: Lesarten historischer Städtebauforschung zur DDR*, ed. Holger Barth (Berlin: Dietrich Reimer Verlag, 2001), 267–78.

34. On Strumilin's vision of the socialist city, see Michael Frolic, "The Soviet City," *Town Planning Review* 34, no. 4 (1964): 285–306.

35. Introduction and N. V. Baranov, "Planning of Metropolitan Areas and New Towns," in United Nations Department of Economic and Social Affairs, *United Nations Symposium on the Planning and Development of New Towns, Moscow, 24 August–7 September 1964* (New York: United Nations, 1966), 6, 213.

36. The invaluable resource on these details is Kazimierz Dziewonski, "Research for Physical Planning in Poland, 1944–1974," *Geographia Polonica* 32 (1975): 5–22. See also Nicole Haumont, Bohdan Jalowiecki, Moïra Munro, and Viktória Szirmai, *Villes nouvelles et villes traditionnelles: Une comparaison internationale* (Paris: L'Harmattan, 1999).

37. Anna Biedrzycka, *Nowa Huta—architektura i twórcy miasta idealnego; Niezrealizowane projekty* (Crakow: Muzeum Historyczne Miasta Krakowa, 2006), 20.

38. Dagmara Jajeśniak-Quast, "Ein Lokaler 'Rat für gegenseitige Wirtschaftshilfe': Eisenhüttenstadt, Kraków Nowa Huta und Ostrava Kunčice," in *Sozialistische Städte zwischen Herrschaft und Selbstbehauptung: Kommunalpolitik, Stadtplanung und Alltag in der DDR*, ed. Christoph Bernhardt and Heinz Reif (Stuttgart: Franz Steiner, 2009), 101.

39. For conditions in Upper Silesia, see Emil Caspari, *The Working Classes of Upper-Silesia: An Historical Essay* (London: Simpson Low, Marston, 1921). On the regional plan, see Adolf Ciborowski, *L'Urbanisme polonais 1945–1955* (Warsaw: Editions Polonia, 1956), 64–82.

40. On Nowa Huta, see Katherine Lebow, *Unfinished Utopia: Nowa Huta, Stalinism, and Polish Society, 1949–56* (Ithaca, NY: Cornell University Press, 2013). Also good is Alison Stenning, "Placing (Post-) Socialism: The Making and Remaking of Nowa Huta, Poland," *European Urban and Regional Studies* 7, no. 2 (2000): 99–118.

41. Katherine Lebow, "Public Works, Private Lives: Youth Brigades in Nowa Huta in the 1950s," *Contemporary European History* 10, no. 2 (2001): 199–219.

42. *A Poem for Adults* appeared in English translation in the Spring 1962 issue of *Dissent*. See Marci Shore, "Some Words for Grown-up Marxists: 'A Poem for Adults' and the Revolt from Within," *Polish Review* 42, no. 2 (June 1997): 131–54.

43. Bolesław Malisz, *La Pologne construit des villes nouvelles*, trans. Kazimiera Bielawska (Warsaw: Editions Polania, 1961), 96–97.

44. Boleslaw Janus, "Labor's Paradise: Family, Work, and Home in Nowa Huta, Poland, 1950–1960," *East European Quarterly* 33, no. 4 (2000): 453–74.

45. Historian Ruth May describes Stalinstadt as an ideal city in the tradition of Tony Garnier's *cité industrielle* and Magnitogorsk. Ruth May, "Planned City Stalinstadt: A Manifesto of the Early German Democratic Republic," *Planning Perspectives* 18 (2003): 47–78.

46. The new town of Schwedt would also fulfill this function. Ökonomisches Forschungsinstitut der Staatlichen Plankommissions, *Planung der Volkswirtschaft in der DDR* (East Berlin: Verlag Die Wirtschaft, 1970), 191; cited in William H. Berentsen, "Regional Change in the German Democratic Republic," *Annals of the Association of American Geographers* 71, no. 1 (1981): 54.

47. See Kurt Leucht, *Die erste neue Stadt in der Deutschen Demokratischen Republik: Planungsgrundlagen und -ergebnisse von Stalinstadt* (Berlin: VEB Verlag Technik, 1957).

48. On the role of the town center in socialist utopian planning, see Elisabeth Knauer-Romani, *Eisenhüttenstadt und die Idealstadt des 20. Jahrhunderts* (Weimar, Germany: Verlag und Datenbank für Geisteswissenschaften, 2000), 89–108.

49. Jajeśniak-Quast, "Ein Lokaler 'Rat für gegenseitige Wirtschaftshilfe,'" 99.

50. May, "Planned City Stalinstadt," 63.

51. See Katerina Clark, "Socialist Realism and the Sacralizing of Space," in *The Landscape of Stalinism*, ed. Evgeny Dobrenko and Eric Naiman (Seattle: University of Washington Press, 2003), 5. See also Anders Aman, *Architecture and Ideology in Eastern Europe during the Stalin Era: An Aspect of Cold War History* (New York: Architectural History Foundation; Cambridge, MA: MIT Press, 1992).

52. On the influence of the Bauhaus, see Wolfgang Thöner, "The 'Indestructible Idea' of the Bauhaus and Its East German Reception," *GHI Bulletin*, suppl. 2 (2005): 115–37. See also Eric Mumford, "CIAM and the Communist Bloc, 1928–1959," *Journal of Architecture* 14, no. 2 (2009): 237–45, as well as his *Defining Urban Design: CIAM Architects and the Formation of a Discipline, 1937–69* (New Haven, CT: Yale University Press, 2009).

53. Christoph Bernhardt, "Entwicklungslogiken und Legitimationsmechanismen im Wohnungsbau der DDR am Beispiel der sozialistischen Modellstadt Eisenhüttenstadt," in *Schönheit und Typenprojektierung: Der DDR-Städtebau im internationalen Kontext*, ed. Christoph Bernhardt and Thomas Wolfes (Erkner, Germany: Institut für Regionalentwicklung und Strukturplanung, 2005), 356.

54. For a comparison of Stalinstadt and Nowa Huta, see Ingrid Apolinarski and Christoph Bernhardt, "Entwicklungslogiken sozialistischer Planstädte am Beispiel von Eisenhüttenstadt und Nova Huta," in Barth, *Grammatik sozialistischer Architekturen*, 51–65.

55. For a comparison of Eastern European steel towns, see Dagmara Jaješniak-Quast, "In the Shadow of the Factory: Steel Towns in Postwar Eastern Europe," in *Urban Machinery: Inside Modern European Cities*, ed. Mikael Hard and Thomas J. Misa (Cambridge, MA: MIT Press, 2008), 187–210.

56. Heinz Colditz and Martin Lücke, *Stalinstadt: Neues Leben—neue Menschen* (Berlin: Kongress Verlag, 1958), 3.

57. See the chapter on capital city planning in Scandinavia in Pierre Merlin, *New Towns: Regional Planning and Development* (London: Methuen, 1971). See also his "Villes nouvelles en Scandinavie," *Cahiers de l'IAURP* 9 (1967).

58. The MARS Plan was created by architects Arthur Ling and Arthur Korn, engineer Felix Samuely, and landscape architect Christopher Tunnard, and published as "A Master Plan for London," *Architectural Review* 91(January 1942): 143–50. It proposed a wide band in central London for cultural and commercial activities, with two industrial zones placed on either side. Sixteen high-density residential districts were stretched along a rapid-rail system running for a distance of twenty-five miles.

59. J. H. Forshaw and Patrick Abercrombie, *County of London Plan: Prepared for the London County Council* (London: Macmillan, 1943), 9, 28–29.

60. See chapter 11 for the detailed study of Ongar in Patrick Abercrombie, *Greater London Plan 1944* (London: Minister of Town and Country Planning, 1945).

61. See Susanne Elizabeth Cowan, "Democracy, Technocracy and Publicity: Public Consultation and British Planning, 1939–1951" (PhD diss., University of California, Berkeley, 2010).

62. On the adaptation of the Radburn principle and the neighborhood unit in Britain, see Anthony Goss, "Neighborhood Units in British New Towns," *Town Planning Review* 32, no. 1 (1961): 66–82, as well as Kermit C. Parsons, "British and American Community Design: Clarence Stein's Manhattan Transfer, 1924–74," *Planning Perspectives* 7, no. 2 (1992): 182–210.

63. "Stevenage New Town" (Stevenage, UK: Stevenage Development Corporation, n.d.). For early photographs of the Stevenage town center and all the British new towns, see Anthony Burton and Joyce Hartley, eds., *The New Towns Record, 1946–1996: 50 Years of UK New Town Development* (Glasgow: Planning Exchange, 1997). See also *Stevenage Master Plan 1966: A Summarised Report of the Stevenage Master Plan Proposals* (Stevenage, UK: Stevenage Development Corporation, 1966).

64. Harriet Atkinson, *The Festival of Britain: A Land and Its People* (London: I. B. Tauris, 2012), 182–85, and Becky Conekin, *The Autobiography of a Nation: The 1951 Festival of Britain* (Manchester: Manchester University Press, 2003). See also John Westergaard and Ruth Glass, "A Profile of Lansbury," *Town Planning Review* 25, no. 1 (1954): 33–58.

65. On Jaqueline Tyrwhitt, see Ellen Shoshkes, *Jaqueline Tyrwhitt: A Transnational Life in Urban Planning and Design* (Surrey, UK: Ashgate, 2013).

66. On the *folkhemmet* ideal and social democracy, see Eva Rudberg, "Building the Utopia of the Everyday," in *Swedish Modernism: Architecture, Consumption and the Welfare State*, ed. Helena Mattsson and Sven-Olov Wallenstein (London: Black Dog, 2010), 152–59. See also Cecilia Widenheim, ed., *Utopia and Reality—modernity in Sweden 1900–1960* (New Haven, CT: Yale University Press, 2002).

67. Stockholms Stads Stadsplanekontor, *GeneralPlan för Stockholm 1952: Förslag uppraättat under aren 1945–52* (Stockholm: P. A. Norstedt, 1952). *Stockholm in the Future: Principles of the Outline Plan for Stockholm* (Stockholm: K. L. Beckmans) was written by Sven Markelius and published in 1946. See also Yngve Larsson, "Building a City

and a Metropolis: The Planned Development of Stockholm," *Journal of the American Institute of Planners* 28 (1962): 220–28.

68. Stadsplanekontor, *GeneralPlan för Stockholm 1952*, 113. See also Linda Vlassen-rood, "Stockholm 1952: Generalplan för Stockholm," in *Mastering the City: North-European Planning 1900–2000*, ed. Koos Bosma and Helma Hellinga, vol. 2 (Rotterdam: NAI; The Hague: EFL Publications, 1997), 290–97.

69. David Pass, *Vällingby and Farsta—from Idea to Reality; The New Community Development Process in Stockholm* (Cambridge, MA: MIT Press, 1973), 116.

70. Peter Hall, *Cities of Tomorrow: An Intellectual History of Urban Design in the Twentieth Century* (New York: Blackwell, 1998), 373.

71. *Observer* (London), July 27, 1958.

72. Stadsplanekontor, *GeneralPlan För Stockholm 1952*, 118.

73. Ibid., 118, 20–21. Introductory essay by Sven Markelius, "Swedish Land Policy," in *Sweden Builds: Its Modern Architecture and Land Policy Background, Development and Contribution*, by G. E. Kidder Smith (New York: Albert Bonnier, 1950), 28.

74. Stockholm City Property Board, *Vällingby Företagens Framtidsstad*, marketing brochure, 1952. Stockholm City Archive.

75. See Lars Nilsson, "The Stockholm Style: A Model for the Building of the City in Parks, 1930s–1960s," in *The European City and Green Space: London, Stockholm, Helsinki and St. Petersburg, 1850–2000*, ed. Peter Clark (Aldershot, UJK: Ashgate, 2006). The Svenska Bostäder housing was designed by architects Sven Backström and Leif Reinius.

76. Kidder Smith, *Sweden Builds*, 94.

77. *New York Times*, October 24, 1963. See as well "The City: Starting from Scratch," *Time*, July 3, 1969: 33–34, and "Finlande: Cité-jardin de Tapiola," *Techniques et architecture* 17 (September 1957): 99–101.

78. Juhana Lahti, "The Helsinki Suburbs of Tapiola and Vantaanpuisto: Post-war Planning by the Architect Aarne Ervi," *Planning Perspectives* 23 (2008): 157–58. A local newspaper-bulletin was established and provided free to every household to create community spirit.

79. Heikki von Hertzen and Paul D. Spreiregen, *Building a New Town: Finland's New Garden City Tapiola* (Boston: MIT Press, 1971), 198–99. Another excellent source in English is Timo Tuomi, ed., *Life and Architecture: Tapiola* (Espoo, Finland: Housing Foundation and City of Espoo, 2003).

80. On Aalto's regional planning in Finland, see Mikael Sundman, "Urban Planning in Finland after 1850," in *Planning and Urban Growth in the Nordic Countries*, ed. Thomas Hall (London: E. and F. N. Spon, 1991), 86–89.

81. The organizations included the Mannerheim League, the Confederation of Finnish Trade Unions (SAK), the Society of Civil Servants, the Central Association of Tenants, and the Finnish Association of Disabled Civilians and Servicemen.

82. These descriptions can be found in von Hertzen and Spreiregen, *Building a New Town*, especially the section titled "Introduction and Background." Tapiola was designed for a population of 15,000 inhabitants. Three residential neighborhoods of 5,000–6,000 residents each were nestled around the town center.

83. Heikki von Hertzen, "Tapiola Puutarhakaupunki: Tapiola Garden City," offprint from *Arkkitehti-Arkitekten* 1–2 (1956), 27 pp. See the interview with von Hertzen in Ilse Irion and Thomas Sieverts, *Neue Städte: Experimentierfelder der Moderne* (Stuttgart: Deutsche Verlags-Anstalt, 1991), 136–40.

84. Von Hertzen and Spreiregen, *Building a New Town*, 87.

85. On this point, see Lahti, "The Helsinki Suburbs of Tapiola and Vantaanpuisto," 156.

86. Erkko Kivikoski, dir., *Puutarhakaupunki Tapiola—Garden City Tapiola* (Helsinki: Filmiryhmä Oy, 1967).

87. See chapter 3 in Karl Popper, *The Poverty of Historicism*, 3rd ed. (London: Routledge and Kegan Paul, 1961), as well as his *The Open Society and Its Enemies*, vol. 1, 5th ed. (Princeton, NJ: Princeton University Press, 1966), 168.

88. For the vicissitudes of the neighborhood unit, see Nicholas N. Patricios, "The Neighborhood Concept: A Retrospective of Physical Design and Social Interaction," *Journal of Architectural and Planning Research* 19, no. 1 (2002): 70–90.

89. Christopher Alexander, "A City Is Not a Tree," was originally published in *Architectural Forum* in two parts: vol. 122, no. 1 (April 1965): 58–61 and vol. 122, no. 2 (May 1955): 58–62. It was republished widely, including in *Ekistics*, and is freely available on the Internet.

90. Wolf Von Eckardt, "The Case for Building 350 New Towns," *Harper's Magazine*, December 1965, 91–92.

CHAPTER THREE

1. Arturo Escobar, *Encountering Development: The Making and Unmaking of the Third World* (Princeton, NJ.: Princeton University Press, 1995), 22. The term *Third World* was coined by French demographer Alfred Sauvy in 1952 to describe countries that were associated with neither Soviet communism nor Western capitalism. See also Michael E. Latham, *The Right Kind of Revolution: Modernization, Development, and U.S. Foreign Policy from the Cold War to the Present* (Ithaca, NY: Cornell University Press, 2011).

2. For a discussion of these dynamics, see Sibel Bozdogan, *Modernism and Nation-Building: Turkish Architectural Culture in the Early Republic* (Seattle: University of Washington Press, 2001), as well as Arturo Almandoz, *Modernization, Urbanization and Development in Latin America, 1900s–2000s* (New York: Routledge, 2015). Lastly, see Joe Nasr and Mercedes Volait, eds., *Urbanism, Imported or Exported?* (New York: John Wiley and Sons, 2003), and Anthony King, *Spaces of Global Cultures: Architecture, Urbanism, Identity* (London: Routledge, 2004).

3. Don Price, quoted in Dwight Macdonald, *The Ford Foundation: The Men and the Millions* (New York: Reynal, 1956), 64. See also Inderjeet Parmar, *Foundations of the American Century: The Ford, Carnegie, and Rockefeller Foundations and the Rise of American Power* (New York: Columbia University Press, 2012).

4. Ford Foundation, "Tapestry for Tomorrow: The Ford Foundation Program in the Middle East," in *Ford Foundation Report* (February 1964), 6. See also Eugene S. Staples, *Forty Years: A Learning Curve; The Ford Foundation Programs in India, 1952–1992* (New Delhi: Ford Foundation, 1992).

5. Vernon Z. Newcombe, "Housing in the Federation of Malaya," *Town Planning Review* 27, no. 1 (1956): 14. For details on Petaling Jaya, see Lee Boon Thong, "Petaling Jaya: The Early Development and Growth of Malaysia's First New Town," *Journal of the Malaysian Branch of the Royal Asiatic Society* 79, no. 2 (2006): 1–22.

6. T. G. McGee and W. D. McTaggart, *Petaling Jaya: A Socio-Economic Survey of a New Town in Selangor, Malaysia*, Pacific Viewpoint Monograph, no. 2 (Wellington, NZ: Victoria University of Wellington, 1967), 3.

7. By then, the New Villages close to the capital had themselves been amalgamated into the suburbs of Kuala Lumpur. See Lee Boon Thong, "New Towns in Malaysia: Development and Planning Policies," in *New Towns in East and South-east Asia,*

ed. David R. Phillips and Anthony G. O. Yeh (New York: Oxford University Press, 1987), 156.

8. See the interesting article by Tadd Graham Fernée, "Modernity and Nation-Making in India, Turkey and Iran," *International Journal of Asian Studies* 9, no. 1 (2012): 71–97.

9. Arieh Sharon, "Planning in Israel," *Town Planning Review* 23, no. 1 (1952): 72. See also the outstanding website on Arieh Sharon: http://www.ariehsharon.org. Accessed June 17, 2015. Also good is Elisha Efrat, *The New Towns of Israel* (New York: Minerva, 1989). On Israeli development towns and regional planning, see S. Ilan Troen, *Imagining Zion: Dreams, Designs, and Realities in a Century of Jewish Settlement* (New Haven, CT: Yale University Press, 2003). See also Haim Yacobi, ed., *Constructing a Sense of Place: Architecture and the Zionist Discourse* (Aldershot, UK: Ashgate, 2004), and Eyal Weizman, *Hollow Land: Israel's Architecture of Occupation* (Verso: London, 2007).

10. See Roy Kozlovsky, "Temporal States of Architecture: Mass Immigration and Provisional Housing in Israel," in *Modernism and the Middle East: Architecture and Politics in the Twentieth Century*, ed. Sandy Isenstadt and Kishwar Rizvi (Seattle: University of Washington Press, 2008), 139–60.

11. S. Ilan Troen, "New Departures in Zionist Planning: The Development Town," in *Israel: The First Decade of Independence*, ed. S. Ilan Troen and Noah Lucas (Albany: State University of New York, 1995), 447.

12. *Song of the Negev* (also known as *Ein Breira*), United Palestine Appeal (Josef Leytes Productions, 1950). YouTube video, 30:16, https://www.youtube.com/watch?v=1aKH-HrcMk.

13. Dana Adams Schmidt, "Israel's Epic: Men against Desert," *New York Times*, April 27, 1952.

14. Jacob Dash, *National Planning for the Redistribution of the Population and the Establishment of New Towns in Israel*, International Federation for Housing and Planning, 27th World Congress (Jerusalem: Planning Department, Ministry of the Interior, 1964), 9, 21.

15. Joan Ash, "The Progress of New Towns in Israel," *Town Planning Review* 45, no. 4 (1974): 388–89. See also Dash, *National Planning for the Redistribution of the Population and the Establishment of New Towns in Israel*, 20–21.

16. E. Brutzkus, "The New Cities in the Framework of National and Regional Planning," *Journal of the Association of Architects and Engineers in Israel* 12 (April–May 1956): 7–9.

17. Helga Keller, dir., *Ashdod Yuli 1961* (Israel: Sherut ha-serat□im ha-Yisśre'eli, 1961), YouTube video, 14:25, https://www.youtube.com/watch?v=eQWpgkT6T4A. See also Keren Filman, "Israeli New-Towns and Propaganda Films in the 1950s," *International Journal of the Arts in Society* 4, no. 2 (2009): 229–37.

18. For a study of the success and failures of Israel's new towns, see Raphael Levy and Benjamin Hanft, eds., *21 Frontier Towns* (New York: United Jewish Appeal; New York: Jewish Agency for Israel, 1971). See also Max Neufeld, *Israel's New Towns: Some Critical Impressions*, Wyndham Deedes Scholars no. 31 (London: Anglo-Israel Association, June 1971).

19. L. W. Jones, "Demographic Review: Rapid Growth in Baghdad and Amman," *Middle East Journal* 23 (Spring 1969): 209.

20. Development Board, Ministry of Development, Government of Iraq, and prepared by Doxiadis Associates, *The Housing Program in Iraq* ([Athens, Greece?]: Ministry of Development, Government of Iraq and Doxiadis Associates, 1957).

21. The contract between the Ford Foundation, Harvard University, and the Government of Pakistan began in 1954 and continued into the early 1960s. Ford Foundation, *The Ford Foundation and Pakistan* (New York, 1959); the reference to Harvard is from page 32. See also the "Request for Allocations," May 27, 1957, and the Center for International Affairs, Harvard University, *Planning in Pakistan: Progress Report of the Development Advisory Service*, September 1960–October 1962. Ford Foundation Archives, Rockefeller Archive Center, Sleepy Hollow, New York, Grant Files PA 54–1. For a good early overview of Ford's programs in Pakistan, see George Gant, "The Ford Foundation in Pakistan," *Annals of the American Academy of Political and Social Science* 323 (1959): 150–59.

22. Mian M. Nazeer, "Urban Growth in Pakistan," *Asian Survey* 6, no. 6 (1966): 312.

23. Ministry of Rehabilitation, Government of Pakistan, and prepared by Doxiadis Associates, *Development of Korangi Area: A Model Community for 500,000 People* ([Athens, Greece?]: Ministry of Rehabilitation, Government of Pakistan and Doxiadis Associates, 1960), unpaginated. For an interesting analysis of Korangi within the context of Pakistan's development, see Steve Inskeep, *Instant City: Life and Death in Karachi* (New York: Penguin, 2011), chapter 6.

24. See chapter 7, "A Different Modernity," in Gyan Prakash, *Another Reason: Science and the Imagination of Modern India* (Princeton, NJ: Princeton University Press, 1999). See also Balkrishna Doshi, "The Modern Movement in India," in *Back from Utopia: The Challenge of the Modern Movement*, ed. Hubert-Jan Henket and Hilde Heynen (Rotterdam: 010 Publishers, 2002). Also see chapter 6, "Nation and Imagination," in Dipesh Chakrabarty, *Provincializing Europe: Postcolonial Thought and Historical Difference* (Princeton, NJ: Princeton University Press, 2000).

25. Letter from Jawaharlal Nehru to Albert Mayer, June 17, 1946. Albert Mayer Etawah Papers, box 1, folder 7, Archives and Manuscripts Division, New York Public Library, New York.

26. See the interesting pamphlet overviews produced by the Government of India's Ministry of Information and Broadcasting, *Housing the Displaced* and *Housing in India* (Delhi: Publications Division, 1951 and 1954).

27. Ford Foundation, "Roots of Change: The Ford Foundation in India," *Ford Foundation Report* (November 1961), 50. The American planning team included sociologists George Goetschius and Gerald Breese, the latter from Princeton University; Bert Hoselitz, a development theorist from the University of Chicago; urban geographer Britton Harris of the University of Pennsylvania; Edward Echeverria and Walter Hedden, both land-use and transportation experts; and Arch Doston, an international development expert. Many became highly influential to the development of systems analysis. Gordon Cullen also contributed urban design studies.

28. Delhi Development Authority, *Delhi Master Plan, 1962* (Delhi: Delhi Development Authority, 1962). Much information can also be found in Gerald Breese, *Urban and Regional Planning for the Delhi-New Delhi Area: Capital for Conquerors and Country* (Princeton, NJ: Gerald Breese, 1974). A Princeton University sociologist, Breese acted as coordinator for the master plan committee. See also Robert Gardner-Medwin, "United Nations and Resettlement in the Far East," *Town Planning Review* 22, no. 4 (1952): 283–98. For a general overview, see Nilendra Bardiar, *A New Delhi—urban, Cultural, Economic and Social Transformation of the City 1947–65* (CreateSpace Independent Publishing Platform, 2014: https://www.createspace.com).

29. On the three regional capitals, see Ravi Kalia, "Modernism, Modernization and Post-colonial India: A Reflective Essay," *Planning Perspectives* 21 (2006): 133–56.

30. On Chandigarh, see Norma Evenson, *Chandigarh* (Berkeley: University of California Press, 1966). Also Ravi Kalia, *Chandigarh: The Making of an Indian City* (Delhi: Oxford University Press, 1999), and Vikramaditya Prakash, *Chandigarh's Le Corbusier: The Struggle for Modernity in Postcolonial India* (Seattle: University of Washington Press, 2002). On the confrontation with Western ideals of urban modernity, see also Ravi Kalia, *Gandhinagar: Building National Identity in Postcolonial India* (Columbia: University of South Carolina Press, 2004).

31. On the Indian steel towns, see the excellent article by Srirupa Roy, "Urban Space, National Time, and Postcolonial Difference: The Steel Towns of India," in *Urban Imaginaries: Locating the Modern City*, ed. Thomas Bender and Alev Cinar (Minneapolis: University of Minnesota Press, 2007), 182–207.

32. Ibid., 191.

33. Albert Mayer, "Working with the People," speech given at Swarthmore College, Swarthmore, Pennsylvania, February 24, 1952, pp. 16–17. Albert Mayer Etawah Papers, box 2, folder 11, Archives and Manuscripts Division, New York Public Library, New York. See also Albert Mayer, *Pilot Project, India: The Story of Rural Development in Etawah, Uttar Pradesh* (Berkeley: University of California Press, 1958). Mayer consulted on a wide range of urban projects in India, including plans for Calcutta, Bombay, Gujarat University in Ahmedabad, Faridabad, and Chandigarh. On his work, see Thomaï Serdari, "Albert Mayer, Architect and Town Planner: The Case for a Total Professional" (PhD diss., New York University, 2005). On India's influence on Mayer's ideas, see Andrew Friedman, "The Global Postcolonial Moment and the American New Towns: India, Reston, Dodoma," *Jounal of Urban History* 38, no. 3 (2012): 553–76.

34. See three examples among many in these English-language Indian newspapers: the *Statesman*, October 11, 1949; the *Pioneer*, May 4, 1952; *Times of India* (Mumbai), August 15, 1952.

35. Ford Foundation, "The Ford Foundation and Foundation Supported Activities in India" (New Delhi, January 1955), 28–31. See also Nicole Sackley, "Passage to Modernity: American Social Scientists, India, and the Pursuit of Development, 1945–1961" (PhD diss., Princeton University, 2004).

36. Albert Mayer, "Community Projects in U.P.: Observations and Recommendations," April 26, 1953, p. 14. Albert Mayer Etawah Papers, box 1, folder 5.

37. S. K. Dey, *Nilokheri* (London: Asia Publishing House, 1962), 19, 21.

38. Speech by Jawaharlal Nehru, February 22, 1950; cited in Syresh K. Sharma, *Haryana: Past and Present* (New Delhi: Mittel, 2005), 248. See also Srirupa Roy, *Beyond Belief: India and the Politics of Postcolonial Nationalism* (Durham, NC: Duke University Press, 2007).

39. Mulk Raj Anand, "Planning and Dreaming," *MARG* 1, no. 1 (1946): 1–2. On the history of *MARG* magazine, see the excellent article by Rachel Lee and Kathleen James-Chakraborty, "*Marg* Magazine: A Tryst with Architectural Modernity," *ABE Journal [Online]* 1 (2012). http://dev.abejournal.eu/.

40. Koenigsberger was also a regular contributor to UN missions. For his urban designs and architecture, see Jon Lang, Madhavi Desai, and Miki Desai, eds., *Architecture and Independence: The Search for Identity—India 1880 to 1980* (New Delhi: Oxford University Press, 1997). See also Rachel Lee, "Constructing a Shared Vision: Otto Koenigsberger and Tata & Sons," *ABE Journal [Online]* 2 (2012). http://dev.abejournal.eu/.

41. Otto H. Koenigsberger, "New Towns in India," *Town Planning Review* 23, no. 2 (1952): 96.

42. On internalizing the neighborhood unit concept by means of the "mohalla unit," see Sanjeev Vidyarthi, "Inappropriately Appropriated or Innovatively Indigenized?: Neighborhood Unit Concept in Post-Independence India," *Journal of Planning History* 9, no. 4 (2010): 267–68.

43. See Sanjeev Vidyarthi, "Reimagining the American Neighborhood Unit for India," in *Crossing Borders: International Exchange and Planning*, ed. Patsy Healey and Robert Upton (New York: Routledge Press, 2010), 73–94.

44. Rhodri Windsor-Liscombe, "In-dependence: Otto Koenigsberger and Modernist Urban Resettlement in India," *Planning Perspectives* 21 (2006): 161. See also Vandana Baweja, "Otto Koenigsberger and the Tropicalization of British Architectural Culture," in *Third World Modernism: Architecture, Development and Identity*, ed. Duanfang Lu (New York: Routledge, 2011), 236–54.

45. The Anglo-Persian Oil Company was renamed the Anglo-Iranian Oil Company in 1935, and then British Petroleum after 1954. See James Bamberg, *British Petroleum and Global Oil, 1950–1975: The Challenge of Nationalism* (Cambridge: Cambridge University Press, 2000), as well as Hossein Askari, *Collaborative Colonialism: The Political Economy of Oil in the Persian Gulf* (New York: Palgrave Macmillan, 2013).

46. Quotation is from Mark Crinson, "Abadan: Planning and Architecture under the Anglo-American Oil Company," *Planning Perspectives* 12 (1997): 350. See also Kaveh Eshani, "Social Engineering and the Contradictions of Modernization in Khuzestan's Company Towns: A Look at Abadan and Masjed-Soleyman," *IRSH (Internationaal Instituut voor Sociale Geschiedenis)* 48 (2003): 361–99. Also good is Mohammad A. Chaichian, *Town and Country in the Middle East: Iran and Egypt in the Transition to Globalization, 1800–1970* (London: Lexington Books, 2009), 89–92. Abadan and its oil refinery were bombed and destroyed in 1980 during the Iran-Iraq War, and then rebuilt.

47. This is discussed for architecture in Gwendolyn Wright, "Building Global Modernisms," *Grey Room* 7 (2002): 124–34.

48. B. S. Hoyle, "New Oil Refinery Construction in Africa," *Geography* 48 (April 1963): 190–91.

49. Henry Tanner, "France Discovers an Oil Oasis," *New York Times*, January 10, 1960.

50. On Hassi-Massaoud, see the Mémoire de thèse by Jean Carail, "Exemple de création urbaine au Sahara pétrolier: Hassi-Massaoud, 1956–1962" (Institut d'urbanisme, Université de Paris, 1962). Quotation is from page II-8. See also Louis Kraft, "The French Sahara and Its Mineral Wealth," *International Affairs* 36 (April 1960): 197–205.

51. See R. I. Lawless, "Uranium Mining at Arlit in the Republic of Niger," *Geography* 59 (January 1974): 45–48. On conditions in Arlit after the collapse of uranium prices and its abandonment, see the excellent documentary film by director Idrissou Mora Kpai, *Arlit, Deuxième Paris* (Niger/France: MKJ Films/Noble Films, 2005), DVD.

52. British Pathé, *Queen's Tour of Nigeria*, February 16, 1956, Film ID 569.18, British Pathé video, 3:04, http://www.britishpathe.com/video/royal-tour-3/query/regatta. Andrew Aptner, *The Pan-African Nation: Oil and the Spectacle of Culture in Nigeria* (Chicago: University of Chicago Press, 2005), 159–62.

53. See C. V. Izeogu, "Urban Development and the Environment in Port Harcourt," *Environment and Urbanization* 59, no. 1 (1989): 59–68. See also L. B. Dangana, "Dynamique urbaine de Port Harcourt, Nigeria," *Annales de Géographie* 89, no. 495 (1980): 605–13, as well as Marc Antoine Pérouse de Montclos, "Port Harcourt: La 'Cité-jardin' dans la Marée Noire," *Politique Africaine* 74 (1999): 42–50. Also A. T. Salau, "The

Oil Industry and the Urban Economy: The Case of Port Harcourt Metropolis," *African Urban Studies*, no. 17 (1983): 75–84. Lastly, Daniel A. Omeweh, *Shell Petroleum Development Company, the State and Underdevelopment in Nigeria's Niger Delta: A Study in Environmental Degradation* (Trenton, NJ: Africa World Press, 2005).

54. For a short overview of Doxiadis's life and work, see Ray Bromley, "Towards Global Human Settlements: Constantinos Doxiadis as Entrepreneur, Coalition-Builder and Visionary," in Nasr and Volait, *Urbanism, Imported or Exported?*, 316–40. See also Ahmed Zaib Khan Mahsud, "Rethinking Doxiadis' Ekistical Urbanism," *Positions* 1 (2010): 6–39. Also useful is the illustrated article "Greek City Planner Constantinos Doxiadis" in *Life*, October 7, 1966.

55. On this point, see Sheila Jasanoff, "Ordering Knowledge, Ordering Society," in *States of Knowledge: The Co-production of Science and Social Order*, ed. Sheila Jasanoff (London: Routledge, 2004), 13–45.

56. Letter to Paul Ylvisaker, October 2, 1959, Doxiadis Associates, General Correspondence 1959, microform. Ford Foundation Archives, Rockefeller Archive Center, Sleepy Hollow, New York. On Doxiadis's connections with the Ford Foundation and American Cold War institutions, see Michelle Provoost, "New Towns on the Cold War Frontier," *Eurozine*, June 28, 2006. Available at www.eurozine.com.

57. Among Doxiadis's most important formulations of ekistics are *Between Dystopia and Utopia* (London: Faber and Faber, 1966) as well as *Ekistics: An Introduction to the Science of Human Settlements* (London: Oxford University Press, 1968), and *Ecumenopolis: The Inevitable City of the Future* (New York: Norton, 1979).

58. Mark Wigley, "Network Fever," *Grey Room* 4 (Summer 2001): 121–22.

59. Memo by Paul Ylvisaker, Ford Foundation, April 28, 1961. Grant Files, PA 60–216, Ford Foundation Archives, Rockefeller Archive Center, Sleepy Hollow, New York.

60. Constantinos Doxiadis, "The City of the Future: Anticipated Research and Planning Programs for the Federal Capital." Grant Files, PA 60–216, Ford Foundation Archives, Rockefeller Archive Center, Sleepy Hollow, New York.

61. Housing and Settlements Agency, Government of West Pakistan, "Housing in West Pakistan: Monthly Report for October 1966" (Lahore 1966), as well as letter to Ford Foundation from Ekistics trainee Tariq Masud Durrani, June 7, 1965. Grant 61–0101, Ford Foundation Archives, Rockefeller Archive Center, Sleepy Hollow, New York.

62. C. A. Doxiadis, *The Federal Capital: A Preliminary Report Prepared for the Government of Pakistan* ([Karachi?]: Doxiadis Associates, 1959), 7–9. See Glenn V. Stephenson, "Two Newly-Created Capitals: Islamabad and Brasilia," *Town Planning Review* 41, no. 4 (1970): 317–32.

63. On this point, see Phillip E. Wegner, *Imaginary Communities: Utopia, the Nation, and the Spatial Histories of Modernity* (Berkeley: University of California Press, 2002), 34–40.

64. See Annie Harper, "Islamabad and the Promise of Pakistan," in *Pakistan: From the Rhetoric of Democracy to the Rise of Militancy*, ed. Ravi Kalia (New Delhi: Routledge, 2011), 64–84.

65. Launched in 1951, the Colombo Plan was a Commonwealth (Australia, Britain, Canada, Ceylon, India, New Zealand, and Pakistan) intergovernmental organization for economic and social development, and provided technical assistance and training for developing countries on a myriad of development projects.

66. C. A. Doxiadis, *The Federal Capital of Pakistan: 3 Principles of Planning. Prepared for the Government of Pakistan* ([Karachi?]: Doxiadis Associates, 1959). For an introduction

to the Master Plan for Islamabad, see "Islamabad: The Creation of a New Capital," *Town Planning Review* 36, no. 1 (1965): 1–28. See also Leo Jamoud, "Islamabad—the Visionary Capital," *Ekistics* 25 (May 1968): 329–33.

67. Orestes Yakas, *Islamabad: The Birth of a Capital* (Oxford: Oxford University Press, 2001), 41.

68. On Hassan Fathy, see Malcolm Miles, "Utopias of Mud? Hassan Fathy and Alternative Modernisms," *Space and Culture* 9 (May 2006): 115–39.

69. Maurice Lee, "Islamabad—the Image," *Architectural Design*, January 1967; reprinted in *Ekistics* 25 (May 1968): 335.

70. These battles and more are recounted in Yakas, *Islamabad*, chapter 4. See also Hasan-Uddin Khan, "The Impact of Modern Architecture on the Islamic World," in Henket and Heynen, *Back from Utopia*, 174–89.

71. Gerard Brigden, "Islamabad: A Progress Report on Pakistan's New Capital City," *Architectural Review* 141 (1967): 211–12.

72. "Blueprint Is Drawn for a 'T.V.A.' in Nigeria," *New York Times*, February 5, 1962. The Kainji Dam project and New Bussa are located in Niger State, Nigeria. Nigeria received full independence from Britain in 1960.

73. On the development of New Bussa, see Brian C. Smith, "New Bussa, a New Town on the Niger," *Urban Studies* 4, no. 2 (1967): 149–64.

74. See the excellent article by Rhodri Windsor Liscombe, "Modernism in Late Imperial British West Africa: The Work of Maxwell Fry and Jane Drew, 1946–56," *Journal of the Society of Architectural Historians* 65, no. 2 (2006): 188–215. See also Iain Jackson and Jessica Holland, *The Architecture of Edwin Maxwell Fry and Jane Drew: Twentieth Century Architecture, Pioneer Modernism and the Tropics* (Farnham, UK: Ashgate, 2014).

75. See chapter 3 of Latham, *The Right Kind of Revolution*.

76. On these points, see Christophe Bonneuil, "Development as Experiment: Science and State Building in Late Colonial and Postcolonial Africa, 1930–1970," *Osiris*, 2nd ser., 15 (2000): 258–81.

77. Keith Jopp, "Tema, Ghana's New Town and Harbour," ed. Development Secretariat (n.p.: Ghana Information Services, 1961), 3.

78. For the image of modern Tema, see Anonymous, "Tema, West Africa's New Town," *West African Review* 33, no. 414 (1962): 4–9.

79. Jopp, "Tema," 23.

80. Government of Ghana and C. A. Doxiadis, *The Town of Tema: Existing Conditions on 1st May 1961* (Accra and Tema, Ghana: Tema Development Organisation, 1961), 20–21.

81. J. Yedu Bannerman, *The Cry for Justice in Tema (Ghana)* (Tema, Ghana: Tema Industrial Mission, 1973), 9.

82. Doxiadis, *Between Dystopia and Utopia*, 50–53. See also C. A. Doxiadis, "Ecumenopolis: The Settlement of the Future," in *ACE Publication Series, Report no. 1* (Athens: Athens Center of Ekistics, 1967), 157–64.

83. Doxiadis, *Between Dystopia and Utopia*, 59.

84. Jean Gottmann and Robert Harper, eds., *Since Megalopolis: The Urban Writings of Jean Gottman* (Baltimore: Johns Hopkins University Press, 1990), 157.

CHAPTER FOUR

1. See the description of Delos III by participant Grady Clay in the *ATI-ACE Newsletter* (Athens, Greece: Athens Technological Institute, Athens Center of Ekistics) 1, no. 10

(April 15, 1965): 4–6. On the Delos Symposia, see also Philip Deane, *Constantinos Doxiadis, Master Builder for Free Men* (Dobbs Ferry, NY: Oceana Publications, 1965), and Pyla Panayiota, "Planetary Home and Garden: Ekistics and Environmental-Developmental Politics," *Grey Room* 36 (2009): 6–35.

2. See Buckminster Fuller, "Worlds Beyond," *Omni* 1, no. 4 (January 1979): 102. Always good is his *Operating Manual for Spaceship Earth* (New York: E. P. Dutton, [ca. 1963], 1972). On American environmentalism and the *Whole Earth Catalog*, see the excellent article by Andrew Kirk, "Appropriating Technology: The *Whole Earth Catalog* and Counterculture Environmental Politics," *Environmental History* 6, no. 3 (2001): 373–94, as well as Fred Turner, *From Counterculture to Cyberculture: Stewart Brand, the Whole Earth Network, and the Rise of Digital Utopianism* (Chicago: University of Chicago Press, 2008).

3. The Declaration of Delos and the reports of the Delos Symposia are available at the World Society for Ekistics archives on the Universidad del Rosario website: http://www.urosario.edu.co/cpg-ri/WorldSocietyforEkistics/archivos/Delos-Declarations-(One-to-Six).pdf. Accessed June 21, 2015.

4. On this point, see Christopher Johnson, "Analogue Apollo: Cybernetics and the Space Age," *Paragraph* 31, no. 3 (2008): 304–26. Also David Mindell, *Digital Apollo: Human and Machine in Spaceflight* (Cambridge, MA: MIT Press, 2008).

5. See Matthew Farish, "Disaster and Decentralization: American Cities and the Cold War," *Cultural Geographies* 10 (2003): 125–48. For a general overview of the period, see Richard Lingeman, *The Noir Forties: The American People from Victory to Cold War* (New York: Nation Books, 2012).

6. Jennifer S. Light, *From Warfare to Welfare: Defense Intellectuals and Urban Problems in Cold War America* (Baltimore: Johns Hopkins University Press, 2003), 6. See Nils Gilman, *Mandarins of the Future: Modernization Theory in Cold War America* (Baltimore: John Hopkins University Press, 2007); Paul Boyer, *By the Bomb's Early Light: American Thought and Culture at the Dawn of the Atomic Age* (New York: Pantheon, 1985). See also Paul N. Edwards, *The Closed World: Computers and the Politics of Discourse in Cold War America* (Cambridge, MA: MIT Press, 1996). And lastly, Walter A. McDougall, *The Heavens and the Earth: A Political History of the Space Age* (New York: Basic Books, 1985).

7. Donald Porter Geddes, ed., *The Atomic Age Opens* (New York: Pocket Books, 1945), 165, 79.

8. Jacob Marshak, Edward Teller, and Lawrence R. Klein, "Dispersal of Cities and Industries," as described in Robert H. Kargon and Arthur P. Molella, "The City as Communications Net: Norbert Wiener, the Atomic Bomb, and Urban Dispersal," *Technology and Culture* 45, no. 4 (2004): 766.

9. Tracy B. Augur, "The Dispersal of Cities as a Defensive Measure," *Journal of the American Planning Association* 14, no. 3 (1948): 32. The many articles on civil defense produced by American scientists and urban planners are also discussed in Michael Quinn Dudley, "Sprawl as Strategy: City Planners Face the Bomb," *Journal of Planning Education and Research* 21 (2001): 52–63.

10. William Fielding Ogburn, "Sociology and the Atom," *Journal of Sociology* 51, no. 4 (1946): 271.

11. Steve Joshua Heims, *The Cybernetics Group* (Cambridge, MA: MIT Press, 1991), 28. The Macy Conferences were held in New York in the late 1940s and early 1950s under the sponsorship of the Josiah Macy Foundation. They were an interdisciplinary forum for the discussion of cybernetics and its applicability to the social sciences.

Among the participants were anthropologist Margaret Mead and sociologists Lawrence Frank and Paul Lazarsfeld.

12. "How U.S. Cities Can Prepare for Atomic War: MIT Professors Suggest a Bold Plan to Prevent Panic and Limit Destruction," and "The Planners Evaluate Their Plan," *Life*, December 18, 1950, 77–86. Quote is from page 85. This article and the unpublished manuscript on which it is based are discussed at length in Kargon and Molella, "The City as Communications Net," 764–65.

13. On the origins of RAND, see David Hounshell, "The Cold War, RAND, and the Generation of Knowledge, 1946–1962," *Historical Studies in the Physical and Biological Sciences* 27, no. 2 (1997): 237–67.

14. David R. Jardini, "Out of the Blue Yonder: The RAND Corporation's Diversification into Social Welfare Research, 1946–1968" (PhD diss., Carnegie Mellon University, 1996), 112. See also "The City Meets the Space Age," *Architectural Forum* 126, no. 1 (January 1967): 60–63, for a description of a 1966 HUD-sponsored seminar at Woods Hole, Massachusetts, that addressed how the US aerospace program could be adapted to urban planning.

15. Ira S. Lowry, "A Model of Metropolis" (Santa Monica, CA: RAND Corporation, August 1964), 133. The Pittsburgh study was part of the US government's Community Renewal Program, which provided funding for the elimination of urban blight. It was carried out for the Pittsburgh Regional Planning Association with support from the RAND Corporation and the Ford Foundation.

16. Ibid., 8.

17. Edward F. R. Hearle, "Information Systems for Urban Planning" (Santa Monica, CA: RAND Corporation, July 1963), 1. See in particular the analysis of Francis Ferguson, *Architecture, Cities and the Systems Approach* (New York: George Braziller, 1975).

18. On this point, see Paul Ricoeur, *Lectures on Ideology and Utopia*, ed. George H. Taylor (New York: Columbia University Press, 1986), 16–17. See also W. J. T. Mitchell, *Iconology: Image, Text, Ideology* (Chicago: University of Chicago Press, 1986).

19. Christopher Alexander, "A City Is Not a Tree," originally published in *Architectural Forum* 122, no. 1 (April 1965): 58–61. See also his *Notes on the Synthesis of Form* (Cambridge, MA: Harvard University Press, 1964).

20. Charles J. Zwick, "Systems Analysis and Urban Planning" (Santa Monica, CA: RAND Corporation, June 1963), 9.

21. Wilfred Owen, *Cities in the Motor Age* (New York: Viking Press, 1959), 145. See as well Owen Gutfreund, *Twentieth-Century Sprawl: Highways and the Reshaping of the American Landscape* (Oxford: Oxford University Press, 2005).

22. On this point, see Sheila Jasanoff, "The Idiom of Co-production," in *States of Knowledge: The Co-production of Science and Social Order*, ed. Sheila Jasanoff (London: Routledge, 2004), 3.

23. The work of Michael Batty is essential to understanding urban modeling. See his *Urban Modelling: Algorithms, Calibrations, Predictions* (Cambridge: Cambridge University Press, 2010), originally published in 1976. See also Batty's *Cities and Complexity: Understanding Cities with Cellular Automata, Agent-Based Models and Fractals* (Cambridge, MA: MIT Press, 2005).

24. Among the many cultural histories of the automobile, see Brian Ladd, *Autophobia: Love and Hate in the Automobile Age* (Chicago: University of Chicago Press, 2008), as well as Cotten Seller, *Republic of Drivers: A Cultural History of Automobility in America* (Chicago: University of Chicago Press, 2008). Lastly, John Heitmann, *The Automobile and American Life* (Jefferson, NC: McFarland, 2009).

25. *National Geographic*, April 1965, 520–21.

26. Office of Metropolitan Development, US Department of Housing and Urban Development, *Tomorrow's Transportation: New Systems for the Urban Future* (Washington, DC: US Department of Housing and Urban Development, 1968). This imagery and the interest in personal rapid transit, or PRT, was reiterated in William F. Hamilton and Dana K. Nance, "Systems Analysis of Urban Transportation," *Scientific American*, July 1969, 19–27.

27. Office of Metropolitan Development, HUD, *Tomorrow's Transportation*, 2.

28. *Life*, December 24, 1965.

29. New Communities Division, US Department of Housing and Urban Development, "Survey and Analysis of Large Developments and New Communities" (Washington, DC: US Department of Housing and Urban Development, February 1969), 4–5. On Los Angeles, see the exhibition catalog edited by Wim De Wit and Christopher James Alexander, *Overdrive: L.A. Constructs the Future, 1940–1990* (Los Angeles: Getty Publications, 2013). On California's new towns and their relationship to highways and water projects, see Edward Eichler and Marshall Kaplan, *The Community Builders* (Berkeley: University of California Press, 1967).

30. *Time*, September 6, 1963.

31. Nathaniel M. Griffin, *Irvine: The Genesis of a New Community* (Washington, DC: Urban Land Institute, 1974), 18.

32. Ann Forsyth and Katherine Crewe, "New Visions for Suburbia: Reassessing Aesthetics and Place-Making in Modernism, Imageability and New Urbanism," *Journal of Urban Design* 14, no. 4 (2009): 430. Parts of the 130-square-mile land grant site had been requisitioned during the war as a Marine Corps airbase, and the aerospace industry exerted a powerful influence over the new town's design and its economy.

33. Melvin M. Webber, "Planning in an Environment of Change: Part I: Beyond the Industrial Age," *Town Planning Review* 39, no. 3 (1968): 181. See also the special issue on Webber in *Access* magazine, published by the University of California Transportation Center (Winter 2006–7).

34. Melvin M. Webber, "The Post-City Age," *Daedalus* 97, no. 4 (1968): 1098.

35. Brian Berry, "Cities as Systems in Systems of Cities," *Papers of the Regional Science Association* 13, no. 1 (1964): 147–48.

36. Lewis Mumford, "Utopia, the City and the Machine," *Daedalus* 94, no. 2 (1965): 278.

37. Britton Harris, "Computers and Urban Planning," in "Computers in the Service of Ekistics," special issue, *Ekistics* 28, no. 164 (1969): 4–7.

38. Louis Edward Alfeld, "Urban Dynamics—the First Fifty Years," *System Dynamics Review* 11, no. 3 (1995): 199–202.

39. See Felicity D. Scott, "Fluid Geographies: Politics and Revolution by Design," in *New Views on R. Buckminster Fuller*, ed. Hsiao-Yun Chu and Roberto G. Trujillo (Palo Alto, CA: Stanford University Press, 2009), 160–75.

40. Webber, "The Post-City Age," 1107.

41. Tony Judt, *Postwar: A History of Europe since 1945* (New York: Penguin, 2005), 17.

42. On postwar urban conditions, see Donald Filtzer, *The Hazards of Urban Life in Late Stalinist Russia* (Cambridge: Cambridge University Press, 2010).

43. See R. W. Davies and Melanie Ilic, "From Khrushchev (1935–1936) to Khrushchev (1956–1964): Construction Policy Compared," in *Khrushchev in the Kremlin: Policy and Government in the Soviet Union, 1953–64*, ed. Jeremy Smith and Melanie Ilic (New York: Routledge Press, 2011), 202–30. See also Steven E. Harris, *Communism*

on Tomorrow Street: Mass Housing and Everyday Life after Stalin (Baltimore: John Hopkins University Press, 2013), and Marina Balina and Evgeny Dobrenko, eds., *Petrified Utopia: Happiness Soviet Style* (London: Anthem, 2011).

44. Stanislav Strumilin, "Family and Community in the Society of the Future," *Soviet Review* 2, no. 2 (1961): 17, 19.

45. Boris Svetlichny, "Les villes de l'avenir," *Recherches internationales à la lumière du marxisme* 7–10, no. 10 (1959): 208–29. See also Svetlichny's "The Future of Soviet Towns," *Town and Country Planning* (1962): 80–82, as well as "City-Planning Processes and Problems of Settlement," *Problems of Economic Transition* 18, no. 3 (1975): 23–30. On the satellite towns of the Soviet Union, see Gary Joseph Hausladen, "Regulating Urban Growth in the USSR: The Role of Satellite Cities in Soviet Urban Development" (PhD diss., Syracuse University, 1983).

46. Statistics on the Soviet new towns fluctuate wildly. The numbers given here are from Ilia Moiseevich Smoliar [Smolyar] of the Central Scientific Research and Design Institute for Town Plannning in Moscow; Smoliar was one of the most important researchers on new towns in the Soviet Union. See I. Smolyar et al., "The Experience of the USSR in the Planning and Construction of New Towns" in United Nations, *Report of the United Nations Seminar on Physical Planning Techniques for the Construction of New Towns: Moscow, USSR, 2–22 September 1968* (New York: United Nations, 1971), 9–16. Also, I. M. Smoliar, ed., *New Towns Formation in the USSR* (Moscow: Central Scientific Research and Design Institute of Town Planning, 1973). See earlier statistics given in Jack A. Underhill, "Soviet New Towns, Planning and National Urban Policy: Shaping the Face of Soviet Cities," *Town Planning Review* 61, no. 3 (1990): 272–73. On the definition of Soviet new towns, see the excellent article by Catherine Chatel and François Moriconi-Ebrard, "Définir les villes nouvelles en Russie," *Revue Regard sur l'Est* 47 (2007). Available online at http://www.regard-est.com.

47. Andrei Ershov and Mikhail Shura-Bura, quoted in Paul Josephson, *New Atlantis Revisited: Akademgorodok, the Siberian City of Science* (Princeton, NJ: Princeton University Press, 1997), 123. See Patrick Major and Rana Mittner, eds., *Across the Blocs: Exploring Comparative Cold War Cultural and Social History* (London: Frank Cass, 2004), as well as Dick Van Lente, ed., *The Nuclear Age in Popular Media: A Transnational History, 1945–1965* (New York: Palgrave Macmillan, 2012).

48. Benjamin Peters, "Normalizing Soviet Cybernetics," *Information and Culture: A Journal of History* 47, no. 2 (2012): 165. See also Paul R. Josephson, "Rockets, Reactors, and Soviet Culture," in *Science and the Soviet Social Order*, ed. Loren R. Graham (Cambridge, MA: Harvard University Press, 1990), 180–85. Lastly, Eva Maurer, ed., *Soviet Space Culture: Cosmic Enthusiasm in Socialist Societies* (New York: Palgrave Macmillan, 2011).

49. Recounted in Willlis H. Ware and Wade B. Holland, "Soviet Cybernetics Technology: I. Soviet Cybernetics, 1959–1962," in *United States Air Force Project RAND* (Santa Monica, CA: RAND Corporation, June 1963), 11.

50. Alexei Gutnov et al., *The Ideal Communist City*, trans. Renée Neu Watkins (New York: George Braziller, 1971), 101. The book was originally published in 1967 as *New Elements of Settling* and then updated in 1968. It was initially translated into Italian in 1968 and then into English and Spanish.

51. "Architecture Soviétique," special issue, *L'Architecture d'aujourd'hui* 147 (December 1969–January 1970).

52. "Cells" and "ganglions" are used by Russian urban geographer Yakov G. Mashbits, "Interdependence of Urbanization and Development of the Territorial Structure of

National Economy," *Geographia Polonica* 44 (1981): 27. "Cybernetic human-machines" is used in Slava Gerovitch, "'New Soviet Man' inside Machine: Human Engineering, Spacecraft Design, and the Construction of Communism," *Osiris* 22 (2007): 143.

53. Dorothy McDonald and Wade B. Holland, "Recent News Items," *Soviet Cybernetics* 3, no. 3 (1969): 3. *Soviet Cybernetics* was published by the RAND Corporation and edited by Wade B. Holland. See also Slava Gerovitch, "InterNyet: Why the Soviet Union Did Not Build a Nationwide Computer Network," *History and Technology* 24, no. 4 (2008): 335–50.

54. Alexei Gutnov, "L'URSS: Vers la ville socialiste; Problèmes actuels de l'urbanisme Sovietique," in *URSS 1917–1978: La Ville, l'architecture*, ed. Jean-Louis Cohen, Marco De Michelis, and Manfredo Tafuri (Paris: L'Equerre, 1979), 356.

55. See for example Y. V. Medvedkov, "Dynamics of Urban Spaces Conditioned by Human Ecology," *Geographia Polonica* 44 (1981): 5–17.

56. I. Smolyar et al., "The Experience of the USSR in the Planning and Construction of New Towns"; United Nations, *Report of the United Nations Seminar on Physical Planning Techniques for the Construction of New Towns. Moscow, USSR, 2–22 September 1968* (New York: United Nations, 1971), 20–21.

57. Zakrytoe administrativno-territorial'noe obrazovanie (Closed Administrative-Territorial Formation), or ZATO.

58. See Richard Rowland, "Russia's Secret Cities," *Post-Soviet Geography and Economics* 37, no. 7 (1996): 426–62. Also Michael Gentile, "Former Closed Cities and Urbanisation in the FSU: An Exploration in Kazakhstan," *Europe-Asia Studies* 56, no. 2 (2004): 263–78.

59. See the remarkable study by Barbara Engel, *Öffentliche Räume in den Blauen Städten Russlands* (Berlin: Wasmuth, 2004), 16–17. See as well Kate Brown, *Plutopia: Nuclear Families, Atomic Cities, and the Great Soviet and American Plutonium Disasters* (New York: Oxford University Press, 2013).

60. Engel, *Öffentliche Räume in den Blauen Städten Russlands*, 83–88.

61. Yevgeny Yevtushenko, *New Works: The Bratsk Station*, trans. Tina Tupikina-Glaessner and Geoffrey Dutton (Melbourne: Sun Books, 1966), 35–36.

62. Andis Cinis, Marija Drémaité, and Mart Kalm, "Perfect Representations of Soviet Planned Space: Mono-industrial Towns in the Soviet Baltic Republics in the 1950s–1980s," *Scandinavian Journal of History* 33, no. 3 (2008): 229.

63. Asif A. Siddiqi, "The Secret Cities," in *ZATO—Soviet Secret Cities during the Cold War*, ed. Xenia Vytuleva (New York: Harriman Institute, Columbia University, 2012), 7.

64. Recounted in Vladislav Zubok, *Zhivago's Children: The Last Russian Intelligensia* (Cambridge, MA: Harvard University Press, 2011), 132.

65. See the excellent analysis by Josephson, *New Atlantis Revisited*. Also excellent is Alexander D'Hooghe, "Science Towns as Fragments of a New Civilisation: The Soviet Development of Siberia," *Interdisciplinary Science Reviews* 31, no. 2 (2006): 135–48. On the Khrushchev years and Akademgorodok, see Francis Spufford, *Red Plenty* (Minneapolis: Graywolf Press, 2012).

66. Walter Sullivan, "Soviet Union's 'Academic Cities' Symbolize New Efforts in Science," *New York Times*, October 16 1967.

67. Ibid. See also Denis Kozlov and Eleonory Gilburd, *The Thaw: Soviet Society and Culture during the 1950s and 1960s* (Toronto: University of Toronto Press, 2014).

68. Robert J. Osborn and Thomas A. Reiner, "Soviet City Planning: Current Issues and Future Perspectives," *Journal of the American Institute of Planners* 28, no. 4 (1962): 240–41.

69. Elke Beyer, "'The Soviet Union Is an Enormous Construction Site,'" in *Soviet Modernism 1955–1991: Unknown History*, ed. Katharina Ritter et al. (Zurich: Park Books, 2012), 258.

70. Alexei Gutnov, "L'URSS," 366.

71. Carol Lubin, community planner, Reston, Virginia, in letter to Bolesław Malisz, October 25, 1966. Carol R. Lubin Papers, 1960–81, Special Collections and Archives, George Mason University Libraries, Fairfax, Virginia.

72. Bolesław Bierut, quoted in Jacek Friedrich, "'. . . A Better, Happier World': Visions of a New Warsaw after World War Two," in *Urban Planning and the Pursuit of Happiness: European Variations on a Universal Theme (18th–21st Centuries)*, ed. Arnold Bartetzky and Marc Schalenberg (Berlin: Jovis, 2009), 112.

73. Bolesław Malisz, *La Pologne construit des villes nouvelles*, trans. Kazimiera Bielawska (Warsaw: Editions Polania, 1961), 5.

74. Bolesław Malisz, "Threshold Analysis as a Tool in Urban and Regional Planning," *Papers in Regional Science* 29, no. 1 (1972): 167–77, as well as his "Physical Planning for the Development of Satellite and New Towns: The Analysis of Urban Development Possibilities" (Warsaw: Institute Papers, Research Institute for Town Planning and Architecture, 1963).

75. On this change, see in particular I. S. Koropeckyj, "Regional Development in Postwar Poland," *Soviet Studies* 29, no. 1 (1977): 108–27.

76. Bolesław Malisz, "Physical Planning for the Development of New Towns," in United Nations Department of Economic and Social Affairs, *Planning of Metropolitan Areas and New Towns*, papers presented at Stockholm, September 14–30, 1961, and Moscow, August 24–September 7, 1964 (New York: United Nations, 1969), 208.

77. Jerzy Kozłowski, "Threshold Theory and the Sub-regional Plan," *Town Planning Review* 39, no. 2 (1968): 99–116.

78. Walter Ulbricht, speech, December 16, 1965. Available at German History in Documents and Images: http://germanhistorydocs.ghi-dc.org/index.cfm. Accessed June 23, 2015. Source: Walter Ulbricht, *Zum Neuen Ökonomischen System der Planung und Leitung [On the New Economic System of Planning and Management]*. Berlin: Dietz, 1966, pp. 668–76.

79. Walter Ulbricht, May 1968 speech at the celebration of the 150th birthday of Karl Marx, quoted in CIA Intelligence Report, "The Prussian Heresy: Ulbricht's Evolving System," Reference Title ESAU XLVI/70, June 29, 1970, 23.

80. Reinhard Sylten, "Zur Prognose und Analyse im Städtebau," *Deutsche Architektur* 17, no. 4 (April 1969): 217.

81. See the explanation of GDR modeling in Gerold Kind, "Modelling of Settlement Systems for Regional Planning," *Geographia Polonica* 44 (1981): 33–43.

82. On the impact of exile on East German architects, see Jay Rowell, "L'exil comme ressource et comme stigmate dans la constitution des réseaux des architectes-urbanistes de la RDA," *Revue d'histoire moderne et contemporaine* 52, no. 2 (2005): 169–91.

83. Richard Paulick, "Hoyerswerda—eine sozialistische Stadt der Deutschen Demokratischen Republik," *Deutsche Architektur* 9, no. 7 (July 1960): 365. On the new town of Schwedt, see Philipp Springer, *Verbaute Träume: Herrschaft, Stadtentwicklung und Lebensrealität in der sozialistischen industriestadt Schwedt* (Berlin: Verlag, 2007).

84. See for example Hans-Georg Heinecke, "Die neuen Typengrundrisse für die Wohnbauten in Neu-Hoyerswerda," *Deutsche Architektur* 5, no. 1 (1956): 27–29, as well as Helmut Mende, "Das Grossplattenwerk von Hoyerswerda," *Deutsche Architektur* 5,

no. 2 (1956): 62–69. And Rudolf Dehmel, "Die neuen Typengrundrisse für Gross-plattenbauweise in Hoyerswerda," *Deutsche Architektur* 5, no. 9 (1956): 410–14.

85. Wolfgang Thöner and Peter Müller, eds., *Bauhaus-Tradition und DDR-Moderne: Der Architekt Richard Paulick* (Berlin: Deutsche Kunstverlag 2006), 126.

86. Paulick, "Hoyerswerda," 357. Excellent material on the design of Hoyerswerda can be found in Thomas Topfstedt, *Städtebau in der DDR 1955–1971* (Leipzig: E. A. Seemann, 1988), 31–36.

87. On these points, see Anthony Vidler, "Diagrams of Diagrams: Architectural Abstraction and Modern Representation," *Representations* 72 (2000): 1–20.

88. "Milton Keynes—eine neue Stadt in England," *Architektur der DDR* 24 (December 1975): 742–45.

89. Paulick, "Hoyerswerda," 366.

90. Brigitte Reimann, *Franziska Linkerhand* (Munich: Kindler, 1974). See also Hunter Bivens, "Neustadt: Affect and Architecture in Brigitte Reimann's East German Novel *Franziska Linkerhand*," *Germanic Review* 83, no. 2 (Spring 2008): 139–66.

91. See for example H. Kowalke et al., "A Study of the Settlement Structure of Agglomeration Regions in the GDR with Special Reference to the Halle-Leipzig Agglomeration," *Geographia Polonica* 4 (1981): 171–78. Jay Rowell, "Les Compétences professionnelles et la production de la ville," *Les Annales de la recherche urbaine* 105 (2008): 149–50.

92. See the visits and quotations in *Architektur der DDR* 23(June 1974): 326.

93. Büro für Städtebau und Architecktur des Rates des Bezirkes Halle, *Halle-Neustadt: Plan und Bau der Chemiearbeiterstadt* (Berlin: VEB Verlag für Bauwesen, 1971), 41.

94. Richard Paulick, "Rationelle Technologie für die Modernisierung von Wohnbauten in den USA," *Deutsche Architektur* 16, no. 2 (February 1967): 117–18.

95. Büro für Städtebau Halle, *Halle-Neustadt*, 143.

96. See the illustrations of Halle-Neustadt in *Deutsche Architektur* 16, no. 4 (April 1967): 196–216, and *Architektur der DDR* 23 (June 1974): 326–73.

CHAPTER FIVE

1. These meetings were sponsored jointly by the government of the nation holding the event (in this case the government of Sweden), the UN, and international organizations such as the World Health Organization and the International Labor Organization.

2. United Nations Department of Economic and Social Affairs, *Planning of Metropolitan Areas and New Towns*, papers presented at Stockholm, September 14–30, 1961, and Moscow, August 24–September 7, 1964 (New York: United Nations, 1969), 1–2, 6.

3. Barbara Ward, "The Processes of World Urbanization," in ibid., 17.

4. United Nations, *Report of the United Nations Seminar on Physical Planning Techniques for the Construction of New Towns. Moscow, USSR, 2–22 September 1968* (New York: United Nations, 1971), 39–40.

5. International Association of New Towns, letter by Michel Boscher, chairman of the ad hoc Working Party, n.d. New Towns Institute Archive, INTA box 1–15, folder 8, Almere, Netherlands.

6. Gordon Cherry, *The Evolution of British Town Planning: A History of Town Planning in the United Kingdom during the 20th Century and of the Royal Town Planning Institute, 1914–74* (New York: Wiley, 1974), 161, as cited in H. W. E. Davies, "Continuity and Change: The Evolution of the British Planning System, 1947–97," *Town Planning Review* 69, no. 2 (1998): 142.

7. Quotation and discussion of the second generation of British new towns are found in Anthony Alexander, *Britain's New Towns: Garden Cities to Sustainable Communities* (London: Routledge, 2009), 41–43. Britain enacted its first national plan in 1965.

8. *Traffic in Towns: A Study of the Long Term Problems of Traffic in Urban Areas; Reports of the Steering Group and Working Group Appointed by the Minister of Transport* (London: Ministry of Transport, 1963). The abridged edition was published as Colin Buchanan, *Traffic in Towns* (Harmondsworth, UK: Penguin, 1964). *Traffic in Towns* was republished by Routledge in 2015 with an introduction by historian Simon Gunn.

9. See the excellent discussion of the Buchanan Report in Simon Gunn, "The Buchanan Report, Environment and the Problem of Traffic in 1960s Britain," *Twentieth Century British History* 22, no. 4 (2011): 521–42. Also useful is Gunn's "People and the Car: The Expansion of Automobility in Urban Britain, c. 1955–70," *Social History* 38, no. 2 (May 2013): 220–37.

10. G. A. Jellicoe, *Motopia: A Study in Evolution of the Urban Landscape* (New York: Frederick A. Praeger, 1961). The short newsreel *Glass City of the Future* was released by British Pathé in 1959 (1:04, Film ID 97.21; available at http://www.britishpathe.com/video/glass-city-of-the-future).

11. On the CES and support by the Ford Foundation, see Mark Clapson, *Anglo-American Crossroads: Urban Research and Planning in Britain, 1940–2010* (London: Bloomsbury, 2013), 41–48.

12. The Llewelyn-Davies family had its roots in Britain's intellectual elite and was close to Beatrice Webb, Bertrand Russell, and the Bloomsbury Group. Richard Llewelyn-Davies was also a member of the Apostles secret society at Cambridge University, whose membership included Leonard Woolf, Julian Bell, and John Maynard Keynes. For his biography, see Noel Annan, *Richard Llewelyn-Davies and the Architect's Dilemma* (Princeton, NJ: Institute for Advanced Study, 1987).

13. Marietta Tree worked closely with Llewelyn-Davies until his death in 1981. See Caroline Seebohm, *No Regrets: The Life of Marietta Tree* (New York: Simon and Schuster, 1997). On Llewelyn-Davies's transnational influence, see as well Clapson, *Anglo-American Crossroads*, chapter 6.

14. For the British-American discussions during this conference, see Jack A. Underhill, *General Observations on British New Town Planning: A Report Written following Participation in the United Nations Seminar on New Towns, London, June 4–19, 1973* (Washington, DC: Office of International Affairs, US Department of Housing and Urban Development, 1973).

15. Llewelyn-Davies Weeks and Partners, *A Study of Urban Development in an Area including Newbury, Swindon and Didcot* (London: Ministry of Housing and Local Government, 1966), 85.

16. Freeman, Fox, Wilbur Smith and Associates was the traffic consultant on the Washington New Town project. *Washington New Town Master Plan and Report* (London: Washington Development Corporation, 1966), 115–21.

17. Llewelyn-Davies, Weeks, Forestier-Walker, and Bor, and Ove Arup and Partners, *Motorways in the Urban Environment*, British Road Federation Report (London: British Road Federation, 1971).

18. Interview with Peter McGovern, member of the Lothian Regional Survey and Plan team and assistant chief planner, 1963–67, in *The New Towns Record, 1946–1996: 50 Years of UK New Town Development*, ed. Anthony Burton and Joyce Hartley, CD format (Glasgow: Planning Exchange, 1997).

19. Michael Alexander, dir., Films of Scotland and Livingston Development Corpora-

tion, producers, *Livingston—a Plan for Living* (Glasgow: Pelicula Films, 1976), celluloid, 20:13.

20. Ibid.

21. Derek Diamond, "New Towns in Their Regional Context," in *New Towns: The British Experience*, ed. Hazel Evans (London: Charles Knight, 1972), 63.

22. See Ministry of Housing and Local Government, *The South East Study, 1961–1981* (London: Her Majesty's Stationery Office, 1964), and the South East Joint Planning Team, *Strategic Plan for the South East* (London: Her Majesty's Stationery Office, 1970). See also Pat Blake, "South East: Strategy or Hypothesis?," *Town and Country Planning* 36 (March 1968): 179–82. The site for Milton Keynes encompassed the existing villages of Bletchley, Stony Stratford, and Wolverton.

23. Stephen Gardiner, "Villages for Thousands: Milton Keynes, a Look at the Plans," *Observer* (London), April 26, 1970.

24. Ibid.

25. "Professor Parkinson wants a 'British Brasília,'" *Daily Telegraph* clipping quoted in Guy Ortolano, "Planning the Urban Future in 1960s Britain," *Historical Journal* 54, no. 2 (2011): 478.

26. Robert Maxwell, "Milton Keynes, the Beautiful City," *Architectural Association Quarterly* 6, no. 3/4 (1974): 6.

27. This imagery is recounted in Richard Williams, *The Anxious City* (New York: Routlege, 2004), 58–59.

28. Reyner Banham et al., "Non-Plan: An Experiment in Freedom," *New Society* 20 (March 20, 1969): 435, 39, 42. On the impact of "Non-Plan," see Jonathan Hughes and Simon Sadler, eds., *Non-Plan: Essays on Freedom, Participation and Change in Modern Architecture and Urbanism* (Oxford: Architectural Press, 2000).

29. Richard Llewelyn-Davies, "Town Design," *Town Planning Review* 37, no. 3 (1966): 165–66.

30. Derek Walker, "New Towns," *Architectural Design* 111 (1995): 7. For the full array of images and descriptions of the designs for Milton Keynes, see Walker's *The Architecture and Planning of Milton Keynes* (London: Architectural Press, 1982).

31. The most well known of the Milton Keynes television advertisements (both quoted here) are the "Red Balloon" (Cogent Elliot Advertising Agency) and "Central Milton Keynes." They were produced for the Milton Keynes Development Corporation. Both are available on the MK Gallery website: http://www.mkgallery.org/information/about/milton_keynes/. Accessed June 24, 2015.

32. Clapson, *Anglo-American Crossroads*, 79. See also Clapson's *The Plan for Milton Keynes* (London: Routledge, 2013) as well as his *A Social History of Milton Keynes: Middle England/Edge City* (London: Frank Cass, 2004).

33. This phrase, for example, appears in the memoir by Michel Mottez, architect in charge of planning the new town of Evry: Michel Mottez, *Carnets de campagne: Evry, 1965–2007* (Paris: L'Harmattan, 2003).

34. On the political battles over the Schéma directeur and Delouvrier's role in de Gaulle's government, see Eric Lengereau, *L'Etat et l'architecture 1958–1981: Une politique publique?* (Paris: Picard, 2001), 119–21. On the oral history of the Schéma directeur and the new towns, see Alessandro Giacone, ed., *Les Grands Paris de Paul Delouvrier* (Paris: Descartes, 2010).

35. For the most recent and authoritative history of the French new towns, see Loïc Vadelorge, *Retour sur les villes nouvelles: Un Histoire urbaine du XXe Siècle* (Paris: CREAPHIS Editions, 2014). See also Kenny Cupers, *The Social Project: Housing Post-*

war France (Minneapolis: University of Minnesota Press, 2014). On the new town of Marne-la-Vallée, see Antoine Picon and Clément Orillard, *De la ville nouvelle à la ville durable, Marne-la-Vallée* (Paris: Parentheses, 2012). Another four new town projects were laid out in the provinces under the auspices of the French state's most powerful and authoritarian planning agency, the Délégation à l'aménagement du terriotoire et à l'action régionale (DATAR).

36. Jean-Louis Voileau, "Team 10 and Structuralism: Analogies and Discrepancies," in *Team 10: 1953–81, in Search of a Utopia of the Present*, ed. Max Risselada and Dirk van den Heuvel (Rotterdam: NAi, 2005), 283.

37. Lion Murard and François Fourquet, eds., *La Naissance des villes nouvelles* (Paris: Presses de l'école nationale des Ponts et Chaussées, 2004), 105–8. See also A. Suquet-Bonnaud, "Les Villes nouvelles en Grande-Bretagne: Progrès récents," *L'Information du bâtiment* 3 (March 1964): 2–14.

38. For one example among many, see Pierre Merlin, *Les Villes nouvelles françaises*, Notes et Etudes Documentaires, nos. 4286–88 (Paris: La Documentation française, May 3, 1976).

39. The architects of Nowa Huta in Poland, for example, studied the new town of Dunaújváros in Hungary through Merlin's research. Anna Biedrzycka, *Nowa Huta—architektura i twórcy miasta idealnego; Niezrealizowane projekty* (Crakow: Muzeum Historyczne Miasta Krakowa, 2006), 17. See also Pierre Merlin, "Aménagement du territoire et villes nouvelles un Hongrie," *Cahiers de l'IAURP* 20 (1970).

40. Claude Ponsard, *Histoire des théories économiques spatiales* (Paris: Armand Colin, 1958).

41. Pierre Merlin, *Géographie humaine* (Paris: Presses universitaires de France, 1997), 20–28, and his *Modèles d'urbanisation* (Paris: IAURP, August 1967). Philippe Pinchemel, preface to Peter Haggett, *L'Analyse spatiale en géographie humaine* (Paris: Armand Colin, 1973), 5–7.

42. Maurice Piau, "Programmation urbaine et immobilière," in *Vingt-cinq ans de villes nouvelles en France*, ed. Annick Jaouen, Chantal Guillet, and Jean-Eudes Roullier (Paris: Economica, 1989), 276.

43. Etablissement public d'aménagement de la ville nouvelle de Cergy-Pontoise, "Dossier: Inauguration du centre ville de Cergy-Pontoise" (Cergy-Pontoise, France: EPA, April 17, 1984), 8.

44. Ibid., 10.

45. Jean-Luc Bodiguel and Jean-Louis Faure, "Les Villes du Schéma directeur de la région parisienne," in "Les origines des villes nouvelles de la région parisienne (1919–1969)," *Cahiers de l'Institut d'histoire du temps présent* 17 (1990): 83.

46. An excellent discussion of this communal rebellion is given in Murard and Fourquet, *La Naissance des villes nouvelles*, 159–64.

47. Thierry Paquot, "Villes nouvelles une utopie de droite," *Espaces et sociétés* 22–23 (1977): 3–23.

48. Jean-Eudes Roullier, ed., *Cergy-Pontoise: "Inventer une ville."* Actes du colloque du 5 septembre 2002 (Lyon: CERTU, September 2002), 41. See also Gérard Monnier and Richard Klein, eds., *Les années ZUP: Architectures de la croissance, 1960–1973* (Paris: Picard, 2002).

49. Eric Rohmer, dir., *Une Ville nouvelle: Cergy-Pontoise*, Journal télévisé (ORTF, December 16, 1970).

50. On this point, see Raewyn Connell, *Southern Theory: The Global Dynamics of Knowledge in Social Science* (Cambridge, MA: Polity, 2007), 56–68.

51. L. R. Vagale, "Physical Planning and Design Principles in the Development of New and Satellite Towns," in United Nations Economic Commission for Asia and the Far East, *Planning for Urban and Regional Development in Asia and the Far East*, Nagoya, Japan, October 10–20, 1966 (New York: United Nations, 1971), 101. Among his many publications, see *A Critical Appraisal of New Towns in Developing Countries: Policy Framework for Nigeria* (Ibadan, Nigeria: Polytechnic, 1977), and *Structure of Metropolitan Regions in India: Planning Problems and Prospects* (New Delhi: L. R. Vagale, 1964).

52. Vagale, "Physical Planning and Design Principles in the Development of New and Satellite Towns," 115.

53. See chapter 7, "A Different Modernity," in Gyan Prakash, *Another Reason: Science and the Imagination of Modern India* (Princeton, NJ: Princeton University Press, 1999).

54. "Our Cities: A Symposium on the Need for a Rational Urban Development," *Seminar*, no. 79 (March 1966): 17, as quoted in Annapurna Shaw, *The Making of Navi Mumbai* (Hyderabad: Orient Longman, 2004), 73.

55. N. S. Lamba, "Emerging Capitals and New Towns," *Journal of Institute of Town Planners, India* 67 (1971): 25, 28, 32.

56. Ved Prakash, *New Towns in India*, Monograph and Occasional Papers Series, vol. 8 (Durham, NC: Program in Comparative Studies on Southern Asia, Duke University, 1969).

57. Ibid., 16.

58. "Bombay: Planning and Dreaming," special issue, *MARG* 18, no. 3 (June 1965): 35, 52.

59. City and Industrial Development Corporation of Maharashtra Ltd. (CIDCO), "Navi Mumbai," www.cidco.maharashtra.gov.in. Accessed June 25, 2015.

60. Gyan Prakash, *Mumbai Fables* (Princeton, NJ: Princeton University Press, 2010), 267.

61. See for example the environmental descriptions in "Twin City on the Sea" and "Alternative Plan," *Times of India* (Mumbai), March 29, 1964.

62. Charles Correa, interviews with Alain Jacquemin, quoted in Alain R. A. Jacquemin, *Urban Development and New Towns in the Third World: Lessons from the New Bombay Experience* (Aldershot, UK: Ashgate, 1999), 246.

63. Charles Correa, "Transfers and Transformations," in *Design for High-Intensity Development*, ed. Margaret Bentley Sevcenko (Cambridge, MA: Aga Khan Program for Islamic Architecture, 1986), 11. See also Correa's *The New Landscape* (Bombay: Book Society of India, 1985) as well as his article "Housing: Space as a Resource," *Times of India* (Mumbai), June 22, 1975.

64. Charles Correa, "New Bombay: Self-Help City," speech delivered at the United Nations Conference on Population, Resources and the Environment, Stockholm, 1974; *Architectural Design* 44, no. 1 (1974): 48–51.

65. On the multiple development projects at Mumbai, see Florian Urban, *Tower and Slab: Histories of Global Mass Housing* (New York: Routledge, 2012), chapter 6. See also the collection of articles in R. N. Sharma and K. Sita, eds., *Issues in Urban Development: A Case of Navi Mumbai* (Jaipur: Rawat Publications, 2001).

66. Comment by B. K. Boman-Behram in "Mayor Not in Favor of Correa's Modified Plan," *Times of India* (Mumbai), April 2, 1975.

67. Melvin M. Webber, "The Post-City Age," *Daedalus* 97, no. 4 (1968): 1107.

68. Clarence Stein, quoted in Tom Vanderbilt, *Survival City: Adventures among the Ruins of Atomic America* (Chicago: University of Chicago Press, 2010), 76. See Greg Hise, "The Airplane and the Garden City: Regional Transformations during World War II," in *World War II and the American Dream*, ed. David Albrecht (Washington,

DC: National Building Museum; Cambridge, MA: MIT Press, 1995), 150. See also the excellent analysis in Kristin Larson, "Cities to Come: Clarence Stein's Postwar Regionalism," *Journal of Planning History* 4, no. 1 (2005): 33–51.

69. James A. Clapp, *New Towns and Urban Policy: Planning Metropolitan Growth* (New York: Dunellen, 1971), xi. See as well Steven Conn, *Americans against the City: Anti-Urbanism in the Twentieth Century* (Oxford: Oxford University Press, 2014).

70. Carl Feiss, "New Towns for America," *AIA Journal* 33, no. 1 (1960): 86–88.

71. Ada Louise Huxtable, "Western Europe Is Found to Lead U.S. in Community Planning," *New York Times*, November 22, 1965, and "Architecture: Virtues of Planned Community," *New York Times*, October 24, 1963.

72. Albert Mayer, *The Urgent Future* (New York: McGraw-Hill, 1967), 3–12, 15. See also Andrew Friedman, "The Global Postcolonial Moment and the American New Towns: India, Reston, Dodoma," *Journal of Urban History* 38, no. 3 (2012): 553–76. Mayer was also working on the new town of Reston in Virginia, where he applied many of the insights gained from working for over a decade with the Nehru government.

73. "By 1976 What City Pattern?," special issue, *Architectural Forum* 105 (September 1956): 103. This issue was edited by Catherine Bauer and Victor Gruen.

74. Catherine Bauer, "First Job: Control New-City Sprawl," in "By 1976 What City Pattern?," special issue, *Architectural Forum* 105 (September 1956): 105.

75. Ibid., 111–12.

76. Donald Canty, ed., *The New City* (New York: Published for Urban America by Frederick A. Praeger, 1969), 19, 31. These introductory remarks to the volume were probably written by Canty.

77. Webber, "The Post-City Age," 1109.

78. See Fredric Jameson, *Archaeologies of the Future: The Desire Called Utopia and Other Science Fictions* (London: Verso, 2007), 56.

79. US Department of Agriculture, *National Growth and Its Distribution*, Symposium on Communities of Tomorrow, December 11–12, 1967 (Washington, DC: US Government Printing Office, 1968), 26.

80. Ibid., 60.

81. Hubert Humphrey, quoted in Walter Vivrett, "Planning for People: Minnesota Experimental City," in *New Community Development: Planning Process, Implementation, and Emerging Social Concerns*, ed. Shirley F. Weiss, Edward J. Kaiser, and Raymond J. Burby; proceedings, New Towns Research Seminar, February 1971 (Chapel Hill: Center for Urban and Regional Studies, University of North Carolina, 1971), 251, 54–55.

82. *City* magazine was supported by Urban America Inc., which had been formed from an early coalition for urban renewal by James Rouse and like-minded businessmen and housing leaders. Originally called the American Council to Improve Our Neighborhoods, its name was changed to Action Inc.; then, after a merger with Urban Coalition, it became known as Urban America Inc. Much of the organization's work was sponsored by Ford Foundation grants.

83. Canty, *The New City*, 19.

84. William E. Finley, "A Fresh Start," in Canty, *The New City*, 163.

85. Ibid., 171.

86. See for example the cover of *Time* magazine addressing the "population explosion," January 11, 1960.

87. Excerpts, "Urban Renewal in America, 1950–1970: A Symposium," *Urban Design Quarterly* 85 (1972): 31.

88. CBS News Special Report, *The Cities*, part 3: "To Build the Future," June 26, 1968.

89. ABC Television Network, *Man and His Universe: Cosmopolis; The Big City*, January 13, 1969.

90. See for example Albert J. Robinson, *Economics and New Towns: A Comparative Study of the United States, the United Kingdom, and Australia* (New York: Praeger, 1975). Also see Gideon Golany, ed., *Innovations for Future Cities* (New York: Praeger, 1976), and Gideon Golany and Daniel Walden, eds., *The Contemporary New Communities Movement in the United States* (Urbana: University of Illinois Press, 1974).

91. Harvey S. Perloff and Neil C. Sandberg, eds., *New Towns: Why—and for Whom?*, Praeger Special Studies Program (New York: Praeger, 1972). Perloff was dean of UCLA's Graduate School of Urban Planning and eventually chair of the Committee on National Urban Policy. The UCLA conference was supported by the American Jewish Committee.

92. See Jack A. Underhill, Paul Brace, and James Rubenstein, "French National Urban Policy and the Paris Region New Towns: The Search for Community" (Washington, DC: Office of International Affairs, US Department of Housing and Urban Development, 1980). See also Underhill's "Soviet New Towns: Housing and National Urban Growth Policy" (Washington, DC: Office of International Affairs, US Department of Housing and Urban Development, 1976).

93. Irving Lewis Allen, ed., *New Towns and the Suburban Dream: Ideology and Utopia in Planning and Development* (Port Washington, NY: Kennikat Press, 1977), 114.

94. Task Force on New Towns, *Summary Report of the Task Force on New Towns* (Washington, DC: US Department of Housing and Urban Development, October 16, 1967), 1.

95. U.S./U.S.S.R. New Towns Working Group, *Planning New Towns: National Reports of the U.S. and U.S.S.R.* (Washington, DC: Office of International Affairs, US Department of Housing and Urban Development, March 1981), 24.

96. Ada Louise Huxtable, "First Light of New Town Era Is on Horizon," *New York Times*, February 17, 1964.

97. US Congress, Committee of Banking and Currency, Subcommittee on Housing, *Oversight Hearings on HUD New Communities Program*, May 31, 1973, 122.

98. Ibid., 231. For the description of New Franconia, see Gerald C. Finn, "New Franconia, New Hampshire: New Community Planning and Zoning Process," in *Strategies for New Community Development in the United States*, ed. Gideon Golany (Stroudsburg, PA: Dowden, Hutchinson and Ross, 1975), 94, 96.

99. For an analysis of federal policy, see chapter 6 in Carol Corden, *Planned Cities: New Towns in Britain and America* (Beverly Hills, CA: Sage Publications, 1977). More recently, see Roger Biles, "New Towns for the Great Society: A Case Study of Politics and Planning," *Planning Perspectives* 13 (1998): 113–32.

100. On the demise of the federal new towns program, see Nicholas Dagen Bloom, "The Federal Icarus: The Public Rejection of 1970s National Suburban Planning," *Journal of Urban History* 28, no. 1 (2001): 55–71.

101. Office of Policy Studies, "An Evaluation of the Federal New Communities Program" (Washington, DC: US Department of Housing and Urban Development, 1985), iii–v.

102. Edward Eichler and Marshall Kaplan, *The Community Builders* (Berkeley: University of California Press, 1967), chapter 4.

103. William Alonso, "Urban Growth in California: New Towns and Other Policy Alter-

natives" (paper presented at the Symposium on California Population Problems and State Policy, University of California, Davis, May 1971), 2–3.

104. Ann Forsyth, *Reforming Suburbia: The Planned Communities of Irvine, Columbia, and the Woodlands* (Berkeley: University of California Press, 2005), 108.

105. "Ford Grant Aids Urban Renewal," *New York Times*, October 19, 1958.

106. James Rouse, quoted in US Department of Agriculture, *National Growth and Its Distribution*, 67.

107. James Rouse, quoted in "Pleasure Domes with Parking," *Time*, October 15, 1956.

108. James Rouse, "Utopia: Limited or Unlimited," speech given at the National Housing Conference, Inter-Religious Coalition for Housing, Interchurch Center, New York City, November 14, 1979. This quote is recounted in the best biography of Rouse, Nicholas Dagen Bloom, *Merchant of Illusion: James Rouse, American Salesman of the Businessman's Utopia* (Columbus: Ohio State University Press, 2004), 27.

109. This statistic is given in Forsyth, *Reforming Suburbia*, 10. Columbia's successes and failures are particularly well documented in Gurney Breckenfeld, *Columbia and the New Cities* (New York: Ives Washburn, 1971).

110. Donald N. Michael, "The Planning Workgroup," in *Creating a New City: Columbia, Maryland*, ed. Robert Tennenbaum (Columbia, MD: Perry, 1996), 11.

111. James Rouse, quoted in Thomas Goldwasser, "New-Town Builder," *New York Times*, August 28, 1977.

112. James Rouse, "Living in a New Town," in *International New Towns Congress, Tehran, Iran, 9–15 December 1977*, ed. INTA [l'Association Internationale des Villes Nouvelles], 138–40. Carton 16, New Towns Institute Archive, Almere, Netherlands.

113. James Rouse, quoted in US Department of Agriculture, *National Growth and Its Distribution*, 67.

114. Tennenbaum, *Creating a New City*, 3.

115. Community Research and Development Inc., "Columbia: A New Town for Howard County; A Presentation for the Officers and Citizens of Howard County," Baltimore, November 11, 1964.

116. *Architectural Record* 160 (March 1972): 113–21.

117. Donald Canty, "A New Approach to New Town Planning," *Architectural Forum* 121 (1964): 194.

118. Interview 0131, quoted in Forsyth, *Reforming Suburbia*, 140.

CHAPTER SIX

1. Constant Nieuwenhuys, "Another City for Another Life," *Internationale Situationniste* #3 (December 1959): 37–40.

2. Peter Smithson, "Capital Cities," *Architectural Design* 28 (November 1958): 437–41; quoted in Jean-Louis Voileau, "Team 10 and Structuralism: Analogies and Discrepancies," in *Team 10: 1953–81, in Search of a Utopia of the Present*, ed. Max Risselada and Dirk van den Heuvel (Rotterdam: NAi, 2005), 281.

3. See Hadas A. Steiner, *Beyond Archigram: The Structure of Circulation* (New York: Routledge, 2009). An excellent discussion of the avant-garde can also be found in Jonathan Hughes and Simon Sadler, eds., *Non-Plan: Essays on Freedom, Participation and Change in Modern Architecture and Urbanism* (Oxford: Architectural Press, 2000). See also Neil Spiller, *Visionary Architecture: Blueprints of the Modern Imagination* (New York: Thames and Hudson, 2006), and Dubravka Djurié and Misko Suvakovié, eds., *Impossible Histories: Historical Avant-Gardes, Neo-Avant-Gardes, and Post-Avant-Gardes in Yugoslavia, 1918–1991* (Cambridge, MA: MIT Press, 2015).

4. Manfredo Tafuri, *Architecture and Utopia: Design and Capitalist Development* (Boston, MA: MIT Press, 1996), 114.

5. Reyner Banham, *Megastructure: Urban Futures of the Recent Past* (London: Thames and Hudson, 1976), 7–9. See also Louis Wilkins, "Dinosaurs of the Modern Movement," *Building Design*, no. 327 (December 1976): 14–15.

6. Paolo Soleri and Jules Noel Wright, "Utopie e o Revoluzione: Utopia and/or Revolution," *Perspecta* 13/14 (1971): 283. See as well the articles in Sarah J. Montross, ed., *Past Futures: Science Fiction, Space Travel, and Postwar Art in America* (Cambridge, MA: MIT Press, 2015).

7. The comparison with Doxiadis was reported in two articles about the exhibit of Soleri's work at the Corcoran Gallery in Washington, DC: Ada Louise Huxtable, "Profit in the Desert," *New York Times*, March 15, 1970, and again in "Soleri Thinks Very Big . . . ," July 26, 1970.

8. Paolo Soleri, *Arcology: The City in the Image of Man* (Cambridge, MA: MIT Press, 1969), XXX. A fourth edition was released by Cosanti Press in 2006.

9. The best source on this forgotten utopian experiment is the doctoral thesis of Todd Wildermuth, "Yesterday's City of Tomorrow: The Minnesota Experimental City and Green Urbanism" (PhD diss., University of Illinois, 2008). See also James R. Prescott, "The Planning for Experimental City," *Land Economics* 46, no. 1 (1970): 68–75, as well as James A. Alcott (who was an MXC general manager), "Planning of an Innovative Free-Standing City: The Case of Minnesota Experimental City," in *Innovations for Future Cities*, ed. Gideon Golany (New York: Praeger, 1976).

10. See Robert B. Semple, "Experimental City Mapped in Midwest," *New York Times*, February 6, 1967.

11. Athelstan Spilhaus, "Technology, Living Cities, and Human Environment," *American Scientist* 57, no. 1 (1969): 24, 25.

12. See "The Experimental City," *Daedalus* 96, no. 4 (1967): 1129–41.

13. On this point, see Paul N. Edwards, *The Closed World: Computers and the Politics of Discourse in Cold War America* (Cambridge, MA: MIT Press, 1996), 14.

14. Andrew Pickering, *The Cybernetic Brain: Sketches of Another Future* (Chicago: University of Chicago Press, 2010), especially part 2.

15. Reinhald Martin makes this point in "Computer Architectures: Saarinen's Patterns, IBM's Brains," in *Anxious Modernisms: Experimentation in Postwar Architectural Culture*, ed. Sarah Williams Goldhagen and Réjean Legault (Montreal: Canadian Center for Architecture; Cambridge, MA: MIT Press, 2000), 153–54. See also Felicity D. Scott, *Architecture or Techno-utopia: Politics after Modernism* (Cambridge, MA: MIT Press, 2010).

16. A. C. Brothers, "Weapons Systems," and M. E. Drummond, "Computers," in "The Science Side," ed. Reyner Banham, special issue, *Architectural Review* 127, no. 757 (March 1960): 188–89. This series of articles and Banham's response are described in detail in Anthony Vidler, *Histories of the Immediate Present* (Cambridge, MA: MIT Press, 2008), 128–33.

17. Richard Llewelyn-Davies, "Human Sciences," in Banham, "The Science Side," 189.

18. Team 10 architects Alison and Peter Smithson, "An Alternative to the Garden City Idea," *Architectural Design* 26 (July 1956): 229–31. Quote is from Theo Grospor, "Thoughts in Progress: The New Brutalism," *Architectural Design* 27, no. 4 (April 1957): 113.

19. On the plans for Tokyo, see Florian Urban, "Case Study III: Mega-Tokyo—zen versus High Tech," in *Megastructure Reloaded: Visionäre Stadtentwürfe der Sechzigerjahre refle-*

ktiert von zeitgenössischen Künstlern = Visionary Architecture and Urban Design of the Sixties Reflected by Contemporary Artists, ed. Sabrina van der Ley and Markus Richter (Ostfildern, Germany: Hatje Cantz Verlag, 2008), 94–96. On the Metabolists, see also Zhongjie Lin, *Kenzo Tange and the Metabolist Movement: Urban Utopias of Modern Japan* (New York: Routledge, 2010).

20. Reyner Banham, *Megastructure: Urban Futures of the Recent Past* (London: Thames and Hudson, 1976).

21. Walter Schwagenscheidt, *Die Nordweststadt: Idee und Gestaltung = The Nordweststadt: Conception and Design* (Stuttgart: Karl Krämer Verlag, 1964). See also Frank Eckardt, "Germany: Neighbourhood Centres, A Complex Issue," *Built Environment* 32, no. 1 (2006): 53–72.

22. "Stadtbau: Frankfurt; Asche und Trabant," *Der Spiegel* 25 (1962): 72. See also Hans Kampffmeyer, *Die Nordweststadt in Frankfurt am Main* (Frankfurt am Main: Europäische Verlagsanstalt, 1968).

23. Schwagenscheidt, *Die Nordweststadt: Idee und Gestaltung*, 89–90. See also Erich Hanke, "Die Nordweststadt Frankfurt am Main," *Anthos: Zeitschrift für Landschaftsarchitektur = Une revue pour le paysage* 7 (1968): 10–15.

24. See "Nordweststadt Centre, Apel & Beckert," *Architectural Design* 36 (1966): 86–88, as well as "Nordwestzentrum Frankfurt/Main," *Das Werk* 57 (1970): 574–77.

25. Cumbernauld was designated a new town in 1956. Robin Crichton, dir., *Cumbernauld: Town for Tomorrow* (Edinburgh: Edinburgh Films Production for Films of Scotland / Cumbernauld Development Corporation, 1970), video, 25 min.

26. News report, "Princess Margaret in Cumbernauld, Scotland," May 18, 1967; Film ID 2031.04, British Pathé Cinema.

27. Wolf Von Eckardt, "The Case for Building 350 New Towns," *Harper's Magazine*, December 1965, 93.

28. This point is made by John Gold, "The Making of a Megastructure: Architectural Modernism, Town Planning and Cumbernauld's Central Area, 1955–75," *Planning Perspectives* 21 (2006): 112.

29. Reyner Banham, quoted in Nigel Whiteley, *Rayner Banham: Historian of the Immediate Future* (Cambridge, MA: MIT Press, 2002), 287.

30. London County Council, *The Planning of a New Town: Data and Design Based on a Study for a New Town of 100,000 at Hook, Hampshire* (London: Greater London Council, 1965). See also "New Town Development: The Hook Study," *RIBA Journal* 69 (1962): 45–77.

31. David Gosling and Barry Maitland, *Design and Planning of Retail Systems* (London: Architectural Press, 1976), 80.

32. Hugh Wilson and Lewis Womersley, *Irvine New Town: Final Report on Planning Proposals* (Edinburgh: Scottish Development Department, 1967), 76.

33. Murray Grigor, dir., *Cumbernauld Hit* (Cumbernauld, Scotland: Cumbernauld Development Corporation, 1977), video, 45:30.

34. Anette Baldauf and Katharina Weingartner, dirs., *The Gruen Effect: Victor Gruen and the Shopping Mall* (Vienna: Pooldoks Filmproduktion KG, 2009), video, 54 min. Alex Wall, *Victor Gruen, from Urban Shop to New City* (Barcelona: Actar, 2005), has undoubtedly the best description and images of Gruen's projects. See also M. Jeffrey Hardwick, *Mall Maker: Victor Gruen, Architect of an American Dream* (Philadelphia: University of Pennsylvania Press, 2004).

35. Victor Gruen, "How to Handle This Chaos of Congestion, This Anarchy of Scatteration," *Architectural Forum* 105 (September 1956): 130–35. Quote is from page 132.

36. Ibid., 132, 135.
37. Victor Gruen, *The Heart of Our Cities: The Urban Crisis; Diagnosis and Cure*, 2nd ed. (London: Thames and Hudson, 1965), 292.
38. "A Break-Through for Two-Level Shopping Centers," *Architectural Forum* 105 (December 1956): 114–26. Quote is from page 117. For an interesting Cold War perspective on Southdale Mall, see Timothy Mennel, "Victor Gruen and the Construction of Cold War Utopias," *Journal of Planning History* 3, no. 2 (2004): 116–50.
39. James Rouse, "Must Shopping Centers Be Inhuman?," *Architectural Forum* 116 (June 1962):104–7. Quotation is from page 7.
40. Stephanie Dyer, "Designing 'Community' in the Cherry Hill Mall: The Social Production of a Consumer Space," *Perspectives in Vernacular Architecture* 9 (2003): 269.
41. Gruen, *The Heart of Our Cities*, 286.
42. See the excellent analysis in Ali Madanipour, *Teheran: The Making of a Metropolis* (New York: John Wiley and Sons, 1998). See also Talinn Grigor, *Building Iran: Modernism, Architecture, and National Heritage under the Pahlavi Monarchs* (New York: Periscope, distributed by Prestel, 2009). Also useful is Hooshang Amirahmadi and Ali Kiafar, "The Transformation of Tehran from a Garrison Town to a Primate City: A Tale of Rapid Growth and Uneven Development," in *Urban Development in the Muslim World*, ed. Hooshang Amirahmadi and Salah S. El-Shakhs (New Brunswick, NJ: Rutgers University Press, 1993), 109–36.
43. The new towns were developed at already existing settlement areas: Vardarvard, Latmar Shomali, Kan, Amirabad, Shemiran, Abbasabad, and Tehran Pars along an east–west trajectory in the northern suburbs of Tehran; Doshantepeh and Shahr-e-Rey in the south.
44. Bernard Hourcade, "Urbanisme et crise urbaine sous Mohammad-Reza Pahlavi," in *Téhéran, capitale bicentenaire*, ed. Chahryar Adle and Bernard Hourcade (Paris-Teheran: Institut français de recherche en Iran, 1992), 220.
45. Michel Ragon, "Pour la première fois tous les plans proposés par les plus grands architectes: Le Corbusier, Michel Holley, Le Coeur et Serge Menil," *Arts*, February 14, 1962, 14–15. See also his "Aventure de la cité futur," *Urbanisme*, no. 92 (1966): 79–81.
46. Tafuri, *Architecture and Utopia*, 139. See also Lary Busbea, *Typologies: The Urban Utopia in France, 1960–1970* (Cambridge, MA: MIT Press, 2007).
47. Craig Buckley and Jean-Louis Violeau, *Utopie: Texts and Projects, 1967–1978*, trans. Jean-Marie Clarke (Los Angeles: Semiotext(e); Cambridge, MA: MIT Press, 2011), 102.
48. Ibid., 104.
49. "Villes nouvelles," special issue, *L'Architecture d'aujourd'hui* 146 (October–November 1969).
50. See the conclusion in Thierry Paquot, *Utopies et utopistes* (Paris: Editions La Découverte, 2007).
51. "Villes nouvelles, l-région parisienne," *Techniques et architecture* 301 (1974): 56. The second quote is from the well-known 1993 symposium held at the new town of Cergy-Pontoise, *L'urbanisme de dalles: Cotinuités et ruptures*; Actes du colloque de Cergy-Pontoise, 16–17 September 1993 (Paris: Presses des Ponts et chaussées, 1995), 2.
52. See for example *Techniques et architecture* 5 (October 1958): 105, 108.
53. Shadrach Woods, "Urban Environment—the Search for a System," typed essay (1962); box 8, folder 6, Shadrach Woods Archive, Avery Library, Columbia University.

54. Shadrach Woods, "Le Mirail, a New Quarter for the City of Toulouse," *Washington University Law Review*, no. 1 (January 1965): 13. See Liane Lefaivre and Alexander Tzonis, "Shadrach Woods, Post War Circulatory Rigourism, Mobility and the 'Stem' and the 'Web,'" in *Team 10: Between Modernity and the Everyday*, ed. Tom Avermaete (Delft, Netherlands: Delft University, 2003), 209.

55. Georges Candilis, *Bâtir la vie: Un architect témoin de son temps* (Paris: Stock, 1977), 256, 62, as well as "Entretiens avec Georges Candilis," *SIA—Ingénieurs et architectes suisses* 26 (December 1994): 497.

56. Michel Mottez, *Carnets de campagne: Evry, 1965–2007* (Paris: L'Harmattan, 2003), 63.

57. Victor Gruen, "Les équipements commerciaux dans les agglomérations urbaines et dans les villes nouvelles" (Victor Gruen Associates, n.d.), Médiathèque IAU Ile-de-France, Paris. On the reaction to Gruen's designs, see Alexis Korganow, "L'interaction ville-équipement en ville nouvelle: Réception et adaptation de la formule de l'équipement socioculturel intégré," in *Programme interministériel d'histoire et d'évaluation des villes nouvelles françaises: Atelier 4*, Paris, April 6–7, 2005 (Paris: Ministère de la Culture et de la Communication, 2007), 10–21.

58. André Darmagnac, François Desbruyères, and Michel Mottez, *Créer un centre ville: Evry* (Paris: Moniteur, 1980), 84–85. See also Corin Hughes-Stanton, "Closed Environment for Living Space," *Design* 245(May 1969): 40–49.

59. "Evry, centre urbain nouveau et ville nouvelle," *Cahiers de l'IAURP*, no. 15 (May 1969): 48, 58, and Etablissement public d'aménagement, *"Créer à Evry un centre ville attachant . . . ,"* unpublished brochure, 1978, Médiathèque IAU Ile-de-France, Paris.

60. Pierre Merlin, *Les Villes nouvelles françaises*, Notes et Etudes Documentaires, nos. 4286–88 (Paris: La Documentation française, May 3, 1976), 55.

61. Banham, *Megastructure*, 74.

62. "The New Brutalism," *Architectural Review* 118, no. 708 (December 1955): 361.

63. Henri Lefebvre, "Notes on the New Town (April 1960)," in *Introduction to Modernity: Twelve Preludes*, trans. John Moore (London: Verso, 1995), 124.

CONCLUSION

1. William Alonso, "The Mirage of New Towns," *Public Interest* 19 (Spring 1970): 3.

2. Zygmunt Bauman, *Liquid Modernity* (Cambridge: Polity Press, 2000), chapter 2.

3. Françoise Choay, "Utopia and the Anthropological Status of Built Space," in *Exit Utopia: Architectural Provocations 1956–76*, ed. Van Schaik and Otakar Mácel (New York: Prestel, 2005), 96, 99.

4. See Dona J. Stewart, "Cities in the Desert: The Egyptian New-Town Program," *Annals of the Association of American Geographers* 86, no. 3 (1996): 459–80.

5. Steve Inskeep, *Instant City: Life and Death in Karachi* (New York: Penguin, 2011), 216–17.

6. Matthieu Aikins, "Kabubble: Counting Down to Economic Collapse in the Afghan Capital," *Harper's Magazine*, February 2013, 53.

7. Ellen Barry, "On Black Sea Swamp, Big Plans for Instant City," *New York Times*, April 22, 2012. See also Giorgi Lomsadze, "Georgia: Whither the City of Lazika," Eurasianet.org, October 3, 2012, http://www.eurasianet.org/node/65995.

8. Northrop Frye, "Varieties of Literary Utopias," in *Utopias and Utopian Thought*, ed. Frank Manuel (Boston: Houghton Mifflin, 1966), 25–49.

9. See for example Ernst Bloch, *The Spirit of Utopia*, trans. Anthony A. Nassar (Stanford, CA: Stanford University Press, 2000) and his *The Principle of Hope*, trans. Neville Plaice, Stephen Plaice, and Paul Knight, vol. 1 (Cambridge, MA: MIT Press, 1995).

10. See the recent study by Rachel Keeton, *Rising in the East: Contemporary New Towns in Asia* (Amsterdam: International New Town Institute; Amsterdam: SUN Publishers, 2011).

11. Harry den Hartog, *Shanghai New Towns: Seaching for Community and Identity in a Sprawling Metropolis* (Rotterdam: 010 Publishers, 2010), 30–36. See also Michael Hulshof and Daan Roggeveen, *How the City Moved to Mr. Sun: China's New Mega-cities* (Amsterdam: Martien de Vletter, SUN Publishers, 2010).

12. On The Woodlands, see George Morgan and John King, *The Woodlands: New Community Development* (College Station: Texas A&M University Press, 1987).

13. See for example Robert Reid, "Supersized," *Civil Engineering* 83, no. 5 (May 2013): 46–55.

14. See Anthony Townsend, *Smart Cities: Big Data, Civic Hackers, and the Quest for a New Utopia* (New York: W. W. Norton, 2013), as well as the publication by the editors of *Scientific American*, *Designing the Urban Future: Smart Cities* (New York: Scientific American, 2014).

15. IBM Smarter Cities at http://www.ibm.com/smarterplanet/us/en/smarter_cities. Accessed July 27, 2015.

SELECTED BIBLIOGRAPHY

The publications and resources about new towns are massive. The selection of secondary sources provided below is an introduction to the new town movement and to its study as utopia. Hundreds of documents exist on individual new towns, and they are foundational as an information base. A few bibliographic collections exist, although they deserve updating. Gideon Golany compiled *New Towns Planning and Development: A World-Wide Bibliography* (Bethesda, MD: Washington Land Institute, 1973). Pierre Merlin published *Bibliographie sur les villes nouvelles françaises et étrangères* (Vincennes, France: Presses Universitaires de Vincennes, 1989). Comprehensive bibliographic collections exist for national new town programs such as those in France, Israel, and Canada. The most complete international compilations and archival collections are at the Centre de documentation de l'urbanisme at La Défense outside Paris, and at the International New Town Institute in Almere, the Netherlands.

Abercrombie, Patrick. *Greater London Plan 1944*. London: Minister of Town and Country Planning, 1945.

Abercrombie, Patrick, and R. Matthew. *The Clyde Valley Regional Plan*. Edinburgh: His Majesty's Stationery Office, 1949.

Adle, Chahryar, and Bernard Hourcade, eds. *Téhéran, capitale bicentenaire*. Paris: Institut français de recherche en Iran, 1992.

Albrecht, David, ed. *World War II and the American Dream*. Washington, DC: National Building Museum; Cambridge, MA: MIT Press, 1995.

Alexander, Anthony. *Britain's New Towns: Garden Cities to Sustainable Communities*. London: Routledge, 2009.

Allen, Irving Lewis, ed. *New Towns and the Suburban Dream: Ideology and Utopia in Planning and Development*. Port Washington, NY: Kennikat Press, 1977.

Almandoz, Arturo. *Modernization, Urbanization and Development in Latin America, 1900s–2000s*. New York: Routledge, 2015.

Alonso, William. *Location and Land Use: Toward a General Theory of Land Rent*. Cambridge, MA: Harvard University Press, 1964.

——. "Urban Growth in California: New Towns and Other Policy Alternatives." Paper presented at the Symposium on California Population Problems and State Policy, University of California, Davis, May 1971.

Aman, Anders. *Architecture and Ideology in Eastern Europe during the Stalin Era: An Aspect of*

Cold War History. New York: Architectural History Foundation; Cambridge, MA: MIT Press, 1992.

Amirahmadi, Hooshang, and Salah S. El-Shakhs, eds. *Urban Development in the Muslim World*. New Brunswick, NJ: Rutgers University Press, 1993.

Annan, Noel. *Richard Llewelyn-Davies and the Architect's Dilemma*. Princeton, NJ: Institute for Advanced Study, 1987.

Apter, Andrew, *The Pan-African Nation: Oil and the Spectacle of Culture in Nigeria*. Chicago, Ill.: University of Chicago Press, 2005.

Arieh Sharon—Architect website. http://www.ariehsharon.org. Accessed July 27, 2015.

Askari, Hossein. *Collaborative Colonialism: The Political Economy of Oil in the Persian Gulf*. New York: Palgrave Macmillan, 2013.

Atkinson, Harriet. *The Festival of Britain: A Land and Its People*. London: I. B. Tauris, 2012.

Avermaete, Tom, ed. *Team 10: Between Modernity and the Everyday*. Delft, Netherlands: Delft University, 2003.

Balina, Marina, and Evgeny Dobrenko, eds. *Petrified Utopia: Happiness Soviet Style*. London: Anthem, 2011.

Bamberg, James. *British Petroleum and Global Oil, 1950–1975: The Challenge of Nationalism*. Cambridge: Cambridge University Press, 2000.

Banham, Reyner. *Megastructure: Urban Futures of the Recent Past*. London: Thames and Hudson, 1976.

Bannerman, J. Yedu. *The Cry for Justice in Tema (Ghana)*. Tema, Ghana: Tema Industrial Mission, 1973.

Bardiar, Nilendra. *A New Delhi—Urban, Cultural, Economic and Social Transformation of the City 1947–65*. CreateSpace Independent Publishing Platform, 2014. https://www .createspace.com.

Bartetzky, Arnold, and Marc Schalenberg, eds. *Urban Planning and the Pursuit of Happiness: European Variations on a Universal Theme (18th–21st Centuries)*. Berlin: Jovis, 2009.

Barth, Holger, ed. *Grammatik sozialistischer Architekturen: Lesarten historischer Städtebauforschung zur DDR*. Berlin: Dietrich Reimer Verlag, 2001.

Batty, Michael. *Cities and Complexity: Understanding Cities with Cellular Automata, Agent-Based Models and Fractals*. Cambridge, MA: MIT Press, 2005.

Bauman, Zygmunt. *Liquid Modernity*. Cambridge: Polity Press, 2000.

Bender, Thomas, and Alev Cinar, eds. *Urban Imaginaries: Locating the Modern City*. Minneapolis: University of Minnesota Press, 2007.

Bernhardt, Christoph, and Heinz Reif, eds. *Sozialistische Städte zwischen Herrschaft und Selbstbehauptung: Kommunalpolitik, Stadtplanung und Alltag in der DDR*. Stuttgart: Franz Steiner, 2009.

Bernhardt, Christoph, and Thomas Wolfes, eds. *Schönheit und Typenprojektierung: Der DDR-Städtebau im internationalen Kontext*. Erkner, Germany: Institut für Regionalentwicklung und Strukturplanung, 2005.

Biedrzycka, Anna. *Nowa Huta—architektura i twórcy miasta idealnego; Niezrealizowane projekty*. Krakow: Muzeum Historyczne Miasta Krakowa, 2006.

Bloch, Ernst. *The Principle of Hope*. Translated by Neville Plaice, Stephen Plaice, and Paul Knight. Vol. 1. Cambridge, MA: MIT Press, 1995.

———. *The Spirit of Utopia*. Translated by Anthony A. Nassar. Stanford, CA: Stanford University Press, 2000.

———. *The Utopian Function of Art and Literature*. Translated by Jack Zipes and Frank Mecklenburg. Cambridge, MA: MIT Press, 1988.

Bloom, Nicholas Dagen. *Merchant of Illusion: James Rouse, American Salesman of the Businessman's Utopia.* Columbus: Ohio State University Press, 2004.

Bosma, Koos, and Helma Hellinga, eds. *Mastering the City: North-European Planning 1900–2000.* Vol. 2. Rotterdam: NAI; The Hague: EFL Publications, 1997.

Boyer, Paul. *By the Bomb's Early Light: American Thought and Culture at the Dawn of the Atomic Age.* New York: Pantheon, 1985.

Bozdogan, Sibel. *Modernism and Nation-Building: Turkish Architectural Culture in the Early Republic.* Seattle: University of Washington Press, 2001.

Branch, Melville Campbell. *Aerial Photography in Urban Planning and Research.* Boston: Harvard University Press, 1948.

Breckenfeld, Gurney. *Columbia and the New Cities.* New York: Ives Washburn, 1971.

Breese, Gerald. *Urban and Regional Planning for the Delhi-New Delhi Area: Capital for Conquerors and Country.* Princeton, NJ: Gerald Breese, 1974.

Brown, Kate. *Plutopia: Nuclear Families, Atomic Cities, and the Great Soviet and American Plutonium Disasters.* New York: Oxford University Press, 2013.

Buchanan, Colin. *Traffic in Towns.* Harmondsworth, UK: Penguin, 1963.

Buckley, Craig, and Jean-Louis Violeau. *Utopie: Texts and Projects, 1967–1978.* Translated by Jean-Marie Clarke. Los Angeles: Semiotext(e); Cambridge, MA: MIT Press, 2011.

Buder, Stanley. *Visionaries and Planners: The Garden City Movement and the Modern Community.* Oxford: Oxford University Press, 1990.

Bullock, Nicholas. *Building the Post-War World: Modern Architecture and Reconstruction in Britain.* London: Routledge, 2002.

Burton, Anthony, and Joyce Hartley, eds. *The New Towns Record, 1946–1996: 50 Years of UK New Town Development.* Glasgow: Planning Exchange, 1997.

Busbea, Larry. *Typologies: The Urban Utopia in France, 1960–1970.* Cambridge, MA: MIT Press, 2007.

Candilis, Georges. *Bâtir la vie: Un architect témoin de son temps.* Paris: Stock, 1977.

Canty, Donald, ed. *The New City.* New York: Published for Urban America by Frederick A. Praeger, 1969.

Caprotti, Federico. *Mussolini's Cities: Internal Colonialism in Italy, 1930–1939.* Youngstown, NY: Cambria Press, 2007.

Carail, Jean. "Exemple de création urbaine au Sahara pétrolier: Hassi-Massaoud, 1956–1962." Mémoire de these, Institut d'urbanisme, Université de Paris, 1962.

Caspari, Emil. *The Working Classes of Upper-Silesia: An Historical Essay.* London: Simpson Low, Marston, 1921.

Centre de documentation sur l'urbanisme. "Villes nouvelles françaises 2001–2005." http://www.cdu.urbanisme.developpement-durable.gouv.fr/villes-nouvelles-francaises -2001–2005-r8243.htmlcaises. Accessed October 13, 2015.

Chaichian, Mohammad A. *Town and Country in the Middle East: Iran and Egypt in the Transition to Globalization, 1800–1970.* London: Lexington Books, 2009.

Chakrabarty, Dipesh. *Provincializing Europe: Postcolonial Thought and Historical Difference.* Princeton, NJ: Princeton University Press, 2000.

Choay, Françoise. *The Rule and the Model: On the Theory of Architecture and Urbanism.* Cambridge, MA: MIT Press, 1997.

———. *L'urbanisme, utopies et réalités: Une anthologie.* Paris: Seuil, 1965.

Chu, Hsiao-Yun, and Roberto G. Trujillo, eds. *New Views on R. Buckminster Fuller.* Palo Alto, CA: Stanford University Press, 2009.

Ciborowski, Adolf. *L'Urbanisme polonais 1945–1955.* Warsaw: Editions Polonia, 1956.

City and Industrial Development Corporation of Maharashtra Ltd. (CIDCO), "Navi Mumbai." www.cidco.maharashtra.gov.in. Accessed July 27, 2015.

Clapp, James A. *New Towns and Urban Policy: Planning Metropolitan Growth*. New York: Dunellen, 1971.

Clapson, Mark. *Anglo-American Crossroads: Urban Research and Planning in Britain, 1940–2010*. London: Bloomsbury, 2013.

———. *The Plan for Milton Keynes*. London: Routledge, 2013.

———. *A Social History of Milton Keynes: Middle England/Edge City*. London: Frank Cass, 2004.

Clark, Peter, ed. *The European City and Green Space: London, Stockholm, Helsinki and St. Petersburg, 1850–2000*. Aldershot, UK: Ashgate, 2006.

Cohen, Jean-Louis. *Architecture in Uniform: Designing and Building for the Second World War*. Montreal: Canadian Centre for Architecture; Paris: Hazan, 2011.

———. *The Future of Architecture since 1889*. New York: Phaidon Press, 2012.

———. *Scenes of the World to Come: European Architecture and the American Challenge, 1893–1960*. Paris: Flammarion; Montreal: Canadian Center for Architecture, 1995.

Cohen, Jean-Louis, Marco De Michelis, and Manfredo Tafuri, eds. *URSS 1917–1978: La Ville, l'architecture*. Paris: L'Equerre, 1979.

Colditz, Heinz, and Martin Lücke. *Stalinstadt: Neues Leben—Neue Menschen*. Berlin: Kongress Verlag, 1958.

Conekin, Becky. *The Autobiography of a Nation: The 1951 Festival of Britain*. Manchester: Manchester University Press, 2003.

Conn, Steven. *Americans against the City: Anti-Urbanism in the Twentieth Century*. Oxford: Oxford University Press, 2014.

Connell, Raewyn. *Southern Theory: The Global Dynamics of Knowledge in Social Science*. Cambridge, MA: Polity, 2007.

Corden, Carol. *Planned Cities: New Towns in Britain and America*. Beverly Hills, CA: Sage Publications, 1977.

Cowan, Susanne Elizabeth. "Democracy, Technocracy and Publicity: Public Consultation and British Planning, 1939–1951." Phd diss., University of California, Berkeley, 2010.

Crowley, David, and Jane Pavitt, eds. *Cold War Modern: Design 1945–1970*. London: V and A, 2008.

Cupers, Kenny. *The Social Project: Housing Postwar France*. Minneapolis: University of Minnesota Press, 2014.

Dahir, James. *The Neighborhood Unit Plan: Its Spread and Acceptance*. New York: Russell Sage Foundation, 1947.

De Wit, Wim, and Christopher James Alexander, eds. *Overdrive: L.A. Constructs the Future, 1940–1990*. An exhibition catalog. Los Angeles: Getty Publications, 2013.

Deane, Philip. *Constantinos Doxiadis, Master Builder for Free Men*. Dobbs Ferry, NY: Oceana Publications, 1965.

DeHaan, Heather. *Stalinist City Planning: Professionals, Performance, and Power*. Toronto: University of Toronto Press, 2013.

Den Hartog, Harry. *Shanghai New Towns: Seaching for Community and Identity in a Sprawling Metropolis*. Rotterdam: 010 Publishers, 2010.

Dey, S. K. *Nilokheri*. London: Asia Publishing House, 1962.

Djurié, Dubravka, and Misko Suvakovié, eds. *Impossible Histories: Historical Avant-Gardes, Neo-Avant-Gardes, and Post-Avant-Gardes in Yugoslavia, 1918–1991*. Cambridge, MA: MIT Press, 2015.

Dobrenko, Evgeny, and Eric Naiman, eds. *The Landscape of Stalinism*. Seattle: University of Washington Press, 2003.

Domhardt, Konstanze Sylva. *The Heart of Our City: Die Stadt in den transatlantischen Debatten der CIAM, 1933–1951*. Zurich: GTA Verlag, 2012.

Dorrian, Mark, and Frédéric Pousin, eds. *Seeing from Above: The Aerial View in Visual Culture*. London: B. Tauris, 2013.

Doxiadis, Constantinos A. *Between Dystopia and Utopia*. London: Faber and Faber, 1966.

———. *Ecumenopolis: The Inevitable City of the Future*. New York: Norton, 1979.

———. *Ekistics: An Introduction to the Science of Human Settlements*. London: Oxford University Press, 1968.

Dufaux, Frédéric, and Annie Fourcaut, eds. *Le Monde des grands ensembles*. Paris: Editions CREAPHIS, 2004.

Eaton, Ruth. *Ideal Cities: Utopianism and the (Un)Built Environment*. Oxford: Thames and Hudson, 2001.

Edwards, Paul N. *The Closed World: Computers and the Politics of Discourse in Cold War America*. Cambridge, MA: MIT Press, 1996.

Eichler, Edward, and Marshall Kaplan. *The Community Builders*. Berkeley: University of California Press, 1967.

Engel, Barbara. *Öffentliche Räume in den Blauen Städten Russlands*. Berlin: Wasmuth, 2004.

Escobar, Arturo. *Encountering Development: The Making and Unmaking of the Third World*. Princeton, NJ: Princeton University Press, 1995.

Estèbe, Philippe, and Sophie Gonnard. *Les Villes nouvelles et le système politique en Ile-de-France*. Paris: Ministère de l'Equipement, du Transport et du Logement, 2005.

Evans, Hazel, ed. *New Towns: The British Experience*. London: Charles Knight, 1972.

Ferguson, Francis. *Architecture, Cities and the Systems Approach*. New York: George Braziller, 1975.

Filtzer, Donald. *The Hazards of Urban Life in Late Stalinist Russia*. Cambridge: Cambridge University Press, 2010.

Fishman, Robert, ed. *The American Planning Tradition, Culture and Policy*. Washington, DC: Woodrow Wilson Center Press, 2000.

———. *Urban Utopias in the Twentieth Century: Ebenezer Howard, Frank Lloyd Wright, Le Corbusier*. Cambridge, MA: MIT Press, 1982.

Forshaw, J. H., and Patrick Abercrombie. *County of London Plan: Prepared for the London County Council*. London: Macmillan, 1943.

Forsyth, Ann. *Reforming Suburbia: The Planned Communities of Irvine, Columbia, and the Woodlands*. Berkeley: University of California Press, 2005.

Foucault, Michel. *Security, Territory, Population: Lectures at the Collège de France, 1977–1978*. Translated by G. Burchell. Basingstoke, UK: Palgrave Macmillan, 2009.

Freedman, Carl. *Critical Theory and Science Fiction*. Middletown, CT: Wesleyan University Press, 2000.

Freestone, Robert. *Urban Nation: Australia's Planning Heritage*. Clayton, Victoria, Australia: Csiro, 2010.

Gaonkar, Dilip Parameshwar, ed. *Alternative Modernities*. Durham, NC: Duke University Press, 2001.

Geddes, Donald Porter, ed. *The Atomic Age Opens*. New York: Pocket Books, 1945.

Geertse, Michel. *Defining the Universal City: The International Federation for Housing and Town Planning and Transnational Planning Dialogue, 1913–1945*. Amsterdam: Vrije Universiteit, 2012.

Ghirardo, Diane. *Building New Communities: New Deal America and Fascist Italy*. Princeton, NJ: Princeton University Press, 1989.

Giacone, Alessandro, ed. *Les Grands Paris de Paul Delouvrier*. Paris: Descartes, 2010.

Gillette, Howard. *Civitas by Design: Building Better Communities, from the Garden City to the New Urbanism*. Philadelphia: University of Pennsylvania Press, 2010.

Gilman, Nils. *Mandarins of the Future: Modernization Theory in Cold War America*. Baltimore: Johns Hopkins University Press, 2007.

Golany, Gideon, ed. *Innovations for Future Cities*. New York: Praeger, 1976.

————, ed. *Strategies for New Community Development in the United States*. Stroudsburg, PA: Dowden, Hutchinson and Ross, 1975.

Golany, Gideon, and Daniel Walden, eds. *The Contemporary New Communities Movement in the United States*. Urbana: University of Illinois Press, 1974.

Gold, John R. *The Experience of Modernism: Modern Architects and the Future City, 1928–53*. London: E. and F. N. Spon, 1997.

————. *The Practice of Modernism: Modern Architects and Urban Transformation, 1954–1972*. London: Routledge, 2007.

Goldhagen, Sarah Williams, and Réjean Legault, eds. *Anxious Modernisms: Experimentation in Postwar Architectural Culture*. Montreal: Canadian Center for Architecture; Cambridge, MA: MIT Press, 2000.

Gordon, Alastair. *Naked Airport: A Cultural History of the World's Most Revolutionary Structure*. Chicago: University of Chicago Press, 2008.

Gordon, Eric. *The Urban Spectator: American Concept Cities from Kodak to Google*. Lebanon, NH: Dartmouth University Press, 2010.

Gosling, David, and Barry Maitland. *Design and Planning of Retail Systems*. London: Architectural Press, 1976.

Gottmann, Jean. *Essais sur l'aménagement de l'espace habité*. Paris: Mouton, 1966.

Gottmann, Jean, and Robert Harper, eds. *Since Megalopolis: The Urban Writings of Jean Gottmann*. Baltimore: Johns Hopkins University Press, 1990.

Graham, Loren R., ed. *Science and the Soviet Social Order*. Cambridge, MA: Harvard University Press, 1990.

Griffin, Nathaniel M. *Irvine: The Genesis of a New Community*. Washington, DC: Urban Land Institute, 1974.

Grigor, Talinn. *Building Iran: Modernism, Architecture, and National Heritage under the Pahlavi Monarchs*. New York: Periscope, distributed by Prestel, 2009.

Gruen, Victor. *The Heart of Our Cities: The Urban Crisis; Diagnosis and Cure*. 2nd ed. London: Thames and Hudson, 1965.

Gutnov, Alexei, A. Baburov, G. Djumenton, S. Kharitonova, I. Lezava, and S. Sadovskij. *The Ideal Communist City*. Translated by Renée Neu Watkins. New York: George Braziller, 1971.

Gutreund, Owen. *Twentieth-Century Sprawl: Highways and the Reshaping of the American Landscape*. Oxford: Oxford University Press, 2005.

Haffner, Jeanne. *The View from Above: The Science of Social Space*. Cambridge, MA: MIT Press, 2013.

Hall, Peter. *Cities of Tomorrow: An Intellectual History of Urban Design in the Twentieth Century*. New York: Blackwell, 1998.

Hall, Peter, and Colin Ward. *Sociable Cities: The Legacy of Ebenezer Howard*. 3rd ed. Chichester, UK: John Wiley and Sons, 2002.

Hall, Thomas, ed. *Planning and Urban Growth in the Nordic Countries*. London: E. and F. N. Spon, 1991.

Hard, Mikael, and Thomas J. Misa, eds. *Urban Machinery: Inside Modern European Cities*. Cambridge, MA: MIT Press, 2008.

Hardwick, M. Jeffrey. *Mall Maker: Victor Gruen, Architect of an American Dream*. Philadelphia: University of Pennsylvania Press, 2004.

Hardy, Dennis. *From New Towns to Green Politics: Campaigning for Town and Country Planning, 1946–1990*. London: E. and F. N. Spon, 1991.

———. *Utopian England: Community Experiments 1900–1945*. New York: Routledge, 2012.

Harris, Steven E. *Communism on Tomorrow Street: Mass Housing and Everyday Life after Stalin*. Baltimore: Johns Hopkins University Press, 2013.

Harvey, David. "The Right to the City." *New Left Review* 53 (September–October 2008): 23–40.

Hatherley, Owen. *Landscapes of Communism: A History through Buildings*. London: Allen Lane, 2015.

Haumont, Nicole. *Les Villes nouvelles d'Europe à la fin du 20ème siècle: Recherche comparative internationale*. 2 vols. Paris: Ecole d'Architecture de Paris-la-Défense, 1997.

Haumont, Nicole, Bohdan Jalowiecki, Moïra Munro, and Viktória Szirmai. *Villes nouvelles et villes traditionnelles: Une comparaison internationale*. Paris: L'Harmattan, 1999.

Hausladen, Gary Joseph. "Regulating Urban Growth in the USSR: The Role of Satellite Cities in Soviet Urban Development." PhD diss., Syracuse University, 1983.

Hayden, Dolores. *Building Suburbia: Green Fields and Urban Growth 1820–2000*. New York: Pantheon, 2003.

Healey, Patsy, and Robert Upton, eds. *Crossing Borders: International Exchange and Planning*. New York: Routledge, 2010.

Heims, Steve Joshua. *The Cybernetics Group*. Cambridge, MA: MIT Press, 1991.

Heitmann, John. *The Automobile and American Life*. Jefferson, NC: McFarland, 2009.

Henderson, Susan R. *Building Culture: Ernst May and the New Frankfurt Initiative, 1926–1931*. New York: Peter Lang International Publishers, 2013.

Henket, Hubert-Jan, and Hilde Heynen, eds. *Back from Utopia: The Challenge of the Modern Movement*. Rotterdam: 010 Publishers, 2002.

Hertzen, Heikki von, and Paul D. Spreiregen. *Building a New Town: Finland's New Garden City Tapiola*. Boston: MIT Press, 1971.

Higgott, Andrew. *Mediating Modernism: Architectural Cultures in Britain*. London: Routledge, 2007.

Hodge, Gerald, and David Gordon, eds. *Planning Canadian Communities*. 5th ed. Toronto: Nelson, 2007.

Hughes, Jonathan, and Simon Sadler, eds. *Non-Plan: Essays on Freedom Participation and Change in Modern Architecture and Urbanism*. Oxford: Architectural Press, 2000.

Hulshof, Michael, and Daan Roggeveen. *How the City Moved to Mr. Sun: China's New Megacities*. Amsterdam: Martien de Vletter, SUN Publishers, 2010.

Inskeep, Steve. *Instant City: Life and Death in Karachi*. New York: Penguin, 2011.

Irion, Ilse, and Thomas Sieverts. *Neue Städte: Experimentierfelder der Moderne*. Stuttgart: Deutsche Verlags-Anstalt, 1991.

Isard, Walter. *History of Regional Science and the Regional Science Association International: The Beginnings and Early History*. Berlin: Springer, 2003.

———. *Location and Space-Economy: A General Theory Relating to Industrial Location, Market Areas, Land Use, Trade, and Urban Structure*. Cambridge, MA: Technology Press, 1965.

Isenstadt, Sandy, and Kishwar Rizvi. *Modernism and the Middle East: Architecture and Politics in the Twentieth Century*. Seattle: University of Washington Press, 2008.

Jackson, Iain, and Jessica Holland. *The Architecture of Edwin Maxwell Fry and Jane Drew:*

Twentieth Century Architecture, Pioneer Modernism and the Tropics. Farnham, UK: Ashgate, 2014.

Jacquemin, Alain R. A. *Urban Development and New Towns in the Third World: Lessons from the New Bombay Experience.* Aldershot, UK: Ashgate, 1999.

Jameson, Fredric. *Archaeologies of the Future: The Desire Called Utopia and Other Science Fictions.* London: Verso, 2007.

Jaouen, Annick, Chantal Guillet, and Jean-Eudes Roullier, eds. *Vingt-cinq ans de villes nouvelles en France.* Paris: Economica, 1989.

Jardini, David R. "Out of the Blue Yonder: The RAND Corporation's Diversification into Social Welfare Research, 1946–1968." PhD diss., Carnegie Mellon University, 1996.

Jasanoff, Sheila, ed. *States of Knowledge: The Co-production of Science and Social Order.* London: Routledge, 2004.

Jellicoe, G. A. *Motopia: A Study in Evolution of the Urban Landscape.* New York: Frederick A. Praeger, 1961.

Johnson, David A. *Planning the Great Metropolis: The 1929 Regional Plan of New York and Its Environs.* New York: Routledge, 2015.

Josephson, Paul. *New Atlantis Revisited: Akademgorodok, the Siberian City of Science.* Princeton, NJ: Princeton University Press, 1997.

Judt, Tony. *Postwar: A History of Europe since 1945.* New York: Penguin, 2005.

Kalia, Ravi. *Gandhinagar: Building National Identity in Postcolonial India.* Columbia: University of South Carolina Press, 2004.

——, ed. *Pakistan: From the Rhetoric of Democracy to the Rise of Militancy.* New Delhi: Routledge, 2011.

Kampffmeyer, Hans. *Die Nordweststadt in Frankfurt am Main.* Frankfurt am Main: Europäische Verlagsanstalt, 1968.

Kargon, Robert H., and Arthur P. Molella. *Invented Edens: Techno-Cities of the Twentieth Century.* Cambridge, MA: MIT Press, 2008.

Keeton, Rachel. *Rising in the East: Contemporary New Towns in Asia.* Amsterdam: International New Town Institute; Amsterdam: SUN Publishers, 2011.

Kidder Smith, G. E. *Sweden Builds: Its Modern Architecture and Land Policy Background, Development and Contribution.* New York: Albert Bonnier, 1950.

King, Anthony. *Spaces of Global Cultures: Architecture, Urbanism, Identity.* London: Routledge, 2004.

Knauer-Romani, Elisabeth. *Eisenhüttenstadt und die Idealstadt des 20. Jahrhunderts.* Weimar, Germany: Verlag und Datenbank für Geisteswissenschaften, 2000.

Kozlov, Denis, and Eleonory Gilburd. *The Thaw: Soviet Society and Culture during the 1950s and 1960s.* Toronto: University of Toronto Press, 2014.

Ladd, Brian. *Autophobia: Love and Hate in the Automobile Age.* Chicago: University of Chicago Press, 2008.

Lang, Jon, Madhavi Desai, and Miki Desai, eds. *Architecture and Independence: The Search for Identity—India 1880 to 1980.* New Delhi: Oxford University Press, 1997.

Laqueur, Walter, and Leopold Labedz, eds. *The Future of Communist Society.* New York: Frederick A. Praeger, 1962.

Latham, Michael E. *The Right Kind of Revolution: Modernization, Development, and U.S. Foreign Policy from the Cold War to the Present.* Ithaca, NY: Cornell University Press, 2011.

Lebow, Katherine. *Unfinished Utopia: Nowa Huta, Stalinism, and Polish Society, 1949–56.* Ithaca, NY: Cornell University Press, 2013.

Lenger, Friedrich. *Metropolen der Moderne: Eine europäische Stadtgeschichte seit 1850.* Munich: C. H. Beck, 2013.

Lengereau, Eric. *L'Etat et l'architecture 1958–1981: Une politique publique?* Paris: Picard, 2001.

Le Normand, Brigitte. *Designing Tito's Capital: Urban Planning, Modernism, and Socialism in Belgrade.* Pittsburgh: University of Pittsburgh Press, 2014.

Leucht, Kurt. *Die erste neue Stadt in der Deutschen Demokratischen Republik: Planungsgrundlagen und -ergebnisse von Stalinstadt.* Berlin: VEB Verlag Technik, 1957.

Levy, Raphael, and Benjamin Hanft, eds. *21 Frontier Towns.* New York: United Jewish Appeal; Jewish Agency for Israel, 1971.

Ley, Sabrina van der, and Markus Richter, eds. *Megastructure Reloaded: Visionäre Stadtentwürfe der Sechzigerjahre reflektiert von zeitgenössischen Künstlern = Visionary Architecture and Urban Design of the Sixties Reflected by Contemporary Artists.* Ostfildern, Germany: Hatje Cantz Verlag, 2008.

Light, Jennifer S. *From Warfare to Welfare: Defense Intellectuals and Urban Problems in Cold War America.* Baltimore: Johns Hopkins University Press, 2003.

Lin, Zhongjie. *Kenzo Tange and the Metabolist Movement: Urban Utopias of Modern Japan.* New York: Routledge, 2010.

Lingeman, Richard. *The Noir Forties: The American People from Victory to Cold War.* New York: Nation Books, 2012.

Llewelyn-Davies, Weeks, Forestier-Walker, and Bor, and Ove Arup and Partners. *Motorways in the Urban Environment.* British Road Federation Report. London: British Road Federation, 1971.

Lu, Duanfang, ed. *Third World Modernism: Architecture, Development and Identity.* New York: Routledge, 2011.

Luccarelli, Mark. *Lewis Mumford and the Ecological Region: The Politics of Planning.* New York: Guilford Press, 1995.

L'urbanisme de dalles: Cotinuités et ruptures. Actes du colloque de Cergy-Pontoise, 16–17 September 1993. Paris: Presses des Ponts et chaussées, 1995.

Macdonald, Dwight. *The Ford Foundation: The Men and the Millions.* New York: Reynal, 1956.

Maciuika, John V. *Before the Bauhuas: Architecture, Politics, and the German State, 1890–1920.* Cambridge: Cambridge University Press, 2008.

Madanipour, Ali. *Teheran: The Making of a Metropolis.* New York: John Wiley and Sons, 1998.

Madge, John. *The Rehousing of Britain.* Target for Tomorrow. London: Pilot Press, 1945.

Major, Patrick, and Rana Mitter, eds. *Across the Blocs: Exploring Comparative Cold War Cultural and Social History.* London: Frank Cass, 2004.

Malisz, Bolesław. *La Pologne construit des villes nouvelles.* Translated by Kazimiera Bielawska. Warsaw: Editions Polania, 1961.

Mannheim, Karl. *Ideology and Utopia: An Introduction to the Sociology of Knowledge.* New York: Harcourt, Brace, 1954.

Manuel, Frank, ed. *Utopias and Utopian Thought.* Boston: Houghton Mifflin, 1966.

Mattsson, Helena, and Sven-Olov Wallenstein, eds. *Swedish Modernism: Architecture, Consumption and the Welfare State.* London: Black Dog, 2010.

Maurer, Eva, ed. *Soviet Space Culture: Cosmic Enthusiasm in Socialist Societies.* New York: Palgrave Macmillan, 2011.

Mayer, Albert. *Pilot Project, India: The Story of Rural Development in Etawah, Uttar Pradesh.* Berkeley: University of California Press, 1958.

———. *The Urgent Future.* New York: McGraw-Hill, 1967.

McDougall, Walter A. *The Heavens and the Earth: A Political History of the Space Age.* New York: Basic Books, 1985.

McGee, T. G., and W. D. McTaggart. *Petaling Jaya: A Socio-Economic Survey of a New Town in Selangor, Malaysia*. Pacific Viewpoint Monograph, no. 2. Wellington, NZ: Victoria University of Wellington, 1967.

Mears, F. C. *A Regional Survey and Plan for Central and Southeast Scotland*. Edinburgh: Central and Southeast Scotland Regional Planning Advisory Committee, 1948.

Merlin, Pierre. *Géographie humaine*. Paris: Presses universitaires de France, 1997.

———. *Les Villes nouvelles françaises*. Notes et Etudes Documentaires, nos. 4286–88. Paris: La Documentation française, May 3, 1976.

———. *New Towns: Regional Planning and Development*. London: Methuen, 1971.

Miliutin, N. A. *Sotsgorod: The Problem of Building Socialist Cities*. Cambridge, MA: MIT Press, 1974.

Mindell, David. *Digital Apollo: Human and Machine in Spaceflight*. Cambridge, MA: MIT Press, 2008.

Monnier, Gérard, and Richard Klein, eds. *Les années ZUP: Architectures de la croissance, 1960–1973*. Paris: Picard, 2002.

Montross, Sarah J., ed. *Past Futures: Science Fiction, Space Travel, and Postwar Art in America*. Cambridge, MA: MIT Press, 2015.

Morgan, George, and John King. *The Woodlands: New Community Development*. College Station: Texas A&M University Press, 1987.

Mottez, Michel. *Carnets de campagne: Evry, 1965–2007*. Paris: L'Harmattan, 2003.

Mumford, Eric. *The CIAM Discourse on Urbanism, 1928–1960*. Cambridge, MA: MIT Press, 2000.

———. *Defining Urban Design: CIAM Architects and the Formation of a Discipline, 1937–69*. New Haven, CT: Yale University Press, 2009.

Mumford, Lewis. *City Development: Studies in Disintegration and Renewal*. New York: Harcourt, Brace, 1945.

———. *The Culture of Cities*. New York: Harcourt, Brace, 1938.

———. *The Story of Utopias*. New York: Peter Smith, 1941.

Murard, Lion, and François Fourquet, eds. *La Naissance des villes nouvelles*. Paris: Presses de l'école nationale des Ponts et Chaussées, 2004.

Nash, Gerald D. *The American West Transformed: The Impact of the Second World War*. Bloomington: Indiana University Press, 1985.

Nasr, Joe, and Mercedes Volait, eds. *Urbanism: Imported or Exported?* New York: John Wiley and Sons, 2003.

Neufeld, Max. *Israel's New Towns: Some Critical Impressions*. Wyndham Deedes Scholars no. 31. London: Anglo-Israel Association, June 1971.

Newsome, W. Brian. *French Urban Planning 1940–1968: The Construction and Deconstruction of an Authoritarian System*. New York: Peter Lang, 2009.

Omeweh, Daniel A. *Shell Petroleum Development Company, the State and Underdevelopment in Nigeria's Niger Delta: A Study in Environmental Degradation*. Trenton, NJ: Africa World Press, 2005.

Owen, Wilfred. *Cities in the Motor Age*. New York: Viking Press, 1959.

Paquot, Thierry. *Utopies et utopistes*. Paris: Editions La Découverte, 2007.

Parmar, Inderjeet. *Foundations of the American Century: The Ford, Carnegie, and Rockefeller Foundations and the Rise of American Power*. New York: Columbia University Press, 2012.

Parsons, Kermit C., and David Schuyler, eds. *From Garden City to Green City: The Legacy of Ebenezer Howard*. Baltimore: Johns Hopkins University Press, 2002.

Pass, David. *Vällingby and Farsta—from Idea to Reality; The New Community Development Process in Stockholm.* Cambridge, MA: MIT Press, 1973.

Perloff, Harvey S., and Neil C. Sandberg, eds. *New Towns: Why—And for Whom?* New York: Praeger, 1972.

Peterson, Sarah Jo. *Plannning the Home Front: Building Bombers and Communities at Willow Run.* Chicago: University of Chicago Press, 2013.

Phillips, David R., and Anthony G. O. Yeh, eds. *New Towns in East and South-east Asia.* New York: Oxford University Press, 1987.

Pickering, Andrew. *The Cybernetic Brain: Sketches of Another Future.* Chicago: University of Chicago Press, 2010.

Picon, Antoine, and Clément Orillard. *De la ville nouvelle à la ville durable, Marne-la-Vallée.* Paris: Parentheses, 2012.

Pinder, David. *Visions of the City: Utopianism, Power, and Politics in Twentieth-Century Urbanism.* New York: Routledge, 2005.

Ponsard, Claude. *Histoire des théories économiques spatiales.* Paris: Armand Colin, 1958.

Popper, Karl. *The Open Society and Its Enemies.* Vol. 1. 5th ed. Princeton, NJ: Princeton University Press, 1966.

———. *The Poverty of Historicism.* 3rd ed. London: Routledge and Kegan Paul, 1961.

Prakash, Gyan. *Another Reason: Science and the Imagination of Modern India.* Princeton, NJ: Princeton University Press, 1999.

———. *Mumbai Fables.* Princeton, NJ: Princeton University Press, 2010.

Prakash, Ved. *New Towns in India.* Monograph and Occasional Papers Series, vol. 8. Durham, NC: Program in Comparative Studies on Southern Asia, Duke University, 1969.

Reid, Susan E., and David Crowley, eds. *Style and Socialism: Modernity and Material Culture in Post-War Eastern Europe.* New York: Berg, 2000.

Reimann, Brigitte. *Franziska Linkerhand.* Munich: Kindler, 1974.

Ricoeur, Paul. *Lectures on Ideology and Utopia.* Edited by George H. Taylor. New York: Columbia University Press, 1986.

Risselada, Max, and Dirk van den Heuvel, eds. *Team 10: 1953–81, in Search of a Utopia of the Present.* Rotterdam: NAi, 2005.

Ritter, Katharina, Ekaterina Shapiro-Obermair, Dietmar Steiner, and Alexandra Wachter, eds. *Soviet Modernism 1955–1991: Unknown History.* Zurich: Park Books, 2012.

Robinson, Albert J. *Economics and New Towns: A Comparative Study of the United States, the United Kingdom, and Australia.* New York: Praeger, 1975.

Robinson, Jennifer. *Ordinary Cities between Modernity and Development.* New York: Routledge, 2006.

Rodgers, Daniel T. *Atlantic Crossings: Social Politics in a Progressive Age.* Cambridge, MA: Harvard University Press, 2000.

Roullier, Jean-Eudes, ed. *Cergy-Pontoise: "Inventer une ville."* Actes du colloque du 5 septembre 2002. Lyon: CERTU, September 2002.

Roy, Srirupa. *Beyond Belief: India and the Politics of Postcolonial Nationalism.* Durham, NC: Duke University Press, 2007.

Saarinen, Eliel. *The City: Its Growth, Its Decay, Its Future.* New York: Reinhold Publishing, 1943.

Sackley, Nicole. "Passage to Modernity: American Social Scientists, India, and the Pursuit of Development, 1945–1961." PhD diss., Princeton University, 2004.

Sargent, Lyman Tower. *Utopianism: A Very Short Introduction.* New York: Oxford University Press, 2010.

Schaik, Van, and Otakar Mácel, eds. *Exit Utopia: Architectural Provocations 1956–76*. New York: Prestel, 2005.

Schneider, Christian. *Stadtgründung im Dritten Reich: Wolfsburg und Salzgitter*. Berlin: Heinz Moos, 1978.

Schwagenscheidt, Walter. *Die Nordweststadt: Idee und Gestaltung = The Nordweststadt: Conception and Design*. Stuttgart: Karl Krämer Verlag, 1964.

Schwarzer, Mitchell. *Zoomscape: Architecture in Motion and Media*. New York: Princeton Architectural Press, 2004.

Scott, Felicity D. *Architecture or Techno-utopia: Politics after Modernism*. Cambridge, MA: MIT Press, 2010.

Scott, James C. *Seeing Like a State: How Certain Schemes to Improve the Human Condition Have Failed*. New Haven, CT: Yale University Press, 1998.

Seebohm, Caroline. *No Regrets: The Life of Marietta Tree*. New York: Simon and Schuster, 1997.

Seller, Cotten. *Republic of Drivers: A Cultural History of Automobility in America*. Chicago: University of Chicago Press, 2008.

Serdari, Thomaï. "Albert Mayer, Architect and Town Planner: The Case for a Total Professional." PhD diss., New York University, 2005.

Sert, José Luis. *Can Our Cities Survive? An ABC of Urban Problems, Their Analysis, Their Solutions*. Cambridge, MA: Harvard University Press, 1942.

Sevcenko, Margaret Bentley, ed. *Design for High-Intensity Development*. Cambridge, MA: Aga Khan Program for Islamic Architecture, 1986.

Shanken, Andrew M. *194X: Architecture, Planning, and Consumer Culture on the American Home Front*. Minneapolis: University of Minnesota Press, 2009.

Sharma, R. N., and K. Sita, eds. *Issues in Urban Development: A Case of Navi Mumbai*. Jaipur: Rawat Publications, 2001.

Sharma, Syresh K. *Haryana: Past and Present*. New Delhi: Mittel, 2005.

Shaw, Annapurna. *The Making of Navi Mumbai*. Hyderabad: Orient Longman, 2004.

Shoshkes, Ellen. *Jaqueline Tyrwhitt: A Transnational Life in Urban Planning and Design*. Surrey, UK: Ashgate, 2013.

Siegfried, Klaus-Jörg. *Wolfsburg—zwischen Wohnstadt und Erlebnisstadt*. Wolfsburg, Germany: Stadt Wolfsburg, 2002.

Smoliar, I. M., ed. *New Towns Formation in the USSR*. Moscow: Central Scientific Research and Design Institute of Town Planning, 1973.

Soleri, Paolo. *Arcology: The City in the Image of Man*. Cambridge, MA: MIT Press, 1969.

Spiller, Neil. *Visionary Architecture: Blueprints of the Modern Imagination*. New York: Thames and Hudson, 2006.

Springer, Philipp. *Verbaute Träume: Herrschaft, Stadtentwicklung und Lebensrealität in der sozialistischen industriestadt Schwedt*. Berlin: Links, 2007.

Spufford, Francis. *Red Plenty*. Minneapolis: Graywolf Press, 2012.

Stadsplanekontor, Stockholms Stads. *General Plan För Stockholm 1952: Förslag uppraättat under aren 1945–52*. Stockholm: P. A. Norstedt, 1952.

Stallabrass, Julian. *Gargantua: Manufactured Mass Culture*. London: Verso, 1996.

Staples, Eugene S. *Forty Years: A Learning Curve; The Ford Foundation Programs in India, 1952–1992*. New Delhi: Ford Foundation, 1992.

Stein, Clarence S. *The Writings of Clarence S. Stein: Architect of the Planned Community*. Edited by Kermit C. Parsons. Baltimore: Johns Hopkins University Press, 1998.

Steiner, Hadas A. *Beyond Archigram: The Structure of Circulation*. New York: Routledge, 2009.

Stern, Robert A. M., David Fishman, and Jacob Tilove. *Paradise Planned: The Garden Suburb and the Modern City*. New York: Monacelli Press, 2013.

Sussman, Carl, ed. *Planning the Fourth Migration: The Neglected Vision of the Regional Planning Association of America*. Cambridge, MA: MIT Press, 1976.

Tafuri, Manfredo. *Architecture and Utopia: Design and Capitalist Development*. Cambridge, MA: MIT Press, 1996.

Tennenbaum, Robert, ed. *Creating a New City: Columbia, Maryland*. Columbia, MD: Perry, 1996.

Thaler, Wolfgang, Maroje Mrduljas, and Vladimir Kulic. *Modernism In-Between: The Mediatory Architectures of Socialist Yugoslavia*. Berlin: Jovis, 2012.

Thöner, Wolfgang, and Peter Müller, eds. *Bauhaus-Tradition und DDR-Moderne: Der Architekt Richard Paulick*. Berlin: Deutsche Kunstverlag, 2006.

Topfstedt, Thomas. *Städtebau in der DDR 1955–1971*. Leipzig: E. A. Seemann, 1988.

Townsend, Anthony. *Smart Cities: Big Data, Civic Hackers, and the Quest for a New Utopia*. New York: W. W. Norton, 2013.

Troen, S. Ilan. *Imagining Zion: Dreams, Designs, and Realities in a Century of Jewish Settlement*. New Haven, CT: Yale University Press, 2003.

Troen, S. Ilan, and Noah Lucas, eds. *Israel: The First Decade of Independence*. Albany: State University of New York, 1995.

Tuomi, Timo, ed. *Life and Architecture: Tapiola*. Espoo, Finland: Housing Foundation and City of Espoo, 2003.

Turner, Fred. *From Counterculture to Cyberculture: Stewart Brand, the Whole Earth Network, and the Rise of Digital Utopianism*. Chicago: University of Chicago Press, 2008.

Urban, Florian. *Tower and Slab: Histories of Global Mass Housing*. New York: Routledge, 2012.

Vadelorge, Loïc. *Retour sur les villes nouvelles: Un Histoire urbaine du XXe Siècle*. Paris: CREAPHIS Editions, 2014.

Vagale, L. R. *A Critical Appraisal of New Towns in Developing Countries: Policy Framework for Nigeria*. Ibadan, Nigeria: Polytechnic, 1977.

———. *Structure of Metropolitan Regions in India: Planning Problems and Prospects*. New Delhi: L. R. Vagale, 1964.

Vanderbilt, Tom. *Survival City: Adventures among the Ruins of Atomic America*. Chicago: University of Chicago Press, 2010.

Van der Cammen, Hans, and Len De Klerk. *The Selfmade Land: Culture and Evolution of Urban and Regional Planning in the Netherlands*. Antwerp: Spectrum, 2012.

Van der Wal, Coen. *In Praise of Common Sense: Planning the Ordinary; A Physical Planning History of the New Towns in the Ijsselmeerpolders*. Rotterdam: 010 Publishers, 1997.

Van Lente, Dick, ed. *The Nuclear Age in Popular Media: A Transnational History, 1945–1965*. New York: Palgrave Macmillan, 2012.

Venis, Bernard. *Le Sahara—Afrique du nord, Algérie*. http://alger-roi.fr/Alger/sahara/sahara.htm. Accessed July 27, 2015.

Vidler, Anthony. *Histories of the Immediate Present*. Cambridge, MA: MIT Press, 2008.

Vytuleva, Xenia, ed. *ZATO: Soviet Secret Cities during the Cold War*. New York: Harriman Institute, Columbia University, 2012.

Wagenaar, Cor, ed. *Happy Cities and Public Happiness in Post-War Europe*. Rotterdam: NAi Publishers / Architecturalia, 2004.

Walker, Derek. *The Architecture and Planning of Milton Keynes*. London: Architectural Press, 1982.

Wall, Alex. *Victor Gruen, from Urban Shop to New City*. Barcelona: Actar, 2005.

Ward, Stephen V., ed. *The Garden City: Past, Present, and Future*. London: Routledge, 1992.

———. *Planning the Twentieth-Century City: The Advanced Capitalist World*. West Sussex, UK: John Wiley and Sons, 2002.

Webber, Melvin M. *Explorations into Urban Structure*. Philadelphia: University of Pennsylvania Press, 1964.

Wegner, Phillip E. *Imaginary Communities: Utopia, the Nation, and the Spatial Histories of Modernity*. Berkeley: University of California Press, 2002.

Whiteley, Nigel. *Rayner Banham: Historian of the Immediate Future*. Cambridge, MA: MIT Press, 2002.

Widenheim, Cecilia, ed. *Utopia and Reality—modernity in Sweden 1900–1960*. New Haven, CT: Yale University Press, 2002.

Wiener, Norbert. *Cybernetics or Control and Communication in the Animal and the Machine*. Cambridge, MA: MIT Press, 1948.

———. *The Human Use of Human Beings*. Boston: Houghton Mifflin, 1950.

Wildermuth, Todd. "Yesterday's City of Tomorrow: The Minnesota Experimental City and Green Urbanism." PhD diss., University of Illinois, 2008.

Williams, Richard. *The Anxious City*. New York: Routledge, 2004.

Yakas, Orestes. *Islamabad: The Birth of a Capital*. Oxford: Oxford University Press, 2001.

Yelavich, Susan, ed. *The Edge of the Millennium: An International Critique of Architecture, Urban Planning, Product and Communication Design*. New York: Watson-Guptill Publications, 1993.

Yevtushenko, Yevgeny. *New Works: The Bratsk Station*. Translated by Tina Tupikina-Glaessner and Geoffrey Dutton. Melbourne: Sun Books, 1966.

Zarecor, Kimberly Elman. *Manufacturing a Socialist Modernity: Housing in Czechoslovakia, 1945–1960*. Pittsburgh: University of Pittsburgh Press, 2011.

Zimmermann, Clemens, ed. *Industrial Cities: History and Future*. Frankfurt am Main: Campus Verlag GmbH, 2013.

Zubok, Vladislav. *Zhivago's Children: The Last Russian Intelligentsia*. Cambridge, MA: Harvard University Press, 2011.

INDEX

Page numbers in italics refer to illustrations.